The Open Dynamics of Braitenberg Vehicles

The Open Dynamics of Braitenberg Vehicles

Scott Hotton and Jeff Yoshimi

The MIT Press
Cambridge, Massachusetts
London, England

This book was set in Times New Roman by Scott Hotton and Jeff Yoshimi. Printed and bound in the United States of America.

Library of Congress Cataloging-in-Publication Data

Names: Hotton, Scott, author. | Yoshimi, Jeffrey, author.
Title: The open dynamics of Braitenberg vehicles / Scott Hotton &
    Jeff Yoshimi.
Description: Cambridge, Massachusetts : The MIT Press, [2023] | Includes
    bibliographical references and index.
Identifiers: LCCN 2023030189 (print) | LCCN 2023030190 (ebook) |
    ISBN 9780262548199 (paperback) | ISBN 9780262378956 (epub) |
    ISBN 9780262378949 (pdf)
Subjects: LCSH: Robots–Dynamics–Mathematical models. | Cognitive
    science–Mathematical models. | Braitenberg, Valentino.
Classification: LCC TJ211.4 .H65 2023 (print) | LCC TJ211.4 (ebook) |
    DDC 629.8/93–dc23/eng/20230912
LC record available at https://lccn.loc.gov/2023030189
LC ebook record available at https://lccn.loc.gov/2023030190

10 9 8 7 6 5 4 3 2 1

Contents

Preface

This book is the result of the authors' long-standing collaboration at the intersection of mathematics, philosophy, cognitive science, and biology. Our initial goal was to provide a mathematical foundation for the use of dynamical systems theory in cognitive science (Thelen and Smith 1994; Port and Van Gelder 1995), in particular by those who take an embodied approach that emphasizes the role of the environment in cognition (Wilson and Foglia 2021; Newen, De Bruin, and Gallagher 2018). Theorists in these areas sometimes draw on dynamical systems theory in ways that overstep the relevant mathematical definitions. For example, they sometimes refer to orbits or trajectories "changing course" or to "attractor layouts" changing even though trajectories and attractors are, strictly speaking, unchanging features of a classical dynamical system.[1]

We understood the intuitions motivating such statements—an attractor , for example, can in some sense "move," and the state of a system can follow this "moving attractor" when the environment changes. We sought a way to express these ideas in a mathematically rigorous way. Thus was born the concept of an *open dynamical system*, a mathematical structure in which it is possible to separately describe the intrinsic dynamics of an agent and the way those intrinsic dynamics change in the presence of an environment.[2] A dynamical system has a single state space, while an open dynamical system has two state spaces: one corresponding to the agent by itself and the other corresponding to the agent in the context of an environment. Each state of the total system (which encompasses the agent and its environment) projects to a unique state for the agent by itself. A host of new concepts emerged from these efforts as

---

1. References to such usages are cataloged in Hotton and Yoshimi 2010.

2. Others had formulated similar ideas—for example, open loop control systems or open systems in thermodynamics—but not in the context of a comprehensive mathematical definition that allows intrinsic and open dynamics to be formally defined and then compared.

well as a language for describing the dynamics of an agent's states under the simultaneous influence of its intrinsic structure and that of its environment. In this framework, it becomes possible to describe how some states of an agent can remain fixed for a time, in one environmental situation, but then shift and change as the agent moves through its environment.

We call the states that unfold in an agent state space *paths*—these are the open analogues of orbits in a dynamical system. Unlike orbits, paths in an agent space can cross each other, forming complicated tangles, making them more difficult to visualize than the neat partitioning of a state space into dynamical system orbits. We developed tools for studying these overlapping paths, classifying them into different types, considering how they attract nearby states, and analyzing the bifurcations they go through. Since the states of an agent state space can often be interpreted as representations of an environment, paths in an agent space can often be interpreted as representational processes. From that standpoint, collections of paths corresponds to the possible representational dynamics of an agent in different circumstances. These ideas provide a balanced approach to embodied cognition, one that recognizes the legitimacy of classical cognitive science and its internal representations while also recognizing the crucial role played by environmental interactions.

An example we studied with an open dynamical systems framework was a Hopfield attractor network in a circle world, an environment with objects that repeatedly pass by the network's sensors (Hotton and Yoshimi 2010, 2011). The system was successful at capturing psychological phenomena such as masking, perceptual ambiguity, apparent motion, and priming. We also showed how open dynamical systems could be used to describe a range of examples in embodied cognition—insect locomotion, agents relying on notebooks (and other external artifacts to solve problems), Scrabble players consulting tiles, ship navigation, and so on—in a more satisfactory way than is possible in either a purely classical or purely embodied approach. The framework was also used to describe physical phenomena such as hysteresis in a more thorough way.

Although we had provided a rigorous mathematical foundation for embodied cognition, our analysis was not complete since our examples only involved one-way interactions from an environment to an agent. Our next task was to include more realistic two-way interactions between an agent and its environment. We chose Braitenberg vehicles because they involve two-way interactions but are also conceptually very simple. To keep things mathematically tractable, we focused on the case of exactly two Braitenberg vehicles interacting with each other.

One of Braitenberg's aims was to show how simple systems are capable of generating surprisingly complicated behaviors. This idea was repeatedly confirmed as we studied pairs of interacting vehicles. Even though the system was simple in terms of its basic mathematical description, it revealed a great wealth of interesting behaviors. Some of our analyses are purely theoretical, like the use of equivariant dynamical systems theory to analyze sophisticated attracting sets such as "relative equilibria" and "relative periodic orbits." Some of the behaviors are too complicated to study analytically, and so numerical analysis was performed using a high-performance computing cluster.[3] From these numerical analyses we found that the system has a propensity for quasiperiodic behavior, that is, its behavior is often a combination of two periodic motions with incommensurate periods.

As we pursued this analysis, we discovered connections to a range of related topics in geometry, biology, and cognitive science. For example, even when the states of the Braitenberg vehicles (i.e. , their internal representations) vary in a strictly periodic manner, their two-way interactions produce quasiperiodic motions associated with "meandering paths" in the physical plane. In some ways, these meandering behaviors resemble the classical mechanics of celestial bodies. Sometimes the vehicles meander about each other along graceful curves that resemble the epicyclic motions performed by some moons in our solar system. At other times they revolve around each other like two stars in a binary star system. A common feature of both epicyclic motions and the meandering paths of the Braitenberg vehicles is that their curvature varies in a periodic manner.

The behaviors of the Braitenberg vehicles also resemble other processes in the natural world, such as the movement of spiral wave tips in the BZ reaction (the Belousov–Zhabotinsky chemical reaction, which has become a model system for studying nonequilibrium dynamics) and the movement of bacterial cells such as *Listeria monocytogenes*. Here again the curvature of the paths traversed along a flat surface varies in a periodic manner. This allowed us to adapt mathematical techniques developed to study these natural systems (Barkley 1994; Barkley and Kevrekidis 1994; Hotton 2010) to the study of the Braitenberg vehicles. Although quasiperiodic dynamical systems are generally regarded as being less complicated than chaotic dynamical systems, the manner in which quasiperiodicity arises with two Braitenberg vehicles is quite complicated. For example, there is a small region of parameter space where they end up meandering about each other at one spatial scale while predominately avoiding each other at a smaller scale.

---

3. https://www.xsede.org

The Braitenberg vehicles also rewarded our approach to embodied cognition. One notable feature of the Braitenberg vehicles is the absence of any substantive internal dynamics so that their behavior is almost completely the result of their interactions. And yet, even in this extreme case, it is helpful to consider both their internal representations and their interactions. On the one hand, we analyzed the "embodied complexity" of the two interacting agents as they undergo circular, straight-line, and meandering motions. On the other hand, we analyzed how each agent represents the other as they move in these different ways. For example, when they rotate about each other indefinitely, each agent represents the other as being in a specific fixed location in its frame of reference. When they meander about each other, each agent represents the other as moving along a simple closed curve. These representational processes in turn ended up being helpful in understanding the bifurcations of the system. For example, within the agent spaces, the complicated saddle node-like and Hopf-like bifurcations occurring in the total dynamical system look like classic saddle node and Hopf bifurcations. We also noted that similar agential representations could guide the behavior of tiger beetles and other organisms during navigation.

One of the authors is a mathematical biologist; the other is a philosopher and cognitive scientist. The two of us have a shared interest in visually rich explanations. Our many discussions of Braitenberg's system often focused on how best to present an idea using images and sometimes turned to historical topics as well. The result is a book that not only presents original scholarship on a system whose complications have not hitherto been examined in such detail but also presents the material in a visually and historically rich way.

1 Introduction

In his 1984 book *Vehicles*, Valentino Braitenberg described a series of thought experiments involving "machines with very simple internal structure" (Braitenberg 1984), as in figure 1.1.[1] We give a dynamical systems analysis for pairs of Braitenberg vehicles. The vehicles (or "agents") we consider are slight variations of Braitenberg's, in which the agents sense each other and the two sensors have weighted connections to a motor that turns the vehicle. One of Braitenberg's intentions was to show how surprisingly complicated an agent's behavior can be, even when its engineering is simple. Our dynamical analysis supports this idea: we show how Braitenberg's vehicles can move along complicated paths despite their simple structure.

Although there have been many empirical and numerical studies of Braitenberg vehicles (e.g., Salomon [1999]; Lilienthal and Duckett [2004]; Stolkin, Sheryll, and Hotaling [2007]; Rañó [2009a, 2009b, 2010, 2011]; Salumäe et al. [2012]; Mamduh et al. [2013]; Dvoretskii et al. [2020]; Kasmarik et al. [2020]), there have only been a few attempts at a dynamical systems approach for this class of systems (e.g., Bicho and Schöner [1997]; Rañó [2012a, 2012b]; Rañó, Khamassi, and Wong-Lin [2021]). In this book, we present the first detailed bifurcation analysis for these systems, using a combination of dynamical systems theory and numerical analysis. [2] We also provide several complete online packages for simulating the system, reproducing the main behaviors either in an online Python simulation or in a graphical user interface, Simbrain, where the agents can be directly observed moving with respect to one another (see www.simbrain.net/braitenberg.html).

---

1. We are grateful to Soraya Boza for figures 1.1, 8.1, 8.2, 8.3, 8.4, 8.6 (right panel), and 8.7, some of which were edited by the authors.

2. Braitenberg vehicles usually have long transients, so we rely on dynamical systems theory to help identify attracting sets. The precise form for the higher-dimensional attracting sets is not known, so we rely on digital computations to describe the system's behavior in those cases, including grid searches of the parameter spaces using a high-performance computing cluster.

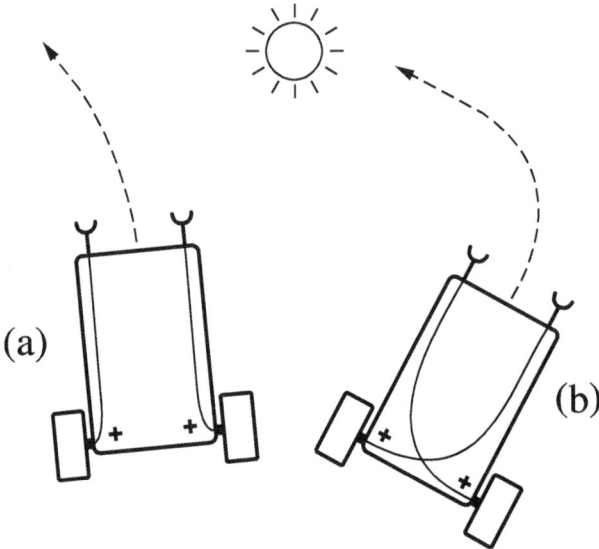

Figure 1.1
Braitenberg vehicles as he originally conceived of them. Shown are (a) an "avoider"
and (b) a "pursuer" relative to a light source. We consider pursuers and avoiders that
are designed slightly differently. Each vehicle emits a signal detectable by the other so
that the vehicles are attracted to or repelled from each other. Adapted from Braitenberg
1984.

The model we present is sufficient to describe an impressively broad range
of behaviors, consistently with Braitenberg's vision. For a sense of this diver-
sity of behavior, see figure 1.2, which shows the main parameter space we
focus on (the full parameter space is discussed in section 2.5), with callouts to
specific regimes of parameter values, indicating what kinds of path the vehicles
follow when their weights are in these regions. As can be seen in the figure,
they pursue and avoid one another in a variety of straight-line, revolving, and
meandering paths. These paths resemble the motions of celestial bodies, the
curves traced out by a spherical pendulum, the moving tip of a spiral wave
in a BZ reaction, actin-based cell motility, and the motions of ascidian larvae.
These are among the cases we consider in this book, though it is likely there
are many other applications of the system. Our work thus unifies a great deal
of work that has been done to date studying these systems while also providing
an explanatory framework that should be applicable in many domains, encompassing naturally occurring systems at multiple scales, from microorganisms
to planets and stars in motion. This is discussed further in chapter 8.

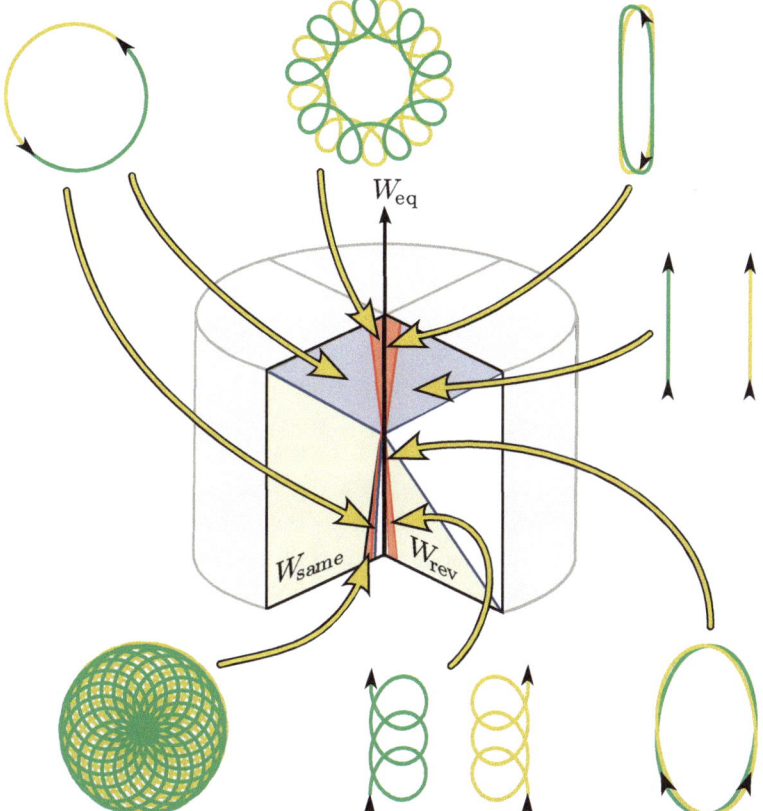

Figure 1.2
The subspaces W_{same} and W_{rev} inside the three-dimensional subspace formed by their direct sum (these subspaces are defined in section 2.5). They are subspaces of the full four-dimensional weight space. We will focus on behaviors of pairs of vehicles when their weight values are in W_{same} (both vehicles have the same weights) and W_{rev} (the weights of the first and second vehicles are reversed). Each panel shows the two physical paths traced by a pair of vehicles. These paths are colored gold and sea green. We will use these color conventions throughout this book in figures showing the physical paths traced by the vehicles. The large arrows from the panels point to the location of the agents' weights in W_{same} or in W_{rev}. The paths shown represent some of the main types of behavior studied in this book, including revolving motion, side-by-side motion, linear and circular meanders, and counter-rotating behaviors.

In addition to the breadth of applications and research threads brought together by the system, several other features of this system make it interesting and worthy of study: (1) the way it combines conservative and dissipative dynamics in a single system, (2) some of its distinctive features as a dynamical system (in particular its bifurcations), and (3) the value of analyzing it as an open dynamical systems, which demonstrates the usefulness of an agent-based perspective in understanding the behavior of complicated systems.

(1) Conservative and dissipative dynamics are associated with different regions of the parameter space of the system. This is especially evident in the two-dimensional subspace of the parameter space that we call W_{same} (shown in figures 1.2, 6.2, and 6.10). The first quadrant of W_{same} corresponds to systems where the two agents pursue each other; the behaviors of the agents resemble conservative systems that produce cyclic or epicyclic paths. The third quadrant corresponds to systems where the two agents avoid each other. The agents in this case, despite avoiding each other, can still interact in complicated and stable ways that resemble dynamics seen in dissipative systems, which tend to have attractors.

There is a certain beauty and elegance to this. Right in the heart of the system we have conservative and dissipative dynamics, which are associated with two distinct long-running research threads in physics. Conservative systems producing cyclic and epicyclic motions are part of the basic history of classical physics, having been studied since the time of Newton, often in an axiomatic way. These are systems whose energy remains constant over time, such as point particles (like celestial bodies) interacting through central forces, frictionless spinning tops moving on flat tables, and spherical pendulums. Dissipative systems, by contrast, require energy to execute their prescribed behaviors. This tradition is less axiomatic and more focused on pragmatic modeling, and is associated with engineering, chemistry, fluid dynamics, thermodynamics, and cybernetics. An important property of dissipative systems, as opposed to conservative systems, is the presence of attracting sets. The analysis of dissipative systems often comes down to identifying the attracting sets of a system and classifying their types. These systems produce, for example, spiral waves whose tips trace out paths that resemble epicyclic orbits. The spiral waves emerge in a saddle node-like bifurcation similar to those observed in the Braitenberg system.

(2) As a dynamical system (section 3), the Braitenberg system has several notable features. First, it is characterized by a basic symmetry. Move the two agents as a pair (either by translating or rotating them), and the paths they follow from their new location and orientation will be be congruent to the paths they would have originally followed. More precisely, the dynamical system is equivariant with respect to the group, $\mathbf{SE}(2)$, of proper congruences of the Euclidean plane.[3] Within this framework, rather than observing attracting fixed

3. The formal definitions of proper congruences and equivariance are in chapter 5.

points and limit cycles, we observe attracting relative equilibria and relative periodic orbits. The different behaviors of the system correspond to different relative equilibria and relative periodic orbits, which in turn correspond to different regions of the system's parameter space. Moving between these regions produces the bifurcations of the system, which are strikingly diverse. There are degenerate saddle node-like bifurcations, degenerate Hopf-like bifurcations, and more exotic bifurcations. For the saddle node-like bifurcations, an attracting and non-attracting pair of relative equilibria emerge together. The attracting relative equilibrium corresponds to the simplest stable motion of the agents, where they revolve around each other in circles or move side by side in straight lines. We also show that the attracting relative equilibria can undergo a degenerate Hopf-like bifurcation, after which the Braitenberg vehicles meander around one another (see figure 1.2).

(3) The system is also noteworthy as an instance of an *open* dynamical system (chapter 4), and in fact that is what initially led us to study this system. Open dynamical systems provide tools to study systems that interact with an environment, as most real systems do (Hotton and Yoshimi 2010, 2011; Yoshimi 2012). Open dynamical systems involve separate dynamical systems to model a total system and one or more agent systems within the total system. While many in the embodied cognition community use environmental couplings to argue against the value of internal representations, we showed how open dynamical systems can be used to merge traditional analysis of representations with analysis of environmental couplings. In particular, we showed how to analyze changing representational dynamics when an agent is embedded in different environments.

In our previous works, we studied how objects influence a stationary agent. Here we show how these ideas work in the more complicated and realistic case of two-way, multi-agent interactions. We will see that internal representational processes continue to be important and revealing in this case. Each vehicle in this system is associated with a pentagonal shaped agent space that contains all of the possible activations of its two sensors (see the right panel of figure 4.1). Points in this space can be interpreted as representations of the position of the other agent. When the agents are out of view of each other, each agent is in its "null state," the state $(0, 0)$. When they come in view of each other, each agent's state enters the interior of its pentagonal shaped state space. We will see that the behaviors and bifurcations of the total six-dimensional system can be understood in terms of changing representations in these two-dimensional agent spaces. In a relative equilibrium, each agent's state settles into a partially fixed point (section 4.1), which represents the other agent as relatively motionless. In a relative periodic orbit, each agent's state (typically) settles into a simple closed curve, which represents the other agent as moving in a periodic fashion. The transitions between these agential representations correspond to the saddle node-like and Hopf-like bifurcations of the total system.

This book is organized as follows. In chapter 2, we give basic specifications for the Braitenberg vehicles and describe the structure of the weight space for pairs of vehicles, focusing on the weight spaces W_{same} and W_{rev}, in which the two agents have the same or reversed weights. In chapter 3, we introduce dynamical systems theory and bifurcation theory, placing an emphasis on techniques we make use of in our study of Braitenberg vehicles. In chapter 4, we describe the theory of open dynamical systems, and formally specify an open dynamical system for pairs of Braitenberg vehicles. In chapter 5, we develop the concepts of a relative equilibrium and relative periodic orbit. In chapters 6 and 7, we describe revolving and translating type relative equilibria and their related relative periodic orbits. We examine each of the weight spaces W_{same} and W_{rev} in some detail. We present a taxonomy of invariant sets that correspond to specific regions in these parameter spaces, and discuss the bifurcations that occur at the boundaries of these regions. These analyses include a detailed discussion of the geometry of the meandering paths that occur in the two weight spaces. We end in chapter 8 by discussing how our results can be applied to a wide range of natural phenomena.

Code and simulations that can be used to study this system and recreate many figures in the book are available at: www.simbrain.net/braitenberg.html.

2 Specifications for the Braitenberg Vehicles

Braitenberg conceived of his vehicles as moving in two dimensions using a pair of wheels driven by motors that are connected to a pair of sensors (see figure 1.1). The speed of each wheel is positively correlated with the activation of the sensor it is connected to. If the left sensor is connected to the left wheel and the right sensor is connected to the right wheel, the vehicle will turn away from the object that it detects (what Braitenberg [1984] calls a type 2a vehicle). If the left sensor is connected to the right wheel and the right sensor is connected to the left wheel, then the vehicle will turn toward the object that it detects (a type 2b Braitenberg vehicle).

We depart from Braitenberg's ideas in a few ways. First, rather than consider vehicles that pursue or avoid stationary objects, we consider pairs of vehicles that pursue or avoid each other. Second, the agents we consider are abstract idealizations of Braitenberg's vehicles. Braitenberg's vehicles are driven by two wheels at the back of a rectangular body. Incorporating the details of such a propulsion system can involve complexities such as time delays in the signals to the motors and the slipping of the wheels as the vehicle moves across a flat surface. Thus we consider idealized vehicles that move rigidly in the Euclidean plane according to precise mathematical rules. In particular, we assume that each vehicle moves forward at a constant linear speed and introduce a simple Braitenberg type rule for how the vehicles turn in each other's presence.

We conceive of Braitenberg vehicles as having simple neural networks that link sensors to motors via weighted connections. The weights allow us to continuously interpolate between different types of Braitenberg vehicles. We have implemented these networks in the computer program Simbrain (see figure 2.1). In the Simbrain implementation, two sensory neurons are connected to two motor neurons that turn the agent. A fifth neuron with a fixed activation moves the agent forward at a fixed linear speed. The angular velocity of the agent is given by the difference between the two motor neurons' activations.

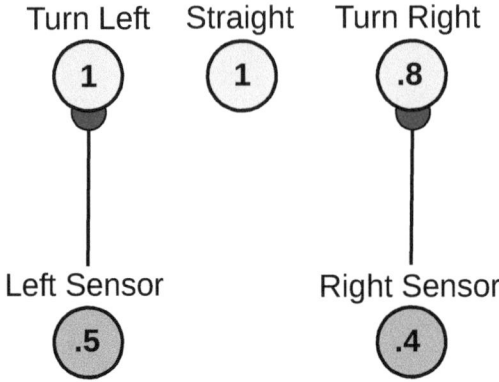

Figure 2.1
A Braitenberg vehicle's neural network as it is modeled in Simbrain (https://simbrain.net). All of the behaviors described here can be recreated using simulations included with Simbrain. The example shown here is a pursuer in pursuit. Both sensor weights have the same value ($w_{(\ell,n)} = w_{(r,n)} = 2$). The other vehicle (not shown) is to the left so the agent receives more activation in its "Left Sensor" ($a_{(\ell,n)} = 0.5$) than in its "Right Sensor" ($a_{(r,n)} = 0.4$). As a result, the activation of the "Turn Left" motor neuron ($w_{(\ell,n)}a_{(\ell,n)} = 2(0.5) = 1$) is greater than the activation of the "Turn Right" motor neuron ($w_{(r,n)}a_{(r,n)} = 2(0.4) = 0.8$). The agent turns to its left, toward the other agent, at a rate equal to the difference between the two motor neuron activations, with angular velocity $0.2 = 2(0.5) - 2(0.4)$.

For the example in figure 2.1, the "Turn Left" motor neuron has more activation than the "Turn Right" motor neuron, so the agent turns left. An advantage of the Simbrain network is that it allows Braitenberg vehicles to be studied in a concrete, intuitively understandable way. For the mathematical analysis presented here, though, we further idealize Braitenberg vehicles as follows.

Braitenberg vehicles, as we conceive of them, are shown in figure 2.2. To distinguish between the two vehicles, they are labeled "1" and "2." The subscript $n = 1, 2$ in the symbol for a quantity indicates which of the two vehicles the quantity refers to. For simplicity we envision the Braitenberg vehicles as circular disks. Also for simplicity we allow the vehicles to overlap in the plane. We let the center of the disk stand for the position of the vehicle in the plane. The position will be represented with a column vector, which for convenience will often be written as the transpose of a row vector: $(x_n, y_n)^T$.

The activation of agent n's left sensor will be denoted by $a_{(\ell,n)}$, and the activation of its right sensor will be denoted by $a_{(r,n)}$. The left and right sensors of agent n are marked in figure 2.2 with two small white dots labeled $a_{(\ell,n)}$ and $a_{(r,n)}$. The activation of each sensor depends on the distance of the sensor to

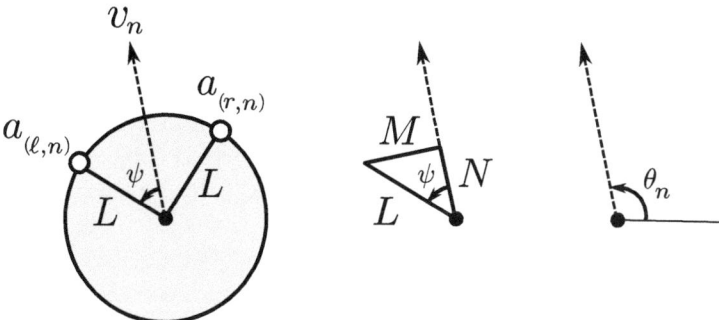

Figure 2.2
(left) An abstract Braitenberg vehicle with a disk-shaped body (shaded light gray) and two sensors (marked with white dots). The activations $a_{(\ell,n)}$ and $a_{(r,n)}$ are associated with its left and right sensors. The agent's position is marked with a black dot, and the stalks are shown as line segments connecting the white dots with the back dot. The parameter L is the radius of the disk and the length of the stalks. This illustrates how the simulation shown in figure 2.1 is implemented in this book. The fixed speed of agent n is v_n, that is, the magnitude of the velocity vector. (middle) The derived parameter $M = L\cos(\psi)$ is the distance from a sensor to the midpoint between the sensors. The derived parameter $N = L\sin(\psi)$ is the distance from the agent to the sensors' midpoint. (right) The velocity vector points in the direction θ_n, the heading of the vehicle in the rest frame.

the other agent's position, which often varies over time. The further away the other agent is, the lower the activation of the sensory neuron. The activation levels will be confined to the unit interval $[0, 1]$. A value of 0 means the other vehicle is beyond the sensor's range, and a value of 1 means the other vehicle is centered on the sensor. The state of agent n will be represented with the column vector: $(a_{(\ell,n)}, a_{(r,n)})^T$.

As with the Simbrain vehicles, we can think of the sensory neurons as being located at the ends of two weighted connections that convey activation to two motor neurons, which induce the agent to turn to its left or right. In our mathematical implementation, the left and right sensor weights for agent n will be denoted by $w_{(\ell,n)}$ and $w_{(r,n)}$, respectively. The activation level of a motor neuron is the product of the activation level of its corresponding sensory neuron and its corresponding weight, that is $w_{(\ell,n)}a_{(\ell,n)}$ for the left motor neuron and $w_{(r,n)}a_{(r,n)}$ for the right motor neuron. Motor neuron activations are not treated as state variables in our analysis since they are just functions of sensor activations and weights.

The rate at which the agent turns (i.e., the agent's angular velocity) is set equal to the difference between the left and right motor neurons' activations:

$$\dot{\theta}_n = w_{(\ell,n)}\, a_{(\ell,n)} - w_{(r,n)}\, a_{(r,n)}. \tag{2.1}$$

If the left motor neuron has more activation than the right motor neuron, then the agent will turn to its left and vice versa.

We distinguish between two sets of parameters. For each vehicle, there are four physical parameters and two neural network parameters. The physical parameters specify properties of each vehicle's physical body. As in Braitenberg's original conceptualization, the physical bodies have bilateral symmetry, that is, their bodies are symmetrical under just one reflection. The neural network parameters are the two sensor weights. Unless the sensor weights are equal, their association with the sensors' locations breaks the bilateral symmetry, that is, the reflectional symmetry of the physical body switches which sensor each weight is associated to. Our analysis focuses on the bifurcations that occur as the neural network parameters—the sensor weights—are varied, while the physical parameters are kept fixed.

2.1 Types of Vehicles

If both sensor weights of an agent are positive and the other vehicle is sufficiently far to the left, the activation will be greater in the left motor neuron than in the right motor neuron, so the agent will turn to its left, towards the other vehicle. The agent will turn to its right if the other vehicle is sufficiently far to the right. We call an agent with this kind of weight pair a *pursuer*. Weight pairs in the first quadrant of the weight space for an individual agent correspond to pursuers, as illustrated in the left panel of figure 2.3.

If both weights are negative and the other vehicle is sufficiently far to its right, then the activation will be greater in the right motor neuron than in the left motor neuron so the agent will turn to its right, away from the other vehicle. The agent will turn to its left if the other vehicle is sufficiently far to its right. We call an agent with this kind of weight pair an *avoider*. Weight pairs in the third quadrant of the weight space for an individual agent correspond to avoiders.

We can also get "lateralized" agents that pursue in one direction and avoid in another. Weights in the second quadrant correspond to vehicles that pursue to the right and avoid on the left. Weights in the fourth quadrant correspond to vehicles that pursue to the left and avoid on the right.

Any type of agent with unequal weights will have a turning bias—that is, it will pursue or avoid more quickly to one side than the other.[1] If $w_{(\ell,n)} < w_{(r,n)}$,

---

1. Such asymmetric behavior is known to occur to a certain extent in humans, and it has been observed throughout the animal kingdom (for reviews see Bisazza, Rogers, and Vallortigara [1998]; Frasnelli [2013]). For instance, mate selection by certain types of finches can be lateralized between the left and right eyes (Workman and Andrew 1986; Templeton et al. 2012), and a right-turning bias has been found in the response of cockroaches to odors (Cooper et al. 2011).

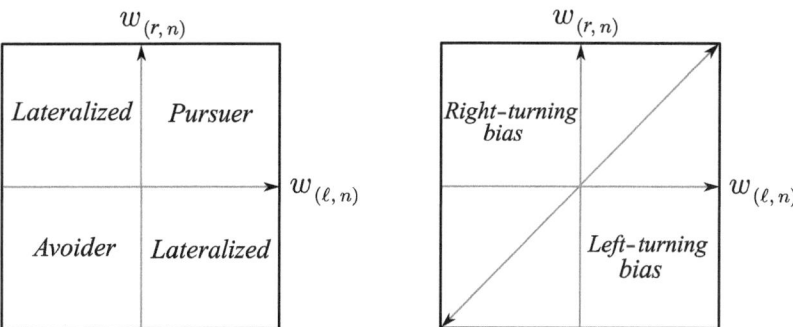

Figure 2.3
Two views for the space of ordered pairs of weights, $(w_{(\ell,n)}, w_{(r,n)})$ for a single vehicle. (left) A vehicle is either a pursuer, an avoider, or lateralized based on whether it tends to pursue the other vehicle (both weights positive), avoid the other vehicle (both weights negative), or pursue on one side and avoid on another (one weight positive, the other negative). (right) A vehicle typically has a bias to turn more strongly in one direction than another. It has a bias to turn right when $(w_{(\ell,n)}, w_{(r,n)})$ is above the diagonal line $w_{(\ell,n)} = w_{(r,n)}$ and to turn left when $(w_{(\ell,n)}, w_{(r,n)})$ is below the diagonal.

the agent has a right-turning bias. If $w_{(\ell,n)} > w_{(r,n)}$, it has a left-turning bias. The diagonal where both weights are the same, $w_{(\ell,n)} = w_{(r,n)}$, separates the weight space into two symmetrical regions, as shown in the right panel of figure 2.3.

It may seem that setting both sensor weights of an agent to be precisely the same is the ideal situation, but we show in subsequent chapters that this actually produces exceptional behavior and that introducing a small turning bias produces more typical behavior for the agents.

2.2 Physical Parameters

The four physical parameters for vehicle n are its speed v_n, the angular offset ψ of its sensors, the radius L of its body, and the range P of its sensors. In almost all of our analysis, the two vehicles will have the same physical parameters, in particular, $v_1 = v_2$.

The speed (i.e., the magnitude of the vehicle's linear velocity) is a fixed positive number. The assumption that the speed remains absolutely constant even as the angular velocity changes is a mathematical idealization that is unlikely to occur in the physical world. In fact, as explained in the introduction and chapter 8, the motion of many naturally occurring systems can be well approximated with the assumption that they maintain their linear speed as they turn.

The vehicles' sensors can be thought of as being located at the ends of two stalks, encasing the weighted connection between the sensors and a turning

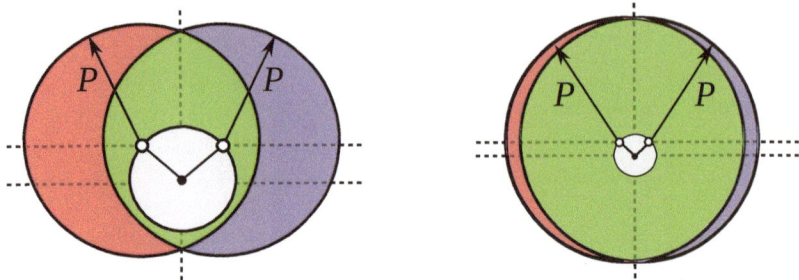

Figure 2.4
The field of view for each sensor is an open disk with radius P centered on the sensor. The sensors' fields of view overlap in the central lune, $D_{(\ell,n)} \cap D_{(r,n)}$, which is colored lime green. The left lune is the portion of the left sensor's field of view that is outside the right sensor's field of view, that is, $D_{(\ell,n)} \backslash D_{(r,n)}$. It is rose colored. The right lune is the portion of the right sensor's field of view that is outside the left sensor's field of view, that is, $D_{(\ell,n)} \backslash D_{(r,n)}$. It is lavender colored. We will use these conventions in color figures showing the sensors' fields of view. (left) The sensor range P is just large enough that the entire vehicle is contained in the central lune. (right) When P is large compared to the agent, the left and right lunes appear much thinner than the central lune. The situation on the left is more conducive to clear pictures; the situation on the right is more characteristic of the cases considered in this book.

mechanism. The stalks have a positive length and are offset by an angle ψ from the vehicle's heading. This angle is a physical parameter of the system. To keep the sensors apart and in front of the agent, the value of ψ has to be strictly between 0 and $\pi/2$ radians (see figure 2.2). For the most part, we set the angular offset to be half-way between these angles, that is, $\psi = \pi/4$.

The length of the stalks is the same as the radius, L, of the disk. It will be convenient to have a derived parameter, $M = L \sin(\psi)$, for the distance between a sensor and the midpoint between the sensors, and a derived parameter, $N = L \cos(\psi)$, for the distance between the center of the agent and the midpoint between the sensors (see figure 2.2). Since M and N are simple functions of the parameters L and ψ they are not treated as independent parameters of the system. They are just used to simplify mathematical expressions.

We assume each agent detects a signal emitted from the center of the other agent. The parameter P specifies the range for the sensors. Each sensor detects the other agent's presence when the center of the other agent is within the distance $P > 0$ from the sensor. The field of view for each vehicle's sensors are two open disks of radius P centered on the sensors' locations (see figure 2.4). We denote the field of view of agent n's left sensor by $D_{(\ell,n)}$ and the field of view of its right sensor by $D_{(r,n)}$.

So long as the sensor range is greater than the distance between a sensor and the midpoint between the sensors (i.e., $P > M$), the two sensors' fields of view overlap. When this occurs the two fields of view generate a partition of the physical plane into four regions. Three of these regions are bounded in size, and the other region is not. The unbounded region is the complement, in the plane, of the union of the two fields of view, that is, the complement of $D_{(\ell,n)} \cup D_{(r,n)}$. The bounded regions are lunes, that is, they are bounded by the arcs of two circles. We call the region where the two fields of view overlap, $D_{(\ell,n)} \cap D_{(r,n)}$, the *central lune*. We call the complement of the right sensor's field of view within the left sensor's field of view, $D_{(\ell,n)} \backslash D_{(r,n)}$, the *left lune*. We call the complement of the left sensor's field of view within the rights sensor's field of view, $D_{(r,n)} \backslash D_{(\ell,n)}$, the *right lune*.

If the fields of view are too small, it will be difficult for one agent to detect the other, and even if it did detect the other agent it would stop turning soon after it started and would only be slightly deflected as it passed by the other agent. If the sensor range were less than or equal to the radius of the vehicles, then the vehicles could even be on top of each without detecting each other, so we require $P > L$. Appendix A explains why we also want $P > 3M$. Succinctly stated we require:

$$P > \max\{3M, L\}. \tag{2.2}$$

The overall size of the vehicles establishes a length scale for the physical paths. Rescaling L, P, v_1, and v_2 by the same factor transforms the physical paths traced by the vehicles to geometrically similar paths. The actual sizes for P, v_1, and v_2 are not especially significant to the behavior of the Braitenberg vehicles. What matters is their sizes relative to L. We can let the radius of a vehicle be the unit of measure for lengths in the plane. From here on, we set $L = 1$. This allows us to think of P as the sensors' range measured in units where the vehicle's radius is 1. We also interpret the speed, v_n, as the distance vehicle n will travel in a unit of time, relative to its radius.

2.3 Reference Frames

We will consider Braitenberg vehicles in several frames of references: a rest frame, a body frame centered on each vehicle, and a rotating frame that turns with a constant angular velocity about a point between the agents. The rotating frame is introduced in section 6.4.

As stated near the beginning of this chapter, the position of agent n in the rest frame will be represented with Cartesian coordinates in the form of a column vector $(x_n, y_n)^T$, and its heading in the rest frame will be denoted by θ_n. This

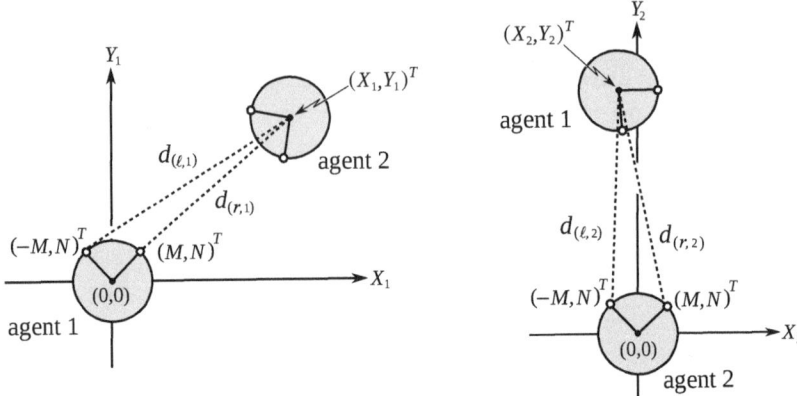

Figure 2.5
The same configuration for both agents is shown in the two body frames. (left) The body
frame for agent 1. In this frame, agent 1 is located at the origin and its heading is up.
Agent 2's location is $(X_1, Y_1)^T$ in agent 1's body frame. The distances from the center
of agent 2 to agent 1's left and right sensors are $d_{(\ell,1)}$ and $d_{(r,1)}$, respectively. (right) The
body frame for agent 2. In this frame agent 2 is located at the origin and its heading
is up. Agent 1's location is $(X_2, Y_2)^T$ in agent 2's body frame. The distances from the
center of agent 1 to agent 2's left and right sensors are $d_{(\ell,2)}$ and $d_{(r,2)}$, respectively. In
both cases, the coordinates for an agent's sensors in its own body frame is $(-M, N)^T$ for
the left sensor and $(M, N)^T$ for the right sensor.

is the angle the vehicle's heading makes with the positive x_n-axis of the rest
frame.

It will be useful to be able to express where one vehicle is in relation to
the other vehicle. Each Braitenberg vehicle will be located at the origin of a
two-dimensional reference frame of its own in which its heading is vertically
upward. We call this its *body frame*.

We denote the coordinates of the other agent in the body frame of agent n
by the column vector $(X_n, Y_n)^T$. Keep in mind that the subscript n in the two
components of the vector $(X_n, Y_n)^T$ refers to the agent in whose body frame we
are working, while the vector itself gives us the coordinates for the *other* agent
in the body frame of agent n. For example, if $X_1 = 4$ and $Y_1 = 3$, then agent 2 is
located at $(4, 3)^T$ in the body frame of agent 1. See figure 2.5.

We can often write general mathematical statements more concisely by
putting n in the subscript of the variables for one agent and $\neg n$ in the subscript of the variables for the other agent. For instance, "agent n detects agent
$\neg n$" is mathematical shorthand for saying that agent n detects the other agent.
We can think of $\neg$ as denoting the function $n \mapsto (3 - n)$ that swaps the numbers

1 and 2. We will use this notation in the formula for the body frames of the two agents that we now derive.

The position of agent $\neg n$ in the body frame of agent n is obtained by positioning the origin of the body frame at the center of agent n and then rotating so that agent n is heading vertically upward. So we subtract the coordinates for agent n in the rest frame from the coordinates for agent $\neg n$ in the rest frame and then rotate around agent n's center so that its heading is vertically upward. This is done by counter-rotating by agent n's heading (i.e., rotating by $-\theta_n$), so that it heads to the right, and then rotating by $\pi/2$, so that it heads upward. Altogether we rotate by $\pi/2 - \theta_n$.

We will use the standard matrix representation for a rotation by any arbitrary angle, Θ, about the origin:

$$R_\Theta = \begin{pmatrix} \cos(\Theta) & -\sin(\Theta) \\ \sin(\Theta) & \cos(\Theta) \end{pmatrix}. \tag{2.3}$$

The equation for the coordinates of agent $\neg n$ in the body frame of agent n is thus:

$$\begin{pmatrix} X_n \\ Y_n \end{pmatrix} = R_{(\pi/2-\theta_n)} \left(\begin{pmatrix} x_{(\neg n)} \\ y_{(\neg n)} \end{pmatrix} - \begin{pmatrix} x_n \\ y_n \end{pmatrix} \right) = R_{(\pi/2-\theta_n)} \begin{pmatrix} x_{(\neg n)} - x_n \\ y_{(\neg n)} - y_n \end{pmatrix} \tag{2.4}$$

for $n = 1, 2$.

2.4 Dynamical Rule for Angular Velocity

We have set the rate of change in agent n's heading (i.e., its angular velocity $\dot\theta_n$) to be equal to the weighted difference of its sensor activations as stated in equation (2.1). It will be useful to have a function that tells us how quickly agent n turns in response to the position of agent $\neg n$ in agent n's body frame. We denote this *turning function* by $\bar\omega_n(X_n, Y_n)^T$ and derive it in this section.

The left and right sensors of a Braitenberg vehicle are located at $(-M, N)^T$ and $(M, N)^T$, respectively, in its body frame (recall figures 2.2 and 2.5). The distances from the left and right sensors of agent n to the center of the other vehicle, $(X_n, Y_n)^T$, will be denoted by:

$$d_{(\ell,n)} = d_{(\ell,n)}\begin{pmatrix} X_n \\ Y_n \end{pmatrix} = \sqrt{(X_n + M)^2 + (Y_n - N)^2}$$

$$d_{(r,n)} = d_{(r,n)}\begin{pmatrix} X_n \\ Y_n \end{pmatrix} = \sqrt{(X_n - M)^2 + (Y_n - N)^2}. \tag{2.5}$$

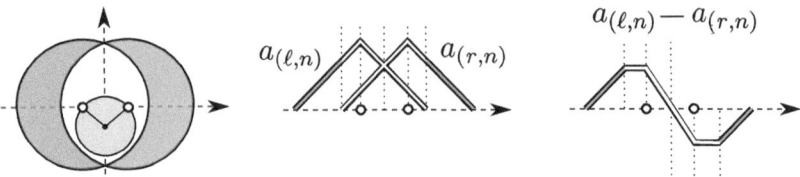

Figure 2.6
The graph of the turning function $\bar{\omega}_n(X_n, Y_n)^T$ restricted to the line $Y_n = N$ through the agent's sensors. The horizontal dashed lines in the three panels correspond to the same line $Y_n = N$ in the body frame. The sensor locations on these dashed lines are marked with $\circ$. The agent here is a pursuer with weights $w_{(\ell,n)} = w_{(r,n)} = 1$. (left) The sensors' fields of view as depicted in figure 2.4. (middle) The graphs of the individual sensor activations, as given by equation (2.7). The peaks are directly above the sensors. (right) The graph of $\bar{\omega}_n(X_n, N)^T$ as given by equation (2.8). When the agent $\neg n$ is to the left of agent n the difference in sensor activations is positive so the agent turns to the left; when agent $\neg n$ is to the right of agent n the difference in sensor activations is negative so the other agent turns to the right. More generally the shape of the graph of $\bar{\omega}_n(X_n, Y_n)^T$ varies dramatically with the sensor weights, which is crucial for the bifurcation analysis in chapter 6. The middle and right panels of this figure show vertical cross sections of the graph of $\bar{\omega}_n(X_n, Y_n)^T$ through the line $Y_n = N$ shown in the left and middle panels of figure 2.7.

The activations of an agent's sensors are computed with a piecewise linear scaling function defined as:

$$f(d) = \begin{cases} 1 - d/P & \text{if } 0 \le d \le P \\ 0 & \text{otherwise} \end{cases} \tag{2.6}$$

for any $d \ge 0$. Substituting the distances between the sensors and the other agent into the scaling function gives us the activations of agent n's sensors.[2]

$$\begin{pmatrix} a_{(\ell,n)} \\ a_{(r,n)} \end{pmatrix} = \begin{pmatrix} f(d_{(\ell,n)}) \\ f(d_{(r,n)}) \end{pmatrix} = \begin{pmatrix} f\left(d_{(\ell,n)}(X_n, Y_n)^T\right) \\ f\left(d_{(r,n)}(X_n, Y_n)^T\right) \end{pmatrix}. \tag{2.7}$$

The middle panel of figure 2.6 shows the graphs of the functions $f(d_{(\ell,n)})$ and $f(d_{(r,n)})$ restricted to the line that passes through the sensors, $Y_n = N$. A sensor achieves its peak activation of 1 only when the other agent is directly on top of it. As the other agent moves away from a sensor its activation drops off linearly to 0 and then remains 0 no matter how much further away the other agent travels since the other agent has gone beyond the sensors' fields of view.

---

2. Note that to improve readability we sometimes drop the parenthesis around the argument to a function when the argument has the form of the transpose of a row vector. For example, $d_{(\ell,n)}\left((X_n, Y_n)^T\right)$ is shortened to $d_{(\ell,n)}(X_n, Y_n)^T$.

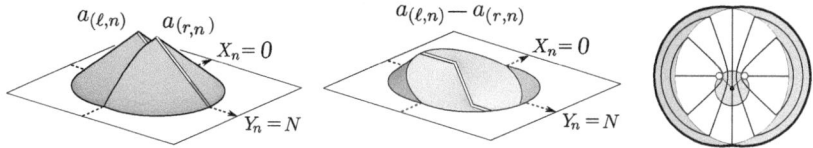

Figure 2.7
(left) The graphs for agent n's sensor activations as functions of $(X_n, Y_n)^T$ in $D_{(\ell,n)} \cup D_{(r,n)}$ are cones, as shown in equation (2.7). (middle) The difference in height between the two cones is the graph of the turning function $\bar{\omega}_n(X_n, Y_n)^T$ when $w_{(\ell,n)} = w_{(r,n)} = 1$. The dark regions are where only one sensor is active; the light region is where both sensors are active. (right) The level curves for $\bar{\omega}_n(X_n, Y_n)^T$ when $w_{(\ell,n)} = w_{(r,n)}$. In the left lune, they are arcs of circles centered at the left sensor. In the right lune, they are arcs of circles centered at the right sensor. In the central lune, they are typically arcs of hyperbolas with their foci located at the sensors.

The turning function $\bar{\omega}_n(X_n, Y_n)^T$ is obtained by substituting equation (2.7) into the right-hand side of equation (2.1):

$$\bar{\omega}_n \begin{pmatrix} X_n \\ Y_n \end{pmatrix} = w_{(\ell,n)} f\left(d_{(\ell,n)}\right) - w_{(r,n)} f\left(d_{(r,n)}\right)$$

$$= w_{(\ell,n)} f\left(d_{(\ell,n)} \begin{pmatrix} X_n \\ Y_n \end{pmatrix}\right) - w_{(r,n)} f\left(d_{(r,n)} \begin{pmatrix} X_n \\ Y_n \end{pmatrix}\right). \tag{2.8}$$

The graph of $\bar{\omega}_n(X_n, Y_n)^T$ in the case $w_{(\ell,n)} = w_{(r,n)} = 1$ (a pursuer) is shown in the middle panel of figure 2.7, and its level curves are shown in the right panel. The value of $\bar{\omega}_n(X_n, Y_n)^T$ is zero on the line segment connecting the vertices of the central lune (where $X_n = 0$). The value of $\bar{\omega}_n(X_n, Y_n)^T$ is positive for negative X_n, so agent n turns to the left, towards the other agent. The value of $\bar{\omega}_n(X_n, Y_n)^T$ is negative for positive X_n, so agent n turns to the right, again towards the other agent.

2.5 Subspaces of the Weight Space

We put all four sensor weights for the two Braitenberg vehicles into a single row vector in the following order: $\left(w_{(\ell,1)}, w_{(r,1)}, w_{(\ell,2)}, w_{(r,2)}\right)$. We allow the sensor weights to have any real number value. We call the vector space for all of these ordered quadruples of sensor weights the *total weight space* and denote it by:

$$W_{\text{total}} = \left\{ \left(w_{(\ell,1)}, w_{(r,1)}, w_{(\ell,2)}, w_{(r,2)}\right) \mid w_{(\ell,1)}, w_{(r,1)}, w_{(\ell,2)}, w_{(r,2)} \in \mathbf{R} \right\}. \tag{2.9}$$

The total weight space is a four-dimensional vector space but we will focus on the two-dimensional subspaces W_{same} and W_{rev} and on their internal direct sum

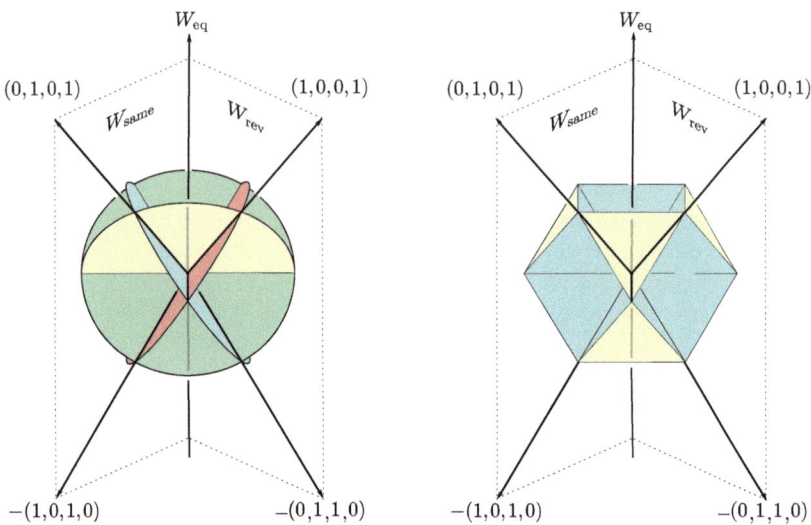

Figure 2.8
Two copies of the subspace $W_{\text{same}} + W_{\text{rev}}$. Basis vectors for W_{same} and W_{rev} are shown in each copy. The left copy shows where the four coordinate 3-spaces of the four-dimensional weight space W_{total} intersect a solid 3-ball centered at the origin of $W_{\text{same}} + W_{\text{rev}}$. The four intersections are disks, each of which is shown in a different color. In the right copy, the circular arcs connecting the vertices shown in the left copy are replaced with line segments, revealing six square cones (blue) and eight triangular cones (yellow). The subspaces W_{same} and W_{rev} bisect the square cones but do not pass through the interiors of the triangular cones. The square cones contain avoider-avoider, pursuer-pursuer, and lateralized-lateralized pairs of vehicles, which we focus on.

$W_{\text{same}} + W_{\text{rev}}$, which is three-dimensional, as illustrated in figures 1.2 and 2.8. In this chapter, we define W_{same} and W_{rev} and describe $W_{\text{same}} + W_{\text{rev}}$.

We will focus on pairs of vehicles that resemble each other so that we can more easily compare their weight spaces to the weight space for a single agent (figure 2.3). We consider two ways that the agents can resemble each other. One way is for the two agents to have the same weights; for instance, both agents could be pursuers with weights such as $(w_{(\ell,n)}, w_{(r,n)}) = (0.7, 0.8)$ or both could be avoiders with weights such as $(w_{(\ell,n)}, w_{(r,n)}) = (-1.5, -1)$. In these cases, there are two exact copies of the same type of Braitenberg vehicle moving in the plane. A convenient way to describe this particular subspace of W_{total} is as the span of a pair of vectors:

$$W_{\text{same}} = \text{span}\{(1, 0, 1, 0), (0, 1, 0, 1)\} = \{(w_\ell, w_r, w_\ell, w_r) \mid w_\ell, w_r \in \mathbf{R}\}.$$
$$(2.10)$$

Recall that the span of a set of vectors is the set of all linear combinations of those vectors; for example, the linear combination

$$(0.7)(1, 0, 1, 0) + (0.8)(0, 1, 0, 1) = (0.7, 0.8, 0.7, 0.8).$$

The vectors $(1, 0, 1, 0)$ and $(0, 1, 0, 1)$ are chosen for convenience. They are linearly independent (neither is a scalar multiple of the other), so they form a basis for the subspace they span. Also note that even though W_{same} is a set of ordered quadruples, because the two agents are constrained to have the same weights, W_{same} is in fact a two-dimensional subspace. This is evident from the fact that a basis for the subspace has two members.

There is a tendency for both agents to circle around each other when they have the same weights. This tendency persists, surprisingly, when they are both avoiders. This is further explained in chapter 6.

Another way for the two agents to resemble each other is for their sensor weights to be reversed. For instance, agent 1 could have weights $(w_{(\ell,1)}, w_{(r,1)}) = (0.7, 0.8)$ and agent 2 could have weights $(w_{(\ell,2)}, w_{(r,2)}) = (0.8, 0.7)$. Although the agents' bodies are bilaterally symmetric, their weights are not. However, they are mirror symmetric copies of each other. This symmetry will be important in several contexts throughout the book.

Again we get a two-dimensional subspace:

$$W_{\text{rev}} = \text{span}\{ (1, 0, 0, 1), (0, 1, 1, 0) \} = \{ (w_\ell, w_r, w_r, w_\ell) \mid w_\ell, w_r \in \mathbf{R} \}. \quad (2.11)$$

When the weights of the two agents are reversed like this—for example, one agent pursues toward the left and the other agent pursues by the same amount toward the right—there is a tendency for them to travel along parallel paths even when they are both avoiders. This is further discussed in chapter 7.

The subspaces W_{same} and W_{rev} intersect in a one-dimensional subspace. The only way the weight pair for one agent can be the same as the weight pair for the other agent and also be reversed from the weight pair for the other agent is for the other agent to have equal weights. And since both agents have the same weights as each other, all four sensor weights must be equal. This gives us a one-dimensional subspace of quadruples of weights for the two agents:

$$W_{\text{eq}} = W_{\text{same}} \cap W_{\text{rev}} = \text{span}\{ (1, 1, 1, 1) \} = \{ (w, w, w, w) \mid w \in \mathbf{R} \}. \quad (2.12)$$

These weights do not break the bilateral symmetry of agents' bodies. Agents with four equal weights engage in exceptional behavior as discussed in chapters 6 and 7.

As we have seen, W_{same} and W_{rev} are two-dimensional despite being sets of quadruples. Moreover, in terms of our chosen bases, the quadrants of the weight space for a single agent are arranged the same way as the quadrants of

W_{same} and W_{rev} (left panel of figure 2.3). The first quadrant in the weight space for an individual agent corresponds to pursuers, and the first quadrants of W_{same} and W_{rev} correspond to pursuer-pursuer pairs. The third quadrant in the weight space for an individual agent corresponds to avoiders, and the third quadrants of W_{same} and W_{rev} correspond to avoider-avoider pairs. The second and fourth quadrants in the weight space for an individual agent correspond to lateralized agents that pursue in one direction and avoid in the other, and the second and fourth quadrants of W_{same} and W_{rev} correspond to pairs of lateralized agents that pursue or avoid in accordance with which side of each other they are on.

In addition to the quadrants of W_{same} and W_{rev} having similar arrangements, the bifurcations that occur in W_{same} and W_{rev} are similar to each other. We provide a detailed analysis of W_{same} in chapter 6 and of W_{rev} in chapter 7. The similarities between the bifurcations can also be seen in figure 1.2.

The diagonal line in the weight space for an individual agent separates the regions of left- and right-turning bias (right panel in figure 2.3). Here again there is an analogy between the individual agent weight space and the subspaces W_{same} and W_{rev}. In both W_{same} and W_{rev} the subspace W_{eq} is the diagonal line with equation $w_\ell = w_r$ (see figures 6.10 and 7.4). It bisects the pursuer-pursuer and avoider-avoider quadrants of W_{same} and W_{rev}. For the region above W_{eq} both agents have a right-turning bias and for the region below W_{eq} both agents have a left-turning bias.

The subspaces W_{same} and W_{rev} intersect along the diagonal line W_{eq}. They intersect orthogonally (using the standard dot product as an inner product). This is shown in figures 1.2 and 2.8.

The internal direct sum, $W_{\text{same}} + W_{\text{rev}}$, is the set of sums over all vectors from W_{same} and W_{rev}. A theorem from linear algebra states that for any two subspaces, V_1 and V_2, of a vector space, the dimension of their internal direct sum is given by:

$$\dim(V_1 + V_2) = \dim(V_1) + \dim(V_2) - \dim(V_1 \cap V_2).$$

Since $\dim(W_{\text{same}}) = 2$, $\dim(W_{\text{rev}}) = 2$, and $\dim(W_{\text{eq}}) = 1$, the subspace $W_{\text{same}} + W_{\text{rev}}$ is three-dimensional.

In the same way that a plane is separated into $4 = 2^2$ quadrants by two coordinate lines, and a three-dimensional space is separated into $8 = 2^3$ octants by three coordinate planes, a four-dimensional vector space is separated into $16 = 2^4$ orthants by four coordinate 3-spaces. To help visualize the orthants of W_{total}, the left panel of figure 2.8 shows where a solid three-dimensional ball centered at the origin of $W_{\text{same}} + W_{\text{rev}}$ intersects the four coordinate 3-spaces of W_{total}. Each intersection of a coordinate 3-space with this ball is a disk. The chosen basis vectors for W_{same} and W_{rev} are also shown.

In the right panel of figure 2.8, circular arcs connecting the vertices have been replaced with line segments. The result is a cuboctahedron, that is, an Archimedean solid formed by truncating or "shaving off" the eight vertices from a cube. A cuboctahedron has six square faces (like a cube) and eight equilateral triangle faces (like an octahedron). These faces are cross sections of six infinite square cones and eight infinite triangular cones in $W_{same} + W_{rev}$. These fourteen cones are where fourteen of the sixteen orthants of W_{total} pass through $W_{same} + W_{rev}$. Each of the subspaces W_{same} and W_{rev} passes through the interiors of the square cones but not the triangular cones. The six square cones contain pursuer-pursuer, avoider-avoider, and lateralized-lateralized pairs of vehicles. The eight triangular cones contain mixed cases with lateralized-pursuer and lateralized-avoider vehicles. The two remaining orthants of W_{total} are outside of $W_{same} + W_{rev}$ and contain pursuer-avoider pairs.

Much of this book focuses on what happens in the subspaces W_{same} and W_{rev}, as shown in figures 1.2 and 2.8. Their union looks like a paddle wheel in which W_{eq} is the axis of the wheel, and the planes W_{same} and W_{rev} form four orthogonal paddles. In the remainder of this section, we briefly describe what happens outside of the "paddle wheel." We separately consider what happens (1) outside the union of W_{same} and W_{rev} but inside their direct sum $W_{same} + W_{rev}$ (that part of figure 1.2 that lies outside the "paddles"), and (2) in the parts of W_{total} that lie altogether outside of the direct sum $W_{same} + W_{rev}$, that is, parts of the total weight space that are outside of what is shown in figure 2.8.

We start with (1), the portion of the direct sum $W_{same} + W_{rev}$ that is outside of W_{same} and W_{rev}. For attracting sets in this region, the two vehicles travel around a common center. To get revolving behavior we need to adjust the speeds of the two vehicles, v_n, to different values. Otherwise we only observe meandering behavior as in W_{same}, except the slower vehicle travels along a smaller path than the other (see figure 2.9).

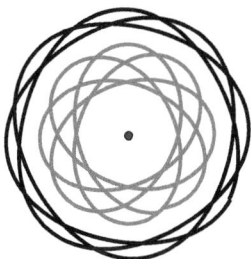

Figure 2.9
The physical paths for two agents whose weights are in $W_{same} + W_{rev}$ but outside of the union of W_{same} and W_{rev}. The path for the slower vehicle is gray; the path for the faster vehicle is black. The common center of their paths is marked by ●.

As the weights are varied along a straight path from W_{same} to W_{rev}, the distances of the vehicles from their common center increases and grows without bound as we get close to W_{rev}. The paths produced by weights in W_{rev} can be thought of as the limiting case in which their distances to the center have grown infinitely large so that they tend to move alongside each other in parallel paths.[3] It is only in W_{rev} that we have translating type behaviors; everywhere else in $W_{\text{same}} + W_{\text{rev}}$ the agents will turn. In this sense the behaviors for W_{rev} are exceptional.

We conclude this chapter by briefly considering (2), what happens outside of the subspace $W_{\text{same}} + W_{\text{rev}}$. Since $W_{\text{same}} + W_{\text{rev}}$ is three-dimensional and W_{total} is four-dimensional the orthogonal complement to $W_{\text{same}} + W_{\text{rev}}$ (i.e., the subspace orthogonal to $W_{\text{same}} + W_{\text{rev}}$) is one-dimensional. We denote this subspace by:

$$W_{\text{opp}} = \text{span}\,\{\,(1, 1, -1, -1)\,\} = \{\,(w, w, -w, -w) \mid w \in \mathbf{R}\,\}\,.$$

For example, W_{opp} includes $(0.5, 0.5, -0.5, -0.5)$, which corresponds to a pursuer-avoider pair. Each agent has the same weight on both of its sensors, but the weights of the two agents are the negative of each other. We can think of these agents as being perfect opposites: under the same circumstances one agent turns to pursue in one direction just as fast as the other agent turns to avoid in the other direction.

We can envision the subspace W_{opp} as pointing in a fourth dimension orthogonal to the three-dimensional subspace $W_{\text{same}} + W_{\text{rev}}$ shown in figure 2.8. The subspace W_{opp} passes through the two orthants that contain all possible pursuer-avoider and avoider-pursuer pairs (see figure 2.10). In general, for these two orthants, the weights of the agents can all differ from one another. On the other hand, for weights in W_{opp} the absolute value is the same for all four weights. For example, $(1.5, 0.5, -0.5, -2.5)$ is in the pursuer-avoider orthant but not in the subspace W_{opp}.

The subspaces W_{opp} and W_{eq} are one-dimensional, and their intersection is the zero-dimensional subspace $\{(0, 0, 0, 0)\}$. So the direct sum $W_{\text{opp}} + W_{\text{eq}}$ is a two-dimensional cross section of W_{total}. The cross sections of the pursuer-pursuer and avoider-avoider orthants of W_{total} are two quadrants of $W_{\text{opp}} + W_{\text{eq}}$. They are shaded dark gray in figure 2.10. The subspace $W_{\text{same}} + W_{\text{rev}}$ also passes through these orthants that are colored light blue in the right panel of figure 2.8.

The cross sections of the "mixed" pursuer-avoider and avoider-pursuer orthants of W_{total} are the other two quadrants of $W_{\text{opp}} + W_{\text{eq}}$. They are shaded

---

3. A line is like a circle with an infinite radius.

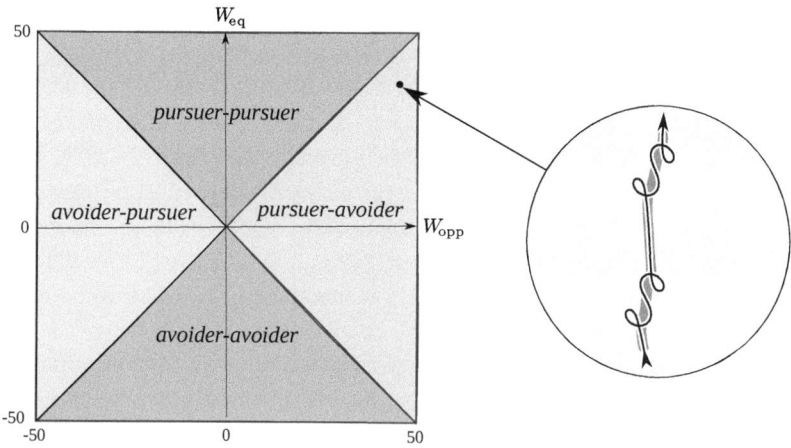

Figure 2.10
The subspace $W_{opp} + W_{eq}$. The circular panel on the right shows the physical paths for a pursuer-avoider pair of agents whose weights in the subspace are marked by •. The paths begin at the bottom of the panel and go upwards. The thin black curve is the path of the pursuer, and the thick gray curve is the path of the avoider. The physical parameters are $v_1 = 9/250$, $v_2 = 5/250$, $\psi = \pi/4$, $L = 1$, and $P = 12$. The weights are $\left(w_{(\ell,1)}, w_{(r,1)}, w_{(\ell,2)}, w_{(r,2)}\right) \approx 45.056 \cdot (1, 1, -1, -1) + 36.864 \cdot (1, 1, 1, 1))$.

light gray in figure 2.10. The pursuer-avoider and avoider-pursuer orthants are the two orthants of W_{total} whose interiors are outside of the subspace $W_{same} + W_{rev}$.

We can also think of $W_{opp} + W_{eq}$ as the subspace of W_{total} in which the agents' weights do not break the bilateral symmetry of their physical bodies. The two sensor weights are the same for each individual agent, but not necessarily the same for both agents. For example, the weights for agent 1 can both be 1.5 and the weights for agent 2 can both be –0.5. The resulting quadruple of weights:

$$(1.5, \ 1.5, \ -0.5, \ -0.5) = (1, \ 1, \ -1, \ -1) + \left(\frac{1}{2}\right) (1, \ 1, \ 1, \ 1)$$

is in $W_{opp} + W_{eq}$. For weights in $W_{opp} + W_{eq}$ the physical bodies and neural networks of the two agents are both bilaterally symmetric , but the agents can still have different neural networks from each other.

Pursuer-avoider and avoider-pursuer agent pairs behave like a pair of predator and prey animals, or like a pair of animals from the same species competing for territory. The pursuer pursues the avoider while the avoider avoids the pursuer. In our implementation of Braitenberg vehicles, the pair can participate in the chase indefinitely. An example is shown in the inset on the right of figure 2.10. In this example, the pursuer moves a little faster than the avoider. As a result, the pursuer repeatedly overshoots the avoider and has to readjust to continue its pursuit. The curvature of both agents varies in a periodic manner, with the average curvature over a single period being close to 0. In the inset of figure 2.10, the pursuer first loops anticlockwise and then clockwise. The pursuer loops like this repeatedly, and on average the curvature of the loops cancel each other. The overall trend of the pursuer is upward with the avoider. On the other hand, the avoider's path veers only slightly to the left and right as it avoids the pursuer. Again the average curvature is close to 0. This behavior has only been observed for pairs of Braitenberg vehicles whose weights are outside the subspace $W_{\text{same}} + W_{\text{rev}}$.

Animals can engage in this kind of behavior although in a less precisely periodic manner. For example, in their seemingly erratic flights, bats often overshoot the small insects they prey on (Ghose et al. 2006). Dragonflies can repeatedly overshoot the intruders they chase from their territories (Lohmann, Corcoran, and Hedrick 2019). In territorial confrontations like this, the goal of the pursuer is not to catch the avoider, but only to drive it from the pursuer's territory. So it is useful to the species for the pursuer to repeatedly overshoot the avoider until it leaves the territory. This is discussed further in section 8.5.

3 Overview of Dynamical Systems Theory

We analyze pairs of Braitenberg vehicles as open dynamical systems, which are defined on the basis of classical or "closed" dynamical systems. We therefore begin with a review of dynamical systems theory and bifurcation theory. This review is selective: we mainly focus on features of dynamical systems theory that we draw on in our study of Braitenberg vehicles. Thus, in some cases, we pass over techniques that are common in introductory treatments; in other cases, we emphasize techniques that are uncommon in introductory treatments. We call attention to these techniques to show how our methods fit in the field of dynamical systems theory. Examples of good introductory texts on dynamical systems theory are Abraham, Abraham, and Shaw (1990); Abraham and Shaw (1992); Izhikevich (2007); and Strogatz (2018).[1]

3.1 The Main Concepts

Classical dynamical systems are abstract mathematical structures that are used to model how things change. They are self-contained deterministic systems: put the system in an initial state and all future states are determined. A dynamical system consists of two sets and a rule:

$$
\begin{array}{lll}
\text{A set of a states:} & S & \\
\text{A set of times:} & T & \text{(3.1)} \\
\text{A function:} & \phi: S \times T \rightarrow S. &
\end{array}
$$

The function ϕ is the rule guiding the temporal evolution of the system. For any given initial state or *initial condition* $s_0 \in S$, and time $t \in T$, the rule ϕ specifies which state $s = \phi(s_0, t)$ will occur at time t. Sometimes ϕ is simply referred to as "*the* dynamical system."

---

1. Other good books on dynamical system theory include Robinson (1995); Perko (2001); Hasselblatt and Katok (2002); Meiss (2007); Hirsch, Smale, and Devaney (2012). Introductory texts on dynamical systems with symmetry are Golubitsky, Schaefer, and Stewart (1988); Golubitsky and Stewart (2003); Field (2020).

Not every function of the form (3.1) is a dynamical system: two conditions must be satisfied. First the function ϕ must take any state to itself in the present moment, that is, at $t = 0$. Second if it takes a state s_0 to s_1 at time t_1, and takes s_1 to s_2 at t_2, then it must take s_0 to s_2 at $t_1 + t_2$.[2]

There are many choices available for the set of states S and the set of times T. The properties of S and T tell us what category of dynamical systems we are working in. We restrict ourselves to cases where the set of times is a subset of the real numbers. The two conditions that the function ϕ must satisfy restricts the possible subsets of $\mathbf{R}$ that T can be. As we have just seen, T must contain 0, which stands for the present moment. Because of the second dynamical system condition we need the sum of any two members of T to be in T. We include the requirement that T must contain positive numbers, which stand for future moments in time. A sequence of moments in time formed by repeatedly adding a positive number increases without bound. So there is no upper bound for the time set T, no limit to how far time goes into the future.

A dynamical system can always make predictions about the future: given a current state s_0 and future time $t > 0$ it tells us what future state $\phi(s_0, t)$ will occur. In fact, so long as we have exact knowledge of the initial condition, there is no limit to how far in the future a dynamical system can make predictions.[3]

The set of times T can also contain negative numbers, which stand for past moments, but T is not required to contain negative numbers. When T does contain negative numbers, the dynamical system is said to be *invertible*, otherwise it is *noninvertible*. For an invertible dynamical system, we can not only predict its future states but can also "retrodict" past states. That is, given a current state s_0 and past time $t < 0$, the system tells us what past state $\phi(s_0, t)$ occurred that led up to the present state.

There are four choices that are usually made for T. They are $\mathbf{R}$, $\mathbf{R}_{\geq 0}$, $\mathbf{Z}$, and $\mathbf{Z}_{\geq 0}$. If $T = \mathbf{R}$, then the dynamical system is called a *flow*. Flows are often obtained as the solutions to a system of ordinary or partial differential equations. For instance, suppose $s(t)$ is a real number that can vary with time t and the rate at which the value of s varies with time is 1. This can be expressed with the differential equation:

$$\dot{s} = 1.$$

---

2. More formally stated: (1) $\forall s_0 \in S$, $\phi(s_0, 0) = s_0$. (2) $\forall s_0 \in S$ and $\forall t_1, t_2 \in T$, $\phi(\phi(s_0, t_1), t_2) = \phi(s_0, t_1 + t_2)$.

3. There are variants of dynamical systems theory that can be applied when the future is uncertain, for example, iterated function systems (Barnsley and Demko 1985), iterated random functions (Diaconis and Freedman 1999), and random dynamical systems (Arnold, Chueshov, and Ochs 2005; Hand 2008).

This has the general solution:

$$s(t) = s_0 + t.$$

This gives us a dynamical system.[4] In this case, the set of states and the set of times are both the set of real numbers. The real number s_0 is the initial condition. The function ϕ: $\mathbf{R} \times \mathbf{R} \to \mathbf{R}$ is:

$$\phi(s_0, t) = s_0 + t. \tag{3.2}$$

States increase continuously with time. This is a flow within the set of real numbers.

If $T = \mathbf{R}_{\geq 0}$ then the dynamical system is called a *semi-flow*. With a semi-flow "orbits" can branch going backwards in time with the different branches merging together in forward time. There have been few applications of semi-flows. Flows and semi-flows are collectively known as *continuous-time* dynamical systems or more commonly just as *continuous* dynamical systems .

If $T = \mathbf{Z}$ or $T = \mathbf{Z}_{\geq 0}$, then the dynamical system is said to be a *discrete time* dynamical system or more commonly a *discrete* dynamical system. Discrete dynamical systems typically arise through the iteration of a function from the set of states to itself. In this case, t corresponds to the number of times the function has been iterated. For instance, suppose $S = \mathbf{R}$. The function f: $\mathbf{R} \to \mathbf{R}$ defined as [5]

$$s_{t+1} = f(s_t) = s_t + 1$$

can be used to define a dynamical system ϕ: $\mathbf{R} \times \mathbf{Z} \to \mathbf{R}$ as:

$$\phi(s_0, t) = f^{[t]}(s_0) = s_0 + t. \tag{3.3}$$

where the $[t]$ in the superscript of f means compose the function f with itself t times. The function f just adds 1 to the current state. Composing f with itself t times just adds t to the current state. In this case, the discrete dynamical system is invertible. For instance, if $s_0 = 2$ then the state at time $t = -10$ must have been $\phi(2, -10) = 2 + (-10) = -8$. This is the only number we can add 1 to ten times to get 2.

Notice that the function in equation (3.3) is the restriction of the function in equation (3.3) to the subset $\mathbf{R} \times \mathbf{Z}$ of $\mathbf{R} \times \mathbf{R}$. We say that the dynamical system defined by equation (3.3) is a *discretization* of the dynamical system defined

---

4. The general solution to a differential equation corresponds to a dynamical system assuming certain conditions are met. Once initial conditions are specified, a particular solution can be obtained from a general solution. These particular solutions are associated with orbits of the dynamical system.

5. This function f is not related to the scaling function f in equation (2.6). The scaling function in equation (2.6) will be used in subsequent chapters.

by equation (3.2). Instead of flowing through the real numbers, we are jumping through them at discrete moments in time.

There are many ways to relate continuous and discrete dynamical systems. One way is to use a numerical integration technique to approximate the solutions to a differential equation. A differential equation can provide us with a continuous dynamical system in an abstract sense, but it can often be difficult to find an exact analytic solution to the differential equation. A numerical integration technique is a discrete dynamical system that approximates a continuous dynamical system. Each iteration of the numerical integrator corresponds to a jump in the set of real numbers.

Ideally a numerical integration technique would give us a discretization of a continuous dynamical system, but even this is often not feasible. Instead, numerical integration techniques usually have the property known as "convergence." In this context, convergence means that as the size of the jumps in the set of times $\mathbf{R}$ goes to 0, the sequence of points in the set of states S produced by the numerical integrator converge to the solution of the differential equation. Such numerical integration techniques can provide accurate approximations to the solutions of differential equations by making the jumps in $T = \mathbf{R}$ sufficiently small. We have made extensive use of convergent numerical integration techniques in our analysis of Braitenberg vehicles.

Whether a dynamical system is continuous or discrete, the set of times T is a topological space, so we call T the *time space* of the dynamical system. We also require the set of states S to be a topological space, so we call it the *state space* of the dynamical system. It is often also called the *phase space*. The structure of the state space tells us about the set of possibilities for a system and how these possibilities are related to each other. The state space is often a vector space like $\mathbf{R}^n$, but it can be other types of continua such as a sphere or cone. The state space can also be a discrete set like the integers or a finite set like the set of states for a computer. It can be a Cartesian product of several distinct spaces, and it can even be infinite-dimensional. For example, when a dynamical system is the solution to a partial differential equation, states are themselves functions, such as curves or surfaces, and the state space is then a set of possible curves or surfaces. Wave equations, for example, can describe the temporal evolution of sound waves and electromagnetic waves.

We also require ϕ to be a continuous function. The dynamical system in this case is called a *topological* dynamical system.[6] We define open dynamical systems in terms of topological dynamical systems. Often the function ϕ is differentiable, in which case we say the dynamical system is *smooth* or *differentiable*. A smooth dynamical system is a type of topological dynamical system. Smooth dynamical systems usually arise as solutions to differential equations. That is the case for our implementation of Braitenberg vehicles.

The state space for a pair of Braitenberg vehicles is a Cartesian product of lines and circles. The Cartesian product $\mathbf{R}^2$ corresponds to the position of a vehicle in a Euclidean plane. The circle $\mathbf{S}^1$ corresponds to the heading of a vehicle in the plane. The Cartesian product $\mathbf{R}^2 \times \mathbf{S}^1$ gives the position and heading of a vehicle in the plane. The Cartesian product $\left(\mathbf{R}^2 \times \mathbf{S}^1\right)^2$ gives the position and heading for two vehicles in the plane. We also say that $\left(\mathbf{R}^2 \times \mathbf{S}^1\right)^2$ is the configuration space for a pair of Braitenberg vehicles in the plane. The time space is $\mathbf{R}$. Our implementation of Braitenberg vehicles is a smooth, continuous-time, invertible dynamical system on the configuration space for two Braitenberg vehicles in a Euclidean plane. It is the solution to differential equation (4.14).

Two important features of a dynamical system are its "orbits" and "invariant sets." Roughly speaking, the orbit of a state $s_0 \in S$ is the set of states that can be reached from s_0 by going forwards and backwards in time.[7] For an invertible dynamical system, the *orbit* of s_0 can be simply defined as the subset of S:

$$\{ \phi(s_0, t) \mid t \in T \}.$$

As explained in (Hotton and Yoshimi 2010), the collection of all orbits for a dynamical system forms a partition of the state space.

For a flow, there are just three types of orbits. An orbit can be a single state, it can have the topology of a circle, or it can have the topology of a line.[8] When an orbit is a single state, it is called a *fixed point*. Fixed points are often also called *equilibria*. When an orbit has the topology of a circle, it is called a *periodic orbit* since each state repeats after the same amount of time.

6. Continuous dynamical systems and discrete dynamical systems are both types of topological dynamical systems. Another choice that is often made is for S to be a measurable space and for ϕ to be a measurable function. This is called a measurable dynamical system.

7. For a noninvertible dynamical system, some or all of the orbits "fork" going backwards in time. So an orbit is defined in general as the union of a forward orbit and a backward orbit. The forward orbit of s_0 is the set of states that occur when we run the dynamical system forward in time. The backward orbit of s_0 is the set all of states that enter its forward orbit after some positive amount of time.

8. More precisely stated, an orbit of a flow is the continuous injection of a point, circle, or line. The continuous injection need not be a topological embedding.

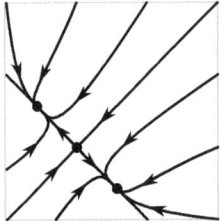 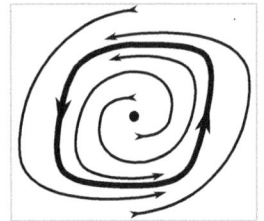 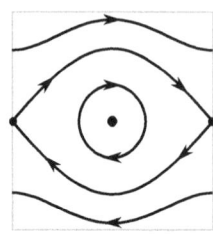

Figure 3.1
Three phase portraits of flows. Fixed points are marked with black dots. (left) Phase portrait for a two-node Hopfield network (we will refer to this as the "Hopfield system"). There are three fixed points: a saddle node and two attracting fixed points. All of the other orbits have the topology of a line. Two orbits converge to the saddle node. The remaining orbits converge to one of the attracting fixed points. There are two heteroclinic orbits connecting the saddle node to the attracting fixed points. (middle) Phase portrait with an attracting periodic orbit (thick curve) surrounding a repelling fixed point. All of the other orbits have the topology of a line and converge to the attracting periodic orbit. (right) Phase portrait for a frictionless pendulum. The state space is a cylinder; the left and right sides are to be glued together. There are two fixed points, one of which is a saddle node with two homoclinic orbits. The other fixed point is nonhyperbolic. The rest of the orbits are periodic. There are no attracting sets for this system.

An *invariant set* is a nonempty subset of S with the property that any state in that subset remains in it for all time, under the dynamics. That is, $S' \subseteq S$ is an invariant set if for all $s \in S'$ and all $t \in T$, we have $\phi(s, t) \in S'$ (this is sometimes called a "strictly invariant set"). A single orbit is an invariant set, as is the state space as a whole. We also explained in (Hotton and Yoshimi 2010) that an invariant set is partitioned by the orbits it contains. As we will see, many important structures in a dynamical system are invariant sets.[9]

Pictures aren't logically essential to dynamical systems, but they are a distinctive feature of the field. We can not draw every single orbit in a state space since that would cover the whole picture, so we draw a representative collection of orbits, or portions of orbits, often with arrows that indicate the future direction of the orbit. This is called a *phase portrait* for a dynamical system. A phase portrait can communicate the prominent behaviors of a system in a way that is not immediately evident just from the equations that define it. Examples of phase portraits are shown in figure 3.1.

9. An invariant set obtained by joining together the members of some arbitrary collection of orbits is generally not very helpful. The invariant sets that are chosen for the analysis of dynamical systems are often closed connected subsets. For chaotic dynamical systems, the attracting set can often be a fractal.

We call a fixed point *attracting* if all of the states in some neighborhood of the fixed point converge to it as time increases without bound. There are two attracting fixed points in the Hopfield system shown in the left panel of figure 3.1. For an invertible dynamical system, a fixed point is *repelling* if all of the states in some neighborhood of the fixed point converge to it as time decreases without bound. The middle panel of figure 3.1 shows a repelling fixed point. A fixed point of an invertible dynamical system is a *saddle node* if in every neighborhood of the fixed point there are states that converge to it as time increases without bound and states that converge to it as time decreases without bound. That is, states move toward the fixed point in some directions, and away from it in others (which is what would happen if we were to place marbles at different locations around the center of a horse's saddle). The left and right panels of figure 3.1 each have one saddle node.

If the states in an orbit converge to the same fixed point as time increases without bound and as time decreases without bound, then it is a *homoclinic* orbit of the fixed point. Roughly speaking, it is an orbit that starts and stops at the same fixed point. The right panel of figure 3.1 shows a fixed point with two homoclinic orbits. If the states in an orbit converge to one fixed point as time increases without bound and converge to a different fixed point as time decreases without bound, then it is a *heteroclinic* orbit of the two fixed points. Roughly speaking, it is an orbit that starts at one fixed point and stops at another. The Hopfield system (figure 3.1, left) shows two *heteroclinic* orbits.

Fixed points are the simplest type of invariant set that a dynamical system can have. The concept of an invariant set allows us to generalize from attracting fixed points, repelling fixed points, and saddle nodes to more complicated variations of these features of a dynamical system. Our study of Braitenberg vehicles involves invariant sets that are more complicated than fixed points.

We defined "attracting sets" and "repelling sets" in (Hotton and Yoshimi 2010, 2011). Roughly speaking, an attracting set is an invariant set with the property that all states sufficiently close to it approach it. Attracting sets correspond to the commonly observed behaviors of a system, and thus they are naturally of interest. They can often be found numerically by running the dynamics from different initial conditions. A repelling set is an invariant set with the property that all states sufficiently close to it move away from it. Repelling sets can be difficult to observe in practice although they can be found in an invertible system by running the dynamics backwards.

The cases where the attracting set is a single orbit are the most prominent and familiar, for example, a fixed point or a periodic orbit. An attracting periodic orbit for a continuous-time dynamical system is often called an attracting *limit cycle*. There are other types of attracting sets such as an attracting torus.

An attractor is an attracting set that contains a dense orbit (i.e., the closure of the orbit is the attracting set). Attracting fixed points and attracting limit cycles are attractors since they consist of a single orbit. Some attracting tori are attractors, and some are not. Chaotic attractors are another well-known type of attractor.

The Braitenberg vehicles studied here do not appear to have any attractors. The attracting sets we observe consist of many orbits, none of which are dense in those attracting sets. These attracting sets are relative equilibria (which are analogous to fixed points) and relative periodic orbits (which are analogous to periodic orbits). These are parallel flows, collections of orbits that correspond to the vehicles revolving around each other or translating together at different locations in the physical plane. None of these orbits is an attractor, nor are any of them dense in an attracting set, but the whole set of orbits is attracting. It is similar for relative periodic orbits, but their topology is different. Much of the book is devoted to the study of relative equilibria, relative periodic orbits, and their bifurcations. The theory of these types of invariant sets is developed in chapter 5 and applied to the study of Braitenberg vehicles in chapters 6 and 7.

3.2 Linearization

Linearization is a useful tool for studying the behavior of dynamical systems, which we use throughout the book. It can be used to understand what is going on in the neighborhood of a fixed point. The technique can also be used to identify bifurcations. On its own it is not a complete method of analysis. It does not guarantee that we will find *all* the important invariant sets of a dynamical system. But it is a well-understood method that can provide us with a qualitative understanding of some of the main behaviors of a dynamical system. It is sometimes useful to find ways to apply the technique even when we are not dealing with fixed points. For example, by suitably changing the frame of reference, we can convert a periodic orbit into a circle of fixed points each of which we can linearize at. A simple example of this method is presented in section 3.6. We use similar methods in chapters 6 and 7.

Linearization is often used to study what is going on in a neighborhood of a fixed point. There are several types of exceptional orbits associated with fixed points and linearization can often help find them. This in turn can help characterize what happens around a fixed point.

Linearizing around fixed points is a stepwise procedure. For the simple examples in this section, we focus on the case where the dynamical system is the solution to a differential equation. A general form for a differential equation on $\mathbf{R}^2$ is

$$\begin{pmatrix} \dot{x} \\ \dot{y} \end{pmatrix} = F \begin{pmatrix} x \\ y \end{pmatrix} = \begin{pmatrix} F_1\,(x,y)^T \\ F_2\,(x,y)^T \end{pmatrix}.$$

To apply the method of linearization:

1. Locate the fixed points of the dynamical system. Fixed points do not move (their time derivative is 0), so to find them in a flow that is the solution of a differential equation, we set the time derivatives of the state variables to 0 and solve the resulting algebraic equation. In two dimensions this looks like:

$$\begin{pmatrix} 0 \\ 0 \end{pmatrix} = \begin{pmatrix} F_1\,(x,y)^T \\ F_2\,(x,y)^T \end{pmatrix}.$$

We let $(x_\odot, y_\odot)^T$ denote a solution to this algebraic equation (the subscript $\odot$ is suggestive of a fixed point in a phase portrait).

Our analysis of the invariant sets for the Braitenberg vehicles will also begin by solving equations in order to find invariant sets, that is, equations (6.1) and (7.2). Since the Braitenberg vehicles are assumed here to move with a constant speed, the dynamical systems for them do not have fixed points but they do have more complicated invariant sets (e.g., relative equilibria), which can still be analyzed in broadly the way we describe in this section.

2. Linearize the differential equation at each of its fixed points. When we linearize a differential equation at a fixed point, we get a new differential equation that approximates the original differential equation in a neighborhood of that fixed point. To linearize a differential equation, we take the total derivative of the function of the state variables in the differential equation, for instance,

$$DF = \begin{pmatrix} \dfrac{\partial F_1}{\partial x} & \dfrac{\partial F_1}{\partial y} \\[2mm] \dfrac{\partial F_2}{\partial x} & \dfrac{\partial F_2}{\partial y} \end{pmatrix}.$$

We evaluate the total derivative at each fixed point $(x_\odot, y_\odot)^T$. This gives us a constant matrix sometimes called the "Jacobian":

$$\mathbf{J} = \begin{pmatrix} \dfrac{\partial F_1}{\partial x}\,(x_\odot, y_\odot)^T & \dfrac{\partial F_1}{\partial y}\,(x_\odot, y_\odot)^T \\[2mm] \dfrac{\partial F_1}{\partial x}\,(x_\odot, y_\odot)^T & \dfrac{\partial F_1}{\partial y}\,(x_\odot, y_\odot)^T \end{pmatrix}.$$

The linear differential equation is defined with this matrix. The state space of the linearized system is a vector space. We write their coordinates with primed variables, for example, $(x', y')^T$. We set the time derivatives of $(x', y')^T$ equal to the product of the constant matrix with $(x', y')^T$.

$$\begin{pmatrix} \dot{x}' \\ \dot{y}' \end{pmatrix} = \mathbf{J} \begin{pmatrix} x' \\ y' \end{pmatrix}.$$

Worked examples of this are in sections 3.5 and 3.6.

3. Find the eigenvalues and eigenvectors for each of the linearized systems. The eigenvalues and eigenvectors of these constant matrices **J** give us important information about a dynamical system. There are two important theorems that tell us how to use this information: the *stable manifold* theorem and the *Hartman-Grobman* theorem. We can use these theorems to reason about the original dynamical system on the basis of the linearized system.

To explain the stable manifold and Hartman-Grobman theorems, we need to introduce some more terminology. An important condition that must hold for these theorems is that the fixed point be "hyperbolic." A fixed point is *hyperbolic* if none of the eigenvalues of its linearization are imaginary, otherwise it is *nonhyperbolic*. The term "hyperbolic" is not intended to suggest "extreme." In fact, it means just the opposite: hyperbolic fixed points are the more typical, well-behaved fixed points. It is the nonhyperbolic fixed points that are unusual.

When a fixed point is hyperbolic, we can gain a qualitative understanding of the system by ignoring the contribution of higher-order terms. The linearization provides us with a good characterization of what happens around a hyperbolic fixed point. We can use the eigenvalues of its linearization to determine whether it is attracting, repelling, or a saddle node[10]:

- If the real part for all of its eigenvalues is negative, then the fixed point is attracting.
- If the real part for all of its eigenvalues is positive, then the fixed point is repelling.
- If the real part is negative for some of its eigenvalues and positive for its other eigenvalues, then the fixed point is a saddle node.

Usually the linearization of a nonhyperbolic fixed point is not as revealing as the linearization of a hyperbolic fixed point. However, in some cases they can be an aid to further analysis. Section 3.6 provides an example of this.

Figure 3.2 illustrates the general idea of using a linearization to understand a dynamical system (the differential equation is from Sfondrini [2013], who analyzes it similarly to the way we do). The left panel shows a phase portrait for a dynamical system that we call the "Dalí flow," based on its resemblance to Salvador Dalí's mustache. To understand the system, we linearize it at the

---

10. A further subclassification is possible using the eigenvalues—for example, we can distinguish attracting nodes from attracting foci based on the eigenvalues—but we do not present this subclassification here.

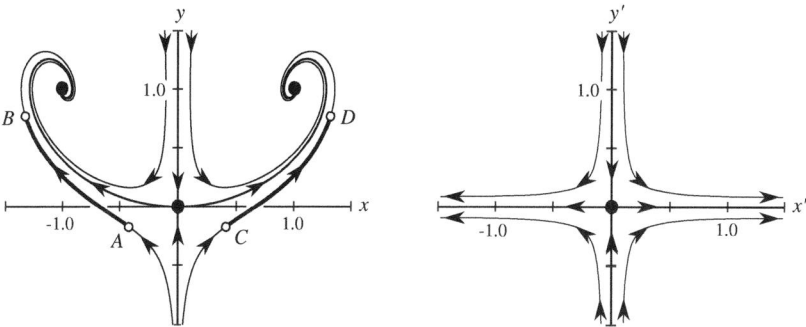

Figure 3.2
(left) Phase portrait for the "Dalí flow." Its differential equation is $(\dot{x}, \dot{y})^T = (x - xy, x^2 - y)^T$. As explained in the main text, linearizing at the three fixed points (marked by black dots) gives a fairly complete qualitative understanding of the system. The fixed point at the origin is a saddle node, and the other two fixed points at $(-1, 1)^T$ and $(1, 1)^T$ (the tips of the mustache) are attracting. This dynamical system is equivariant with respect to the reflection about the y-axis, which we explain (in the main text) in terms of the points A, B, C, and D. (right) Phase portrait for the linearization of the Dalí flow at the origin. Its differential equation is $(\dot{x}', \dot{y}')^T = (x', -y')^T$. All of its orbits outside of the coordinate axes are hyperbolas whose asymptotes are the two axes. It can be seen that the two dynamical systems are qualitatively the same in a neighborhood of the origin even though the linearization straightens the orbits.

origin (this linearization is shown in the right panel). We also linearize it at the two fixed points located at the "tips of the mustache" (these two linearizations are not shown in the figure). When we do this, we gain a fairly complete qualitative understanding of the system. The fixed point at the origin has one positive and one negative eigenvalue, so it is a saddle node. States approach from above and below, but otherwise they move away from it. The other two fixed points have complex conjugate eigenvalues with negative real part, so they are attracting. Two heteroclinic orbits come out of the origin to connect the saddle node with each of the attracting fixed points. Thus, in this system, states near the y-axis approach the origin from above or below and then shoot away toward one of the two attracting fixed points.

We can use the stable manifold and Hartman-Grobman theorems to analyze this system more thoroughly. The stable manifold theorem provides a way to associate stable and unstable eigenspaces in the linearization with stable and unstable manifolds in the original system. We must first define these concepts. An *eigenspace* for an eigenvalue is the space spanned by its eigenvectors.[11] The

11. Or its "generalized eigenvectors" when an eigenvalue has multiplicity greater than 1.

stable eigenspace corresponds to the eigenvalues with negative real part, the *unstable eigenspace* corresponds to the eigenvalues with positive real part, and the *center eigenspace* corresponds to eigenvalues with zero real part, imaginary eigenvalues. In the linearization of the Dalí flow at its saddle node (right panel of figure 3.2), the stable eigenspace is the y'-axis and the unstable eigenspace is the x'-axis. In the linearizations of the Dalí flow at its attracting fixed points (not shown in figure 3.2), the stable eigenspace is the whole plane, and the unstable eigenspace is just a point.

We can associate the stable and unstable eigenspaces of the linearization with the stable and unstable manifolds in the original system. A "manifold" is a topological space that is everywhere like a Euclidean space. Every point in a manifold has an open neighborhood with the same topology as an open ball in Euclidean space.[12] An invariant set of a dynamical system which is a manifold is called an *invariant manifold* The *stable manifold* of a fixed point is an invariant manifold which contains the fixed point, and which is such that all of the other states in it converge to the fixed point as time increases without bound. The *unstable manifold* of a fixed point is an invariant manifold which is such that all of the other states in it converge to the fixed point as time decreases without bound.

The stable manifold theorem asserts that every hyperbolic fixed point has a unique stable manifold and a unique unstable manifold. For the Dalí flow, the theorem implies (and figure 3.2 shows) that at the origin, there is a unique stable manifold (the y-axis), and a unique unstable manifold. The unstable manifold is the union of the saddle node with the heteroclinic orbits (roughly speaking, the mustache). At the mustache tips, the theorem implies (and again, the figure shows) that there is a unique stable manifold for each tip: the half-plane to the left of the y-axis for the left tip, and the half plane to the right of the y-axis for the right tip. The unstable manifolds for the attracting fixed points in each case is trivial, that is, they are the attracting fixed points. Although there are no repelling fixed points in the Dalí flow, it is the case that the stable manifold for a repelling fixed point is just the fixed point itself.

The stable manifold theorem further asserts that the stable and unstable eigenspaces are each tangent to the corresponding stable and unstable manifolds at the point where the dynamical system has been linearized. To visualize this, imagine superimposing the right panel of figure 3.2 on the left panel. The

---

12. Every Euclidean space is a type of manifold. Even a single point is a manifold since a point is a zero-dimensional Euclidean space. Not all manifolds are Euclidean spaces. A one-dimensional manifold either has the topology of a line or the topology of a circle. A two-dimensional manifold is a type of surface such as a plane, a sphere, or a torus. There are many types of manifolds in higher dimension.

x'-axis in the linearization is tangent to the unstable manifold of the saddle node of the original system, and the y'-axis in the linearization is tangent to the stable manifold of the saddle node. So we learn from the linearized system what is happening in these two directions from the saddle node. The linearized system tells us that states in the nearly horizontal direction from the saddle node converge to it as time goes to negative infinity. It also tells us that states in the nearly vertical direction from the saddle node converge to it as time goes to positive infinity. In fact, for the Dalí flow, all states in the exact vertical direction from the saddle node converge to it as time goes to positive infinity, but linearization can only tell us what is going on near the fixed point. In general, the stable manifold of a fixed point can bend away from the stable eigenspace, just as the unstable manifold of the saddle node of the Dalí flow bent away from the unstable eigenspace.

For nonhyperbolic fixed points there is, in addition to the stable and unstable eigenspaces, a center eigenspace. This is the subspace spanned by an eigenbasis corresponding to the imaginary eigenvalues of the nonhyperbolic fixed point. There is also the center manifold theorem which asserts that there is an invariant manifold with the property that the center eigenspace is tangent to it at the nonhyperbolic fixed point. Unlike with the stable and unstable manifold, the center manifold of a nonhyperbolic fixed point need not be unique. Center manifolds don't involve attraction or repulsion internally, but can (for example) be a line or circle of fixed points, or a collection of parallel orbits or concentric periodic orbits. The center manifold theorem is similar to the stable manifold theorem, but it specifies that in addition to the association between stable and unstable eigenspaces and manifolds, there are one or more center manifolds associated to the center eigenspace.

Summarizing these results:

1. Hyperbolic fixed points: Each stable eigenspace in the linearization corresponds to a unique stable manifold in the system being linearized, and the stable eigenspace is tangent to the stable manifold at the fixed point. Each unstable eigenspace of the linearization corresponds to a unique unstable manifold in the original system, and the unstable eigenspace is tangent to the unstable manifold at the fixed point (stable manifold theorem).

2. Nonhyperbolic fixed points: Stable and unstable eigenspaces are associated to tangential stable and unstable manifolds as with the stable manifold theorem, but there is also a nontrivial (positive-dimensional) center eigenspace of the fixed point, which corresponds to one or more center

manifolds in the original system. The center eigenspace in the linearization is tangent to the center manifold or center manifolds at the fixed points (center manifold theorem).

The phase portrait in the right panel of figure 3.1 shows two fixed points, a saddle node and a nonhyperbolic fixed point (in the middle of the panel). The two eigenvalues for the nonhyperbolic fixed point are imaginary. This fixed point's stable and unstable manifolds are just a single point. States near this fixed point do not move closer or further from it. A center manifold for this nonhyperbolic fixed point is the whole two-dimensional state space, which in this case is a cylinder.

The stable manifold theorem can help us identify the stable and unstable manifolds of a fixed point. However, strictly speaking, it does not tell us what is happening near a fixed point outside the stable and unstable manifolds. For that, we turn to the Hartman-Grobman theorem, which asserts that every hyperbolic fixed point has an open neighborhood that can be mapped homeomorphically to the state space of the linear system, and further that the arcs of the orbits near the hyperbolic fixed point get mapped homeomorphically to arcs of the orbits of the linear system. Thus, in a sufficiently zoomed-in neighborhood of the fixed point, the system and its linearization will appear to be the same. This can be observed in figure 3.2, where the arcs near the origin in each panel resemble each other. Though we do not show the linearizations for the attracting fixed points, the situation is the same in those cases: we have orbits spiraling into a fixed point in the linearization, and we would see essentially the same thing if we zoomed in on the mustache tips.

Invariant sets can also have stable, unstable, and center manifolds, just as fixed points do. This is discussed in general in section 5.6 and specifically for the Braitenberg vehicles in chapters 6 and 7.

3.3   Equivalence of Dynamical Systems

The concepts of bifurcation, equivariance, and open dynamical systems are defined in terms of equivalence relations for dynamical systems. There are two main equivalence relations for continuous-time dynamical systems. They are called "topological conjugacy" and "topological equivalence." These equivalence relations basically tell us that two dynamical systems are qualitatively the same even when there are some quantitative differences between them.

Topological conjugacy between dynamical systems is based on the existence of a homeomorphism between their state spaces. This is similar to what the conclusion of the Hartman-Grobman theorem tells us about hyperbolic fixed points, except the homeomorphism of the Hartman-Grobman theorem only applies near a fixed point. Topological equivalence is based on the existence of a homeomorphism between the state spaces as well as an order preserving homeomorphism between the time spaces, that is, a homeomorphism that preserves the order between past and future times (a "reparameterizing" of time). More formally:

- Two dynamical systems are *topologically conjugate* if there is a homeomorphism from the state space of one to the state space of the other such that we get the same result whether we (1) run the first dynamical system from any initial condition for any amount of time and then apply the homeomorphism, or (2) first apply the homeomorphism to the initial condition and then run the second dynamical system for the same amount of time.

- Two dynamical systems are *topologically equivalent* if, in addition to the homeomorphism between their state spaces, there is also a reparameterization of time such that we get the same result whether we (1) run the first dynamical system from any initial condition for any amount of time, then apply the homeomorphism, or (2) first apply the homeomorphism to the initial condition, run the second dynamical system for the same amount of time, and then reparameterize time.

Even though topological equivalence adds a condition to topological conjugacy, it is more broadly applicable since the reparameterization of time allows us to adjust the rate at which states travel through corresponding orbits. Topological conjugacy does not allow this and is thus more restrictive. Topological conjugacy requires that a system end up *in the same state* whether we apply a homeomorphism and run the dynamics or run the dynamics and apply the homeomorphism. On the other hand, if we allow time to be reparameterized, then we can apply a homeomorphism and run the dynamics, or run the dynamics and apply the homeomorphism, and even if we end up in a different state, we could still meet the condition for equivalence as long there is a way to stretch or contract time so that it's in the same state. For example, the Hopfield and Dalí flows are not topologically conjugate, but they are topologically equivalent. Traveling along one of the heteroclinic orbits of the Hopfield system does not proceed at the same rate as traveling along one of the heteroclinic orbits of the Dalí's system even though the two orbits are topologically the same.

Intuitively, these are ways of saying that two dynamical systems are qualitatively the same, with respect to their state spaces, or with respect to both their state spaces and time spaces. The homeomorphism between the state spaces maps the orbits of one dynamical system to the orbits of the other dynamical system. The homeomorphism between the time spaces allows us to adjust how fast states move along the orbits.

The Hopfield system (left panel of figure 3.1) has a phase portrait that is topologically equivalent to that of the Dalí flow, (i.e., left panel of figure 3.2). They only differ in the shape of the orbits. Both systems have a saddle node and two attracting fixed points. They both have heteroclinic orbits connecting the saddle node to the attracting fixed points. In both cases, the unstable manifold of the saddle node is the union of the saddle node with the heteroclinic orbits. In both cases, the stable manifold of the saddle node is a curve passing through the saddle node. In both cases, the stable manifold of each attracting fixed point is the set of points that are on the same side of the stable manifold of the saddle node. The middle panel of figure 3.1 and the top middle panel of figure 3.6 show phase portraits for another pair of topologically equivalent dynamical systems.

An important concept related to topological conjugacy is "equivariance." Any dynamical system is trivially topologically conjugate to itself, using the identity map on the state space as the homeomorphism. Sometimes a dynamical system is topologically conjugate to itself in other nontrivial ways. The Dalí flow shown in the left panel of figure 3.2 provides an example. It can be seen that the phase portrait is symmetrical under a reflection about the y-axis. However the dynamical system is more than merely symmetrical under this reflection; it is also topologically conjugate to itself under this reflection. We can run the dynamical system from any initial condition, such as point A in figure 3.2, for some amount of time to reach point B, and then reflect B about the y-axis to get to point D. Or we can reflect the initial condition A to point C and then run the dynamical system for the same amount of time to get to the same point D. Since this is true for any initial condition in the plane, and any amount of time, the Dalí flow is topologically conjugate to itself under the reflection about the y-axis.

The set of all homeomorphisms that make a dynamical system topologically conjugate to itself forms a transformation group, which we often just call a group. A transformation is a bijection of a set to itself that generally preserves one or more properties of the set. A set of transformations is called a transformation group if the composition of any two members of the set is in the set and the inverse of any member of the set is in the set.

For the Dalí flow the transformation group is the set consisting of the identity map and the reflection about the y-axis. This is the cyclic group of two elements, which is commonly denoted $\mathbf{Z}_2$. We say the dynamical system is "equivariant" with respect to this group. We can run the dynamics and apply any homeomorphism in the group, or apply the same homeomorphism in the group and run the dynamics, and get the same result either way. Thus, the Dalí flow is equivariant with respect to $\mathbf{Z}_2$. Equivariance plays an important role in the dynamical systems for the Braitenberg vehicles. We discuss equivariance with an example system in section 3.6 and discuss the concept in greater detail in chapter 5.

3.4 Bifurcation Theory

To understand the concept of a bifurcation, we begin with the concept of a parameterized family of dynamical systems, that is, a space of dynamical systems associated with a set of parameters. Parameters are often implicit in the specification of the rule ϕ. That is, the quantities used to define ϕ can often be divided into two classes: *state variables* and *parameters*. The state variables provide a coordinate system for the state space and the parameters provide a coordinate system for the space of dynamical systems. The parameters have fixed values as the dynamical system unfolds. The set of all possible parameter values is a *parameter space*. A particular dynamical system is associated to each point in the parameter space, and we can think of the parameter space as a space of dynamical systems.

Bifurcation theory applies the concept of topological equivalence to parameterized families of dynamical systems. Varying one or more parameters of a parameterized family of dynamical systems typically produces topologically equivalent dynamical systems. As the parameters are varied, the orbits shift and move a little, but nothing drastic changes. When this is not the case—when something more extreme happens like a fixed point appears or disappears or a change occurs where one type of invariant set is replaced by another—we say we are at a *bifurcation* in the parameterized family of dynamical systems.

A fairly simple example of a parameterized family of dynamical systems exhibiting a bifurcation is the general solution to the linear differential equation:

$$\dot{x} = \lambda x. \tag{3.4}$$

The members of this parameterized family of dynamical systems have the form:

$$x(t) = x_0 e^{\lambda t}.$$

The state variable is $x \in \mathbf{R}$ (which can vary with time) and the parameter is $\lambda \in \mathbf{R}$ (which does not vary with time). The set of times is also $\mathbf{R}$. Thus, the parameter space is $\mathbf{R}$, and each value of λ specifies a particular one-dimensional continuous-time dynamical system on $\mathbf{R}$.

Notice that in this case we can see how the abstract formalism introduced at the start of the chapter applies. We can express the solution of the differential equation explicitly in terms of ϕ with a parameter λ:

$$\phi(x_0, t; \lambda) = x_0 e^{\lambda t}.$$

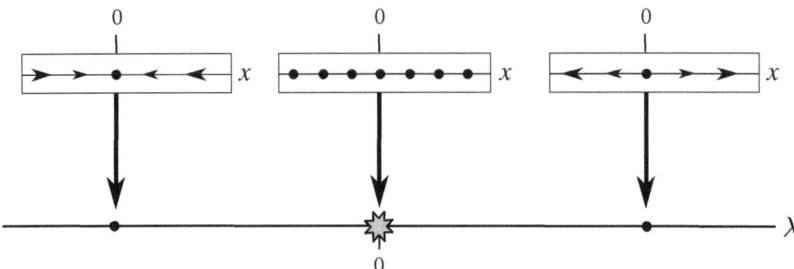

Figure 3.3
The three top panels show one-dimensional phase portraits for differential equation
(3.4). The bottom of the figure shows a one-dimensional bifurcation diagram for the
system. Each panel has an arrow that points to its value for λ in the bifurcation diagram.
(left) When $\lambda < 0$, the origin of the associated phase portrait is an attracting fixed point.
(middle) When $\lambda = 0$, every state of the associated phase portrait is a fixed point. (right)
When $\lambda > 0$, the origin of the associated phase portrait is a repelling fixed point. The
bifurcation occurs at $\lambda = 0$, which is marked by the eight-pointed star.

λ specifies which dynamical system we are working with. To determine what
state the system is in at time t, given initial condition x_0, we simply compute
$x_0 e^{\lambda t}$. Often, however, we don't have solutions to differential equations, and
we cannot express the rule of the dynamical system in this explicit way.

We can now proceed with the bifurcation analysis, considering what the
dynamics are for different values of λ. For $\lambda < 0$, the state variable $x(t)$ undergoes exponential decay toward 0, which is the unique fixed point of the
dynamical system. It is an attracting fixed point, as shown in figure 3.3 (left
panel). For $\lambda = 0$, every point in the state space is fixed. This dynamical system
is called the *identity dynamical system*. For $\lambda > 0$, the state variable $x(t)$ undergoes exponential growth away from 0, which is the unique fixed point of the
system. It is a repelling fixed point.

All of the dynamical systems with $\lambda < 0$ are topologically equivalent to each
other.[13] They have just three orbits: the set of negative numbers, the fixed point,
and the set of positive numbers. The nonzero numbers converge to the fixed
point in positive time. The dynamical systems with $\lambda > 0$ are also topologically
equivalent to each other. Again there are just three orbits: the set of negative
numbers, the fixed point, and the set of positive numbers. The nonzero numbers
go off to infinity in positive time. However the dynamical systems associated
with negative values of λ are not topologically equivalent to the dynamical

---

13. The homeomorphism from the state space to itself is $x \mapsto \text{sign}(x) |x|^{(\lambda_1/\lambda_2)}$ where $\lambda_1, \lambda_2 < 0$
are two choices for the parameter. This also works for $\lambda_1, \lambda_2 > 0$.

systems associated with positive values of λ. We would have to reverse the direction of time to map the two orbits of nonzero numbers for a dynamical system with negative λ to the two orbits of nonzero numbers for a dynamical system with positive λ. The only dynamical system that is topologically equivalent to the identity dynamical system is itself. Thus there are three equivalence classes of dynamical systems in this parameterized family of dynamical systems. They correspond to the parameter λ being negative, zero, or positive.

We say that the parameterized family of dynamical systems undergoes a bifurcation as the value for λ passes through 0. The presence of this bifurcation can be detected just by looking at the fixed point at the origin. For $\lambda < 0$, all states near 0 converge to it. For $\lambda = 0$, all states near 0 neither move closer to 0 nor further from 0. For $\lambda > 0$, all states near 0 diverge from it. This is enough to tell us a bifurcation has occurred. We do not need to know what happens in the state space far from 0. The changes that occur in a small neighborhood of 0 are enough to make the dynamical systems inequivalent.

Since differential equation (3.4) is linear, the parameter λ is in fact the eigenvalue. For all $\lambda \neq 0$, the origin is a hyperbolic fixed point. For $\lambda = 0$, the origin is a nonhyperbolic fixed point. The bifurcation occurs where the fixed point becomes nonhyperbolic. More generally we can detect bifurcations by linearizing at the fixed points of a parameterized family of dynamical systems. One way a bifurcation can occur is when a hyperbolic fixed point becomes nonhyperbolic, that is, when some of the eigenvalues of the linearization become imaginary numbers.

Each eigenvalue for the linearization at a fixed point is a point in the complex plane. As one or more parameters are varied, the eigenvalues move around in the complex plane. We call the set of eigenvalues in the complex plane as a parameter is varied an *eigenvalue trace* . We use eigenvalue traces both in our introductory examples and in our study of Braitenberg vehicles. One way to detect a bifurcation is to check if the eigenvalue trace for a fixed point intersects the imaginary axis. In the case of the family of linear differential equations (3.4), the eigenvalue equals the parameter and a bifurcation occurs as the parameter passes through zero.

Sometimes fixed points come into or out of existence as a parameter is varied. This is another way to detect a bifurcation. This also works for differential equation (3.4). For $\lambda \neq 0$, there is just one fixed point and it is located at the origin. For $\lambda = 0$, all of the states are fixed points. Therefore a bifurcation must occur at $\lambda = 0$.

There are many types of bifurcations, and some of them are common enough that they have their own names (e.g., a pitchfork bifurcation). A standard type of bifurcation, known as a "saddle node" bifurcation, involves a pair of fixed points.[14] In a saddle node bifurcation, a pair of fixed points comes into existence as a parameter is varied. In a Hopf bifurcation an attracting fixed point is replaced by a nonattracting fixed point and an attracting limit cycle.[15] We focus on saddle node bifurcations in section 3.5 and Hopf bifurcations in section 3.6, since analogs of these bifurcations are important in the analysis of the Braitenberg vehicles. We also consider a nonstandard bifurcation in section 3.5. This is an analog of a nonstandard bifurcation that occurs with the Braitenberg vehicles.

While phase portraits are useful for depicting the state space of a dynamical system, "bifurcation diagrams" are useful for depicting a portion of the parameter space for a parameterized family of dynamical systems. Usually some of the equivalence classes in the parameter space are open regions. These regions are sometimes called *dynamical regimes*. Dynamical systems in the same regime have the same types of attracting sets, repelling sets, and so on, just stretched and shifted in different ways. Bifurcations occur along the boundaries separating distinct dynamical regimes. As the parameters cross such a boundary, the dynamical systems change in some way: gaining or losing an attracting set or undergoing a change where one type of invariant set is replaced by another. A *bifurcation diagram* shows where some or all of the dynamical regimes and bifurcations are in the parameter space.

The bottom of figure 3.3 shows a bifurcation diagram for the parameterized family of dynamical systems given by equation (3.4). There is one bifurcation in the diagram, indicated via the eight-pointed star that separates the two dynamical regimes. In the left regime, the system has an attracting fixed point at the origin; in the right regime, the system has a repelling fixed point at the origin. At the bifurcation, the system is a collection of fixed points. This is a nonstandard bifurcation in the intuitive sense that it does not have a common name (as contrasted with standard named bifurcations like saddle node and Hopf bifurcations). We will follow the convention of indicating nonstandard bifurcations by eight-pointed stars.

Another example is shown in figure 3.4 of the next section, which shows a one-dimensional bifurcation diagram for the parameterized family of dynamical systems given by equation (3.5). The bifurcation diagram is associated with phase portraits for the three dynamical regimes between the two bifurcations (a saddle node bifurcation and another nonstandard bifurcation), and also shows

---

14. Note that "saddle node" is used to refer both to a type of fixed point and to a type of bifurcation.

15. Technically this is called a "supercritical" Hopf bifurcation. There is also a "subcritical" Hopf bifurcation, which is less commonly observed.

phase portraits at the bifurcations. Similarly figure 3.6 shows a bifurcation diagram for a system with two dynamical regimes and a single bifurcation (the Hopf bifurcation).

Other types of invariant sets can be involved in bifurcations besides fixed points. For instance, there are analogs of the saddle node bifurcation of fixed points where invariant sets appear in pairs. This can happen whether the invariant sets are fixed points, periodic orbits, or relative equilibria. A saddle node-like bifurcation begins with the appearance of a single invariant set that subsequently splits into two invariant sets with the same topology. In a Hopf-like bifurcation an attracting set ceases to be attracting, and a new attracting set appears that engages in an additional oscillation.

Bifurcation diagrams are a central feature in our work. Figure 1.2 shows a bifurcation diagram for the part of the parameter space we focus on for the Braitenberg system. In this case, the panels show the physical paths of the vehicles in those dynamical regimes. The physical paths are projections of orbits of the attracting sets for the Braitenberg vehicles. The parameter space for the Braitenberg vehicles includes the several physical parameters (discussed in section 2.2) and also the quadruple of sensor weights $(w_{(\ell,1)}, w_{(r,1)}, w_{(\ell,2)}, w_{(r,2)})$ in the total weight space $W_{\text{total}} \cong \mathbf{R}^4$, (discussed in section 2.5). Each choice of parameter value specifies a particular dynamical system for the Braitenberg vehicles: they behave differently as the parameters are varied. We will choose particular values for physical parameters and focus on the bifurcations that occur as the weights of the neural network are varied.

A saddle node-like bifurcation occurs for the Braitenberg vehicles in which a pair of relative equilibria appear, one of which is attracting (this shows that other invariant sets besides fixed points and periodic orbits can be involved in bifurcations). Relative equilibria are explained in chapter 5, and the saddle node-like bifurcation of relative equilibria is covered in detail in chapter 6. A Hopf-like bifurcation also occurs for the Braitenberg vehicles in which an attracting relative equilibrium becomes unstable and an apparent attracting relative periodic orbit begins. This is covered in detail in chapters 6 and 7. In the next few sections, we develop examples that are analogous to the bifurcations that occur with the Braitenberg vehicles.

3.5 A Saddle Node and a Nonstandard Bifurcation

As we have just explained, two ways of identifying bifurcations are by detecting a change in the number of fixed points and by finding where the eigenvalue traces of a linearization intersect the imaginary axis. We now demonstrate both of these techniques in our analysis of a parameterized family of dynamical systems that displays a well-known saddle node bifurcation as well as a non-standard bifurcation. The system is given by the solutions to the differential equation:

$$\begin{pmatrix} \dot{x} \\ \dot{y} \end{pmatrix} = F\begin{pmatrix} x \\ y \end{pmatrix} = \begin{pmatrix} \lambda - x^2 \\ xy - y \end{pmatrix}. \tag{3.5}$$

The state variables are $(x, y)^T$, and the real number λ is the parameter. Although it is possible to obtain a general solution for equation (3.5) in terms of elementary functions, it is very complicated, so we rely on dynamical systems theory and numerical integration to analyze this example.

The first component of the differential equation $\dot{x} = \lambda - x^2$ does not contain the variable y. It is a standard equation for the saddle node bifurcation in one dimension, and it is often used to illustrate the saddle node bifurcation in higher dimensions. We will see that for this example there are two bifurcation values for λ: $\lambda = 0$ (the saddle node bifurcation) and $\lambda = 1$ (a nonstandard bifurcation). This is summarized in figure 3.4.

We will follow the stepwise procedure for linearization outlined in section 3.2. Before we locate the fixed points, there is a simple fact about the system that we can easily deduce. If $y = 0$ at some moment in time, then the rate of change of y is 0 for any value of x. So the x-axis is an invariant set for equation (3.5) regardless of the value for the parameter λ.

We now proceed to locate the fixed points. We set the time derivatives of $(x, y)^T$ equal to $(0, 0)^T$ and solve the algebraic equation:

$$\begin{pmatrix} 0 \\ 0 \end{pmatrix} = \begin{pmatrix} \lambda - x^2 \\ xy - y \end{pmatrix}. \tag{3.6}$$

The set of solutions depends on the parameter λ. There will be five cases to consider:

$$\lambda < 0, \qquad \lambda = 0, \qquad 0 < \lambda < 1, \qquad \lambda = 1, \qquad \text{and} \qquad 1 < \lambda.$$

We begin with cases where $\lambda < 0$ and $\lambda = 0$. It will then be easier to move back and forth through the last three cases, where $\lambda > 0$.

When $\lambda < 0$, the rate of change in x is $\dot{x} = \lambda - x^2$, which is negative for all $x \in \mathbf{R}$. So the value of x is always decreasing, and there are no fixed points. The phase portrait in the bottom left panel of figure 3.4 shows that all of the states of the system move leftward. They also converge to the x-axis as time increases without bound.

Next we consider the case $\lambda = 0$. For $x = 0$ the rate of change is $\dot{x} = 0$, so the y-axis is also an invariant set. The x-axis and the y-axis only intersect at the origin. The only way the origin can remain in both the x and y axes is by being a fixed point. If $x = 0$, then the rate of change in y is $\dot{y} = -y$, which only

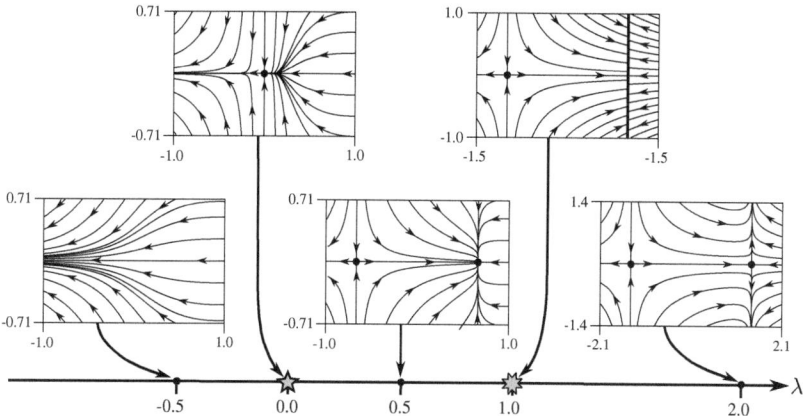

Figure 3.4
The five panels show numerically computed phase portraits for equation (3.5). (top row) Phase portraits at the two bifurcation values. (middle row) Phase portraits for the three regimes between the bifurcations. (bottom row) A one-dimensional bifurcation diagram for equation (3.5). Each phase portrait has an arrow that points to its value for λ in the bifurcation diagram. The saddle node bifurcation occurs at $\lambda = 0$ (marked by the five-pointed star), and the nonstandard bifurcation occurs at $\lambda = 1$ (marked by the eight-pointed star).

equals 0 for $y = 0$. So $(0,0)^T$ is the only fixed point on the y-axis. For $x \neq 0$, the rate of change in x is $\dot{x} = -x^2$, which must be negative, and so in this case x always decreases in value. So there are no fixed points with $x \neq 0$. So the only fixed point when $\lambda = 0$ is at $(0,0)^T$. The number of fixed points has increased from 0 to 1 as λ went from negative to 0, so a bifurcation has taken place. The phase portrait in the top left panel of figure 3.4 shows that all of the states with negative values for x move to the left without bound (just as they did for $\lambda < 0$), while states where x is nonnegative converge to the origin.

To set up the analysis of the bifurcations that occur as we continue to vary λ, we linearize the system at its fixed point. To linearize differential equation (3.5) at the origin, we take the total derivative of the function on the right-hand side of the equation:

$$DF\begin{pmatrix} x \\ y \end{pmatrix} = \begin{pmatrix} \dfrac{\partial}{\partial x}\left(\lambda - x^2\right) & \dfrac{\partial}{\partial y}\left(\lambda - x^2\right) \\ \dfrac{\partial}{\partial x}\left(xy - y\right) & \dfrac{\partial}{\partial y}\left(xy - y\right) \end{pmatrix} = \begin{pmatrix} -2x & 0 \\ y & x - 1 \end{pmatrix}. \quad (3.7)$$

and then substitute the coordinates for the fixed point, $(x, y)^T = (0, 0)^T$ (when $\lambda = 0$) into the matrix to get:

$$\begin{pmatrix} 0 & 0 \\ 0 & -1 \end{pmatrix}.$$

This matrix can be used to define our linearized differential equation when $\lambda = 0$. Since it is a new system, we express it using the primed variables $(x', y')^T$:

$$\begin{pmatrix} \dot{x}' \\ \dot{y}' \end{pmatrix} = \begin{pmatrix} 0 & 0 \\ 0 & -1 \end{pmatrix} \begin{pmatrix} x' \\ y' \end{pmatrix} = \begin{pmatrix} 0 \\ -y' \end{pmatrix}.$$

Linear differential equations are usually easily solved. In this case, it can be checked by substitution that the general solution is:

$$\begin{pmatrix} x'(t) \\ y'(t) \end{pmatrix} = \begin{pmatrix} x'_0 \\ y'_0 e^{-t} \end{pmatrix}$$

for any initial condition $(x'_0, y'_0)^T \in \mathbf{R}^2$. Since the matrix is diagonal, the two variables $x'(t)$ and $y'(t)$ are decoupled from each other. The value of $x'(t)$ at any time only depends on the initial value of x'_0, and the value of $y'(t)$ at any time only depends on the initial value of y'_0.

We can see from the solution to the linear equation that if $x'_0 = 0$ then $x'(t) = 0$ for all $t \in \mathbf{R}$ and that if $y'_0 = 0$ then $y'(t) = 0$ for all $t \in \mathbf{R}$. In other words, the x' and y' axes are invariant sets and the origin is fixed. In fact, for this linear system, all of the points in the x'-axis are fixed. Moreover, the x' coordinate for any state is fixed, that is, $x'(t) = x'_0$ for all $t \in \mathbf{R}$. For any state not on the x'-axis, the y' coordinate decays exponentially to 0. Aside from the fixed points in the x'-axis, all of the orbits for the linearized system are vertical half-lines that converge to the x'-axis as time increases without bound. The linearized system can be conceptualized as a collection of copies of the same one-dimensional dynamical system arranged along the x'-axis and vertical to it.

Since the matrix for the linear system is diagonal, the eigenvalues can be read off the diagonal: they are 0 and -1. Since 0 is an imaginary number, the fixed point $(x, y)^T = (0, 0)^T$ of the nonlinear system is nonhyperbolic. This also indicates that a bifurcation takes place when $\lambda = 0$. We can not apply the stable manifold or Hartman-Grobman theorems to the fixed point since it is nonhyperbolic.

However the center manifold theorem does tell us that there are nontrivial stable and center manifolds for the nonhyperbolic fixed point of the nonlinear system. Because there is one imaginary eigenvalue, 0, the center eigenspace has one dimension. Because the one remaining eigenvalue is negative, the stable eigenspace has one dimension. Since the matrix is diagonal, the members of the standard basis of $\mathbf{R}^2$ are eigenvectors. The eigenvector for the eigenvalue

0 is $(1, 0)^T$. The eigenvector for the eigenvalue -1 is $(0, 1)^T$:

$$\begin{pmatrix} 0 & 0 \\ 0 & -1 \end{pmatrix} \begin{pmatrix} 1 \\ 0 \end{pmatrix} = 0 \begin{pmatrix} 1 \\ 0 \end{pmatrix} \qquad \begin{pmatrix} 0 & 0 \\ 0 & -1 \end{pmatrix} \begin{pmatrix} 0 \\ 1 \end{pmatrix} = -1 \begin{pmatrix} 0 \\ 1 \end{pmatrix}.$$

The x'-axis is the center eigenspace, and the y'-axis is the stable eigenspace for the linearized system.

The $\lambda = 0$ case illustrates a role for the center manifold theorem. By the center manifold theorem, the stable eigenspace is tangent to the stable manifold at the fixed point and the center eigenspace is tangent to one or more center manifolds at the fixed point. We showed that the y-axis is an invariant set. Furthermore $\dot{y} = -y$ when $x = 0$, so the dynamics in the y-axis for the nonlinear system is exactly the same as the dynamics in the y'-axis of the linear system. Every point in the y'-axis exponentially decays to the origin, so the y-axis is the stable manifold of the origin in the nonlinear system. This is shown in the top left panel of figure 3.4.

We also showed that the x-axis is an invariant set of the nonlinear system. The top left panel of figure 3.4 also shows that the x-axis is one of the center manifolds for the origin of the nonlinear system. Each of the other center manifolds for the origin is the union of the negative x-axis, the point $(0, 0)^T$, and any one of the orbits in the half-plane $x > 0$. Also note that the dynamics in the center manifold of the origin is not equivalent to the dynamics in the center eigenspace. For the nonlinear system, all of the points in the center manifold of the origin, aside from the origin, move to the left while for the linear system all of the points in center eigenspace are fixed.

We now proceed to the $\lambda > 0$ cases. We will show that there are always at least two fixed points for $\lambda > 0$:

$$\left(\pm\sqrt{\lambda}, \, 0 \right)^T. \tag{3.8}$$

These fixed points are symmetrically positioned about the origin of the x-axis, as can be seen in the right three panels of figure 3.4. When λ is increased above 0, the origin ceases to be fixed. Instead, the two fixed points in (3.8) emerge from $(0, 0)^T$. There are no fixed points when $\lambda < 0$, one fixed point when $\lambda = 0$, and at least two fixed points when $\lambda > 0$. This is indicative of a saddle node bifurcation occurring at $\lambda = 0$. We will also show that there are no other fixed points for $\lambda > 0$ unless $\lambda = 1$, in which case there are an infinite number of fixed points. This shows that another bifurcation occurs at $\lambda = 1$.

When $\lambda > 0$, the first component of equation (3.6) is $0 = \lambda - x^2$, which has two solutions: $x = \pm\sqrt{\lambda}$. When we substitute them into the second component of (3.6), we get these equations:

$$\text{For} \quad x = -\sqrt{\lambda}: \qquad\qquad \text{For} \quad x = \sqrt{\lambda}:$$
$$0 = (\sqrt{\lambda} + 1)y \qquad\qquad 0 = (\sqrt{\lambda} - 1)y.$$

Both of these equations are satisfied for $y = 0$, which confirms that the points in (3.8) are fixed. We now consider whether these are unique solutions, starting with the case where $x = -\sqrt{\lambda}$. In this case, there are no additional solutions, because it is impossible for $(\sqrt{\lambda} + 1)$ to be 0, which forces $y = 0$. So the only fixed point with $x = -\sqrt{\lambda}$ is $(-\sqrt{\lambda}, 0)^T$. Now consider the case where $x = \sqrt{\lambda}$. In this case, the only way for $\sqrt{\lambda} - 1 = 0$ is for $\lambda = 1$. So long as $\lambda \neq 1$, $(\sqrt{\lambda}, 0)^T$ is the only fixed point with $x = \sqrt{\lambda}$. In the case where $\lambda = 1$, every real value for y satisfies the equation for $x = \sqrt{\lambda}$, so every point in the vertical line $x = 1$ is fixed. These facts can be observed in the right three panels of figure 3.4, where two fixed points are visible for the phase portraits where $\lambda > 0$, except at $\lambda = 1$ where the whole vertical line $x = 1$ is fixed (i.e., every point in it is a fixed point).

We say a fixed point is *isolated* if it has an open neighborhood in which it is the only fixed point, otherwise it is *nonisolated*. When $\lambda = 1$ every point in the vertical line $x = 1$ is a nonisolated fixed point since any open neighborhood of any point in the fixed line will contain other points in the fixed line.

In summary, the only fixed points for the dynamical system given by equation (3.5) when $0 < \lambda < 1$ or $\lambda > 1$ are the two points in (3.8). For $\lambda = 1$, there is an isolated fixed point at $(-1, 0)^T$, and an entire vertical line of nonisolated fixed points at $x = 1$. For all $\lambda > 0$, the x-axis is invariant and there are no fixed points on the x-axis except for the two points in (3.8), which we designate the "left fixed point" and "right fixed point." The vertical line through each of these fixed points is an invariant set since $\dot{x} = 0$ when $x = \pm\sqrt{\lambda}$ regardless of the value for y. The open line segment between the left and right endpoints is a heteroclinic orbit. This is shown in the three panels on the right of figure 3.4. This gives us an overall sense of the dynamical system for $\lambda > 0$.

To understand what is happening in more detail, we first linearize the system, then use the stable manifold theorem, the Hartman-Grobman theorem, and the center manifold theorem. We linearize differential equation (3.5) at the two fixed points in (3.8) by substituting them into the total derivative in equation (3.7). The matrices are:

$$\text{At} \quad (-\sqrt{\lambda}, 0)^T: \qquad\qquad \text{At} \quad (\sqrt{\lambda}, 0)^T:$$

$$\begin{pmatrix} 2\sqrt{\lambda} & 0 \\ 0 & -\sqrt{\lambda} - 1 \end{pmatrix} \qquad\qquad \begin{pmatrix} -2\sqrt{\lambda} & 0 \\ 0 & \sqrt{\lambda} - 1 \end{pmatrix}.$$

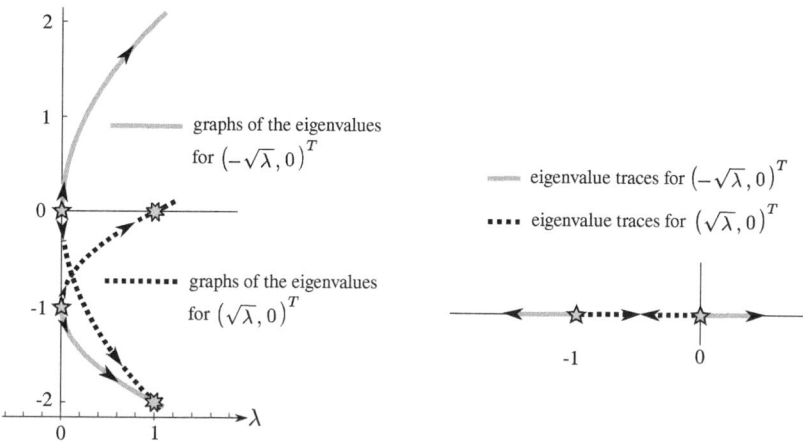

Figure 3.5
(left) The graphs of the eigenvalues as a function of λ. (right) The eigenvalue traces in the complex plane for the two fixed points shortly after the saddle node bifurcation at $\lambda = 0$. The eigenvalues at the saddle node bifurcations are marked by five-pointed stars. As λ increases above 0, the eigenvalues for $(-\sqrt{\lambda}, 0)^T$ move outside of the interval $[-1, 0]$, while the eigenvalues for $(\sqrt{\lambda}, 0)^T$ move inside $[-1, 0]$. A nonstandard bifurcation involving the fixed point $(\sqrt{\lambda}, 0)^T$ occurs at $\lambda = 1$. The eigenvalues for $(\sqrt{\lambda}, 0)^T$ at the nonstandard bifurcation are marked by the eight-pointed stars.

Once again these matrices are diagonal so their eigenvalues are the diagonal entries, and the standard basis of $\mathbf{R}^2$ is an eigenbasis for both fixed points. All of the eigenvalues are real since λ is positive. The graphs of the eigenvalues as a function of the parameter λ are shown in the left panel of figure 3.5. The right panel of the figure shows eigenvalue traces for these eigenvalues.

A few points are worth emphasizing about figure 3.5. First, the stars in the figure correspond to eigenvalues, not fixed points. There is a single fixed point at $\lambda = 0$ with two eigenvalues marked by the two five-pointed stars in figure 3.5. After the bifurcation, the graphs of the two eigenvalues for each fixed point are a pair of half-parabolas, which are indicated by the shading and stroke of the curves in the figure:

- The graphs of the eigenvalues for the left fixed point, $(-\sqrt{\lambda}, 0)^T)$, are the pair of half-parabolas shown with the solid gray lines.
- The graphs of the eigenvalues for the right fixed point, $(\sqrt{\lambda}, 0)^T$, are the pair of half-parabolas shown with dotted black lines.

Second, this is an important transitional figure because it shows how parameter variations are related to eigenvalue traces. In the left panel, we see how the eigenvalues of the fixed points change as the parameter changes. In the eigenvalue trace in the right panel, we simply show how the eigenvalues move in the complex plane; the parameter is not present in the panel. We make use of eigenvalue traces in the rest of the book.

We can apply the stable manifold and Hartman-Grobman theorems to these fixed points when they are hyperbolic, and this will tell us about the behavior of differential equation (3.5) in a neighborhood of each fixed point. For all positive values of λ, the eigenvalues of $(-\sqrt{\lambda}, 0)^T$ are nonzero and this fixed point is hyperbolic. For all positive values of λ except $\lambda = 1$, the eigenvalues of $(\sqrt{\lambda}, 0)^T$ are nonzero and this fixed point is hyperbolic. For $\lambda = 1$ the fixed point $(\sqrt{\lambda}, 0)^T$ is nonhyperbolic. This further confirms that a bifurcation takes place at $\lambda = 1$.

The bifurcation that takes place at $\lambda = 0$ is a saddle node bifurcation. There are no fixed points when $\lambda < 0$, there is one fixed point when $\lambda = 0$, and there are two fixed points when λ is a little greater than 0. This is shown in the left three panels in figure 3.4. The bifurcation value of $\lambda = 0$ is marked by the five-pointed star in the bifurcation diagram. As λ passes through 0 from below, a nonhyperbolic fixed point appears with eigenvalues -1 and 0. These are marked by the five-pointed stars in figure 3.5. When λ becomes positive, the nonhyperbolic fixed point separates into two hyperbolic fixed points. The eigenvalues for the left fixed point $(-\sqrt{\lambda}, 0)^T$ move outward so that one eigenvalue remains negative while the other becomes positive. Therefore the left fixed point is a saddle node. The eigenvalues for the right fixed point $(\sqrt{\lambda}, 0)^T$ move inward so that they are both negative. Therefore the right fixed point is attracting.

In general, a saddle node bifurcation produces a single pair of hyperbolic fixed points that emerge from the nonhyperbolic fixed point that momentarily comes into existence at the beginning of the bifurcation. To be a saddle node bifurcation, there must be one and only one eigenvalue for the nonhyperbolic fixed point that equals 0. In two dimensions, the other eigenvalue for the nonhyperbolic fixed point must be positive or negative. The nonhyperbolic fixed point subsequently splits into a pair of hyperbolic fixed points. The zero eigenvalue for the nonhyperbolic fixed point becomes positive for one of the hyperbolic fixed points and negative for the other. The eigenvalues depend continuously on the parameters, so the sign for the nonzero eigenvalue of the nonhyperbolic fixed point does not change as the parameter is varied only slightly. Therefore the eigenvalues for one of the hyperbolic fixed points produced by the saddle node bifurcation must have opposite signs, and that fixed point must be a saddle node. The eigenvalues for the other hyperbolic fixed point must have the same sign. It can be either repelling or attracting. In example (3.5), it is an attracting fixed point when $0 < \lambda < 1$.

We now consider what happens to the left fixed point after the saddle node bifurcation at $\lambda = 0$. It persists as the same type of fixed point for all $\lambda > 0$. As can be seen in figure 3.5, it is a saddle node with a stable and an unstable manifold. To show this, we linearize differential equation (3.5) at the left fixed point by plugging it in to equation (3.7). In this case, the linear differential equation is:

$$\begin{pmatrix} \dot{x}' \\ \dot{y}' \end{pmatrix} = \begin{pmatrix} 2\sqrt{\lambda} & 0 \\ 0 & -\sqrt{\lambda}-1 \end{pmatrix} \begin{pmatrix} x' \\ y' \end{pmatrix}.$$

It can be checked by substitution that the general solution for this differential equation is:

$$\begin{pmatrix} x'(t) \\ y'(t) \end{pmatrix} = \begin{pmatrix} x'_0 \exp\left(\left(2\sqrt{\lambda}\right)t\right) \\ y'_0 \exp\left(\left(-\sqrt{\lambda}-1\right)t\right) \end{pmatrix}$$

for all initial conditions $(x'_0, y'_0)^T \in \mathbf{R}^2$.

Since the matrix is diagonal, the two variables $x'(t)$ and $y'(t)$ are decoupled from each other. Like before, the standard basis vectors are eigenvectors, and the x' and y' axes are invariant sets.

- Unstable eigenspace: The eigenvalue $2\sqrt{\lambda}$ is positive and $(1, 0)^T$ is one of its eigenvectors. So the x'-axis is the unstable eigenspace for the left fixed point, $(-\sqrt{\lambda}, 0)^T$. Within the x'-axis, the value of $x'(t)$ undergoes exponential growth away from 0.

- Stable eigenspace: The eigenvalue $(-\sqrt{\lambda}-1)$ is negative and $(0, 1)^T$ is one of its eigenvectors. So the y'-axis is the stable eigenspace for the left fixed point, $(-\sqrt{\lambda}, 0)^T$. Within the y'-axis, the value of $y'(t)$ undergoes exponential decay to 0.

Thus, the fixed point of the linear system is a saddle node for all $\lambda > 0$. By the Hartman-Grobman theorem, there is a neighborhood of the left fixed point of the nonlinear system (3.5) in which it is equivalent to its linearization. So the left fixed point, $(-\sqrt{\lambda}, 0)^T$, is a saddle node of (3.5) for all $\lambda > 0$.

By the stable manifold theorem, there are unique stable and unstable manifolds that pass through the left fixed point and that are tangent to the stable and unstable eigenspaces for the left fixed point.

- Unstable manifold: All states on the x-axis to the left of the left fixed point move left without bound. States on the x-axis just to the right of the left fixed point converge to the right fixed point but never reach it in finite time. The unstable manifold of the left fixed point, $(-\sqrt{\lambda}, 0)^T$, is the open interval $(-\infty, \sqrt{\lambda})$ within in the x-axis.

- Stable manifold: If $x = -\sqrt{\lambda}$ then the rate of change of x is 0 regardless of the value for y. So the vertical line through the left fixed point is an invariant manifold containing the fixed point. On this line the sign for the rate of change in y is opposite to the sign for the value of y so states in the line $x = -\sqrt{\lambda}$ converge to the fixed point $(-\sqrt{\lambda}, 0)^T$ as time increases without bound. Therefore the entire line $x = -\sqrt{\lambda}$ is the stable manifold of the left fixed point, $(-\sqrt{\lambda}, 0)^T$.

We now consider the right fixed point $(\sqrt{\lambda}, 0)^T$. The linearized differential equation at the right fixed point is:

$$\begin{pmatrix} \dot{x}' \\ \dot{y}' \end{pmatrix} = \begin{pmatrix} -2\sqrt{\lambda} & 0 \\ 0 & \sqrt{\lambda}-1 \end{pmatrix} \begin{pmatrix} x' \\ y' \end{pmatrix}.$$

The general solution is:

$$\begin{pmatrix} x'(t) \\ y'(t) \end{pmatrix} = \begin{pmatrix} x_0' \exp\left(\left(-2\sqrt{\lambda}\right)t\right) \\ y_0' \exp\left(\left(\sqrt{\lambda}-1\right)t\right) \end{pmatrix}$$

for the initial condition $(x_0', y_0')^T$. There are three cases to consider: $0 < \lambda < 1$, $\lambda > 1$, and $\lambda = 1$.

In the case where $0 < \lambda < 1$, both of the eigenvalues $-2\sqrt{\lambda}$ and $(\sqrt{\lambda}-1)$ are negative and both $x'(t)$ and $y'(t)$ undergo exponential decay. All initial conditions, $(x_0', y_0')^T$ converge to $(0, 0)$ as time increases without bound. This is an attracting fixed point, and the stable eigenspace is the entire $(x', y')^T$ plane. By the Hartman-Grobman theorem, this means the right fixed point is attracting when $0 < \lambda < 1$.[16] By the stable manifold theorem, there is a unique two-dimensional stable manifold for the right fixed point. As can be seen in the bottom middle panel of figure 3.4, it is the half-plane to the right of the stable manifold of the left fixed point, that is, the set of all $(x, y)^T \in \mathbf{R}^2$ with $x > -\sqrt{\lambda}$.

For the right fixed point when $\lambda > 1$, the eigenvalues $-2\sqrt{\lambda}$ and $(\sqrt{\lambda}-1)$ have opposite signs. This fixed point is a saddle node, and the situation is similar to the left fixed point. We can summarize this as follows:

- Stable eigenspace: The eigenvalue $-2\sqrt{\lambda}$ is negative so $x'(t)$ undergoes exponential decay to 0. An eigenvector for $-2\sqrt{\lambda}$ is $(1, 0)^T$, which spans the x'-axis. Thus the x'-axis is the stable eigenspace for the right fixed point.
- Unstable eigenspace: The eigenvalue $(\sqrt{\lambda}-1)$ is positive so $y'(t)$ undergoes exponential growth away from 0. An eigenvector for $(\sqrt{\lambda}-1)$ is $(0, 1)^T$,

---

16. Technically the unstable eigenspace does exist, but it is zero-dimensional, consisting of just the single point $(0, 0)^T$.

which spans the y'-axis. The y'-axis is the unstable eigenspace for the right fixed point.

By the Hartman-Grobman theorem, the right fixed point, $(\sqrt{\lambda}, 0)^T$, is a saddle node. By the stable manifold theorem, the stable and unstable eigenspaces are associated with stable and unstable manifolds. They are:

- Stable manifold: All points on the x-axis to the right of the right fixed point converge to it. All points on the x-axis between the left and right fixed point converge to the right fixed point. States to the left of the left fixed point move left so they can not converge to the right fixed point. The stable manifold for the right fixed point is the open interval $(-\sqrt{\lambda}, \infty)$ in the x-axis.
- Unstable manifold: We also showed that the vertical line through the right fixed point is an invariant set. We showed that there are no other fixed points in this vertical line so it is the unstable manifold for the right fixed point.

In addition to the saddle node bifurcation that occurs at $\lambda = 0$ in equation (3.5), there is a nonstandard bifurcation that occurs at $\lambda = 1$. Standard bifurcations only involve isolated fixed points. At $\lambda = 1$, there is an isolated fixed point $(-1, 0)^T$ on the negative x-axis. However the fixed point $(1, 0)^T$ on the positive x-axis is part of an entire vertical line of fixed points, that is, a fixed line. The points in this vertical line are only fixed when $\lambda = 1$. Otherwise, there are never more than two fixed points for equation (3.5). The number of fixed points going from two to infinity and back down to two confirms that a bifurcation takes place at $\lambda = 1$. A phase portrait for the bifurcation is shown in the top right panel of figure 3.4, and the bifurcation value $\lambda = 1$ is marked by the eight-pointed star in the bifurcation diagram. Also, the eigenvalues for the right fixed point at $\lambda = 1$ are marked by the eight-pointed stars in the left panel of figure 3.5.

We can also use linearization to show that a bifurcation takes place at $\lambda = 1$. We can linearize at any of the points in the fixed line by substituting $(1, y)^T$ into equation (3.7). We get:

$$\begin{pmatrix} -2 & 0 \\ y & 0 \end{pmatrix}.$$

This is a lower triangular matrix, so the eigenvalues are the diagonal entries as in the previous cases. All of the points in the fixed line have the same eigenvalues, which are -2 and 0. Thus, all of the points in the fixed line are nonhyperbolic. A general fact about nonisolated fixed points is that they are always nonhyperbolic and have nontrivial center manifolds. The right fixed point, $(\sqrt{\lambda}, 0)^T$, going from hyperbolic, to nonhyperbolic, and back to hyperbolic as λ passes through 1 confirms that a bifurcation takes place.

For $\lambda = 1$, the linearization at the right fixed point $(1,0)^T$ gives us a diagonal matrix and the standard basis vectors are the eigenvectors. We can use them to identify the stable and center eigenspaces:

- Stable eigenspace: an eigenvector for the eigenvalue -2 is $(1,0)^T$, and the x'-axis is the stable eigenspace.
- Center eigenspace: an eigenvector for the eigenvalue 0 is $(0,1)^T$, and the y'-axis is the center eigenspace.

We cannot use the stable manifold theorem or the Hartman-Grobman theorem for the right fixed point since it is nonhyperbolic when $\lambda = 1$, but by the center manifold theorem, there is a stable manifold and one or more center manifolds. In this case, the stable manifold and center manifolds are:

- Stable manifold: the half-line $(-1, \infty)$ in the x-axis.
- Center manifolds: the fixed vertical line $x = 1$. Any open interval of this line that contains the fixed point $(1, 0)^T$ is also a center manifold.

Note that unlike in the saddle node bifurcation, in this nonstandard bifurcation the dynamics in the center eigenspace and in the center manifold of $(1,0)^T$ are the same: every point in them is fixed.

In this nonstandard bifurcation, where λ increases through 1, the right fixed point, $(\sqrt{\lambda}, 0)^T$, goes from being attracting, to momentarily becoming part of an attracting fixed line, and then becoming a saddle node. Nonstandard bifurcations like this occur for the Braitenberg vehicles. In particular, in chapter 7, a relative equilibrium comes into existence as part of an infinite collection of relative equilibria. All of the relative equilibria in the collection, except for one, only exist at the bifurcation.

3.6   A Hopf Bifurcation

The Braitenberg vehicles undergo a Hopf-like bifurcation involving relative equilibria. We now consider an example of a Hopf bifurcation in which an attracting fixed point is replaced by a non-attracting fixed point and an attracting limit cycle. It turns out that this limit cycle is a simple example of a relative equilibrium. The concept of a relative equilibrium does not do much work in this example (this system is usually analyzed without the concept), but our goal here is to introduce the concepts and methods employed in chapters 6 and 7 in the context of a simple example.

The parameterized family of dynamical systems that exhibits the Hopf bifurcation is the general solution to the differential equation:[17]

$$\begin{pmatrix} \dot{x} \\ \dot{y} \end{pmatrix} = F \begin{pmatrix} x \\ y \end{pmatrix} = \begin{pmatrix} (\lambda - x^2 - y^2) \, x - y \\ (\lambda - x^2 - y^2) \, y + x \end{pmatrix}$$

$$= \left((\lambda - x^2 - y^2) \, R_0 + R_{(\pi/2)} \right) \begin{pmatrix} x \\ y \end{pmatrix}. \qquad (3.9)$$

where R_0 and $R_{(\pi/2)}$ are rotation matrices as in equation (2.3).

The points $(x, y)^T$ are the states of the system, and the real number λ is a fixed parameter. Substituting $(x, y)^T = (0, 0)^T$ on the right-hand side shows that the origin is a fixed point. To see that there are no other fixed points, we set $(\dot{x}, \dot{y})^T = (0, 0)^T$ on the left-hand side to get the equation:

$$\begin{pmatrix} 0 \\ 0 \end{pmatrix} = \begin{pmatrix} (\lambda - x^2 - y^2) \, x - y \\ (\lambda - x^2 - y^2) \, y + x \end{pmatrix}. \qquad (3.10)$$

Note that it follows from the first component of equation (3.10) that $x = 0$ implies $y = 0$. And it follows from the second component that $y = 0$ implies $x = 0$. Thus $x = 0$ or $y = 0$ implies $(x, y)^T = (0, 0)^T$. Or, by the contrapositive, if $(x, y)^T \neq (0, 0)^T$ then $x \neq 0$ and $y \neq 0$. In other words, the origin is the only fixed point in the coordinate axes.

Suppose $(x, y)^T$ is a solution to equation (3.10) and that $(x, y)^T \neq (0, 0)^T$. This implies that $x \neq 0$ and $y \neq 0$, which in turn implies:

$$\begin{pmatrix} y/x \\ -x/y \end{pmatrix} = \begin{pmatrix} \lambda - x^2 - y^2 \\ \lambda - x^2 - y^2 \end{pmatrix} \implies \frac{y}{x} = \lambda - x^2 - y^2 = -\frac{x}{y}$$

$$\implies \frac{y}{x} = -\frac{x}{y} \implies x^2 + y^2 = 0.$$

The last equation is for a circle of radius 0 about the origin, which just is the origin. This contradicts the supposition that $(x, y)^T \neq (0, 0)^T$. Therefore, $(0, 0)^T$ is the only fixed point for equation (3.9).

We cannot detect a bifurcation in differential equation (3.9) by a change in the number of fixed points because it always has exactly one fixed point. However, it does have a bifurcation, which we can detect in other ways. We can linearize the system at the origin and find where the eigenvalue traces cross the imaginary axis, or we can see where there is a change in the number of periodic orbits the system has. We show both methods.

---

17. This differential equation comes from Guckenheimer and Holmes (1983).

To linearize the system at the origin, we take the total derivative of the right-hand side of equation (3.9):

$$DF\begin{pmatrix} x \\ y \end{pmatrix} = \begin{pmatrix} \dfrac{\partial}{\partial x}\left(\left(\lambda-x^2-y^2\right)x-y\right) & \dfrac{\partial}{\partial y}\left(\left(\lambda-x^2-y^2\right)x-y\right) \\ \dfrac{\partial}{\partial x}\left(\left(\lambda-x^2-y^2\right)\right)y+x & \dfrac{\partial}{\partial y}\left(\left(\lambda-x^2-y^2\right)\right)y+x \end{pmatrix}$$

$$= \begin{pmatrix} \lambda-3x^2-y^2 & -1-2xy \\ 1-2xy & \lambda-x^2-3y^2 \end{pmatrix}.$$

Evaluating the total derivative at the origin gives us an easy matrix to deal with:

$$\begin{pmatrix} \lambda & -1 \\ 1 & \lambda \end{pmatrix}.$$

The eigenvalues of this matrix are the complex conjugate numbers $\lambda \pm i$:

$$\begin{pmatrix} \lambda & -1 \\ 1 & \lambda \end{pmatrix}\begin{pmatrix} -1 \\ \pm i \end{pmatrix} = (\lambda \pm i)\begin{pmatrix} -1 \\ \pm i \end{pmatrix}.$$

Varying λ only changes the real part of the eigenvalues; their imaginary parts are always $\pm i$. As the parameter λ passes through 0, one eigenvalue passes through the positive imaginary axis at i while the other eigenvalue passes through the negative imaginary axis at $-i$. The eigenvalue traces are shown in the bottom left and bottom middle panels of figure 3.6. Two complex conjugate eigenvalues passing through the imaginary axis is a characteristic feature of the Hopf bifurcation.

The linearized differential equation for the origin is:

$$\begin{pmatrix} \dot{x}' \\ \dot{y}' \end{pmatrix} = \begin{pmatrix} \lambda & -1 \\ 1 & \lambda \end{pmatrix}\begin{pmatrix} x' \\ y' \end{pmatrix} = \left(\lambda R_0 + R_{(\pi/2)}\right)\begin{pmatrix} x' \\ y' \end{pmatrix}.$$

Linearizing differential equation (3.9) at the origin amounts to just replacing $\left(\lambda-x^2-y^2\right)$ with λ in equation (3.9). The solution to this linearized system is much simpler. Taking the derivative of both sides of:

$$\begin{pmatrix} x'(t) \\ y'(t) \end{pmatrix} = e^{\lambda t}R_t\begin{pmatrix} x'_0 \\ y'_0 \end{pmatrix} \tag{3.11}$$

and using the product rule for matrices gives:

$$\begin{pmatrix} \dot{x}'(t) \\ \dot{y}'(t) \end{pmatrix} = \left(\lambda e^{\lambda t}R_t + e^{\lambda t}\dot{R}_t\right)\begin{pmatrix} x'_0 \\ y'_0 \end{pmatrix}$$

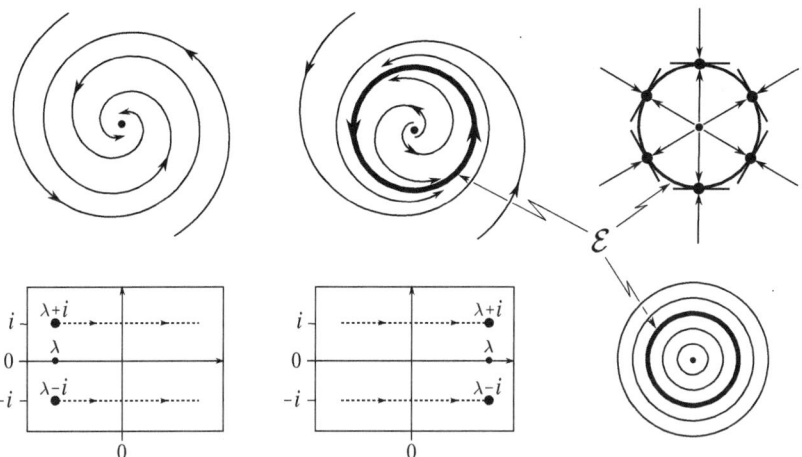

Figure 3.6
(top left) Phase portrait with orbits spiraling toward the origin. (top middle) Phase portrait with orbits spiraling away from the origin and toward the attracting limit cycle $\mathcal{E}$. (top right) The same phase portrait from the top middle panel as seen in a rotating frame. Each orbit is confined within a line through the origin. Three lines through the origin are shown, and they each contain seven orbits. The limit cycle $\mathcal{E}$ now appears as a circle of fixed points. Each of their stable eigenspaces are orthogonal to $\mathcal{E}$, and each of their center eigenspaces are tangent to $\mathcal{E}$. Six tangents are shown. (bottom left) The location of the parameter $\lambda < 0$ is marked by the small dot, and the corresponding location for the complex conjugate eigenvalues $\lambda \pm i$ are marked by the large dots. (bottom middle) The location of $\lambda > 0$ is marked by the small dot, and the corresponding location for the complex conjugate eigenvalues $\lambda \pm i$ are marked by the large dots. A Hopf bifurcation occurs between the left and middle columns when the eigenvalues pass through the imaginary axis. Their traces are marked with dotted lines. (bottom right). The equivalence classes for $SO(2)$ acting on the state space.

Substituting $R_t = R_0\,R_t$ and $\dot{R}_t = R_{(\pi/2)}\,R_t$ into the right-hand side gives:

$$\left(\lambda e^{\lambda t} R_0\,R_t + e^{\lambda t} R_{(\pi/2)}\,R_t\right)\begin{pmatrix} x_0' \\ y_0' \end{pmatrix} = \left(\lambda R_0 + R_{(\pi/2)}\right)\left(e^{\lambda t} R_t \begin{pmatrix} x_0' \\ y_0' \end{pmatrix}\right)$$

$$= \left(\lambda R_0 + R_{(\pi/2)}\right)\begin{pmatrix} x'(t) \\ y'(t) \end{pmatrix}.$$

which confirms that equation (3.11) is the general solution to the linearized differential equation.

We can use the Hartman-Grobman theorem to draw conclusions about the nonlinear system based on its linearization. The origin is the only fixed point for the linearized system; the rest of the orbits are equiangular spirals centered on that fixed point (so long as $\lambda \neq 0$). Because of the rotation factor R_t in solution (3.11), these spirals wind anticlockwise around the origin. Because of the scaling factor $e^{\lambda t}$, they coil inward for $\lambda < 0$ and outward for $\lambda > 0$. By the Hartman-Grobman theorem, the nonlinear system is qualitatively the same as this linearization in a neighborhood of the origin (so long as $\lambda \neq 0$). So for the nonlinear system, the origin is an attracting fixed point when $\lambda < 0$ and a repelling fixed point when $\lambda > 0$. The resemblance of the nonlinear system to the linear system in a neighborhood of the origin can be seen in the phase portraits in the top left and middle panels of figure 3.6.

The eigenvalue traces crossing the imaginary axis at $\lambda = 0$ indicates that a bifurcation occurs. That bifurcation involves the appearance of the following circular periodic orbit:

$$\begin{pmatrix} x \\ y \end{pmatrix} = \sqrt{\lambda} \begin{pmatrix} \cos(t) \\ \sin(t) \end{pmatrix}. \tag{3.12}$$

It can be checked by substitution that this is a solution to differential equation (3.9). We will denote this periodic orbit by $\mathcal{E}$. It is shown in the top middle, top right, and bottom right panels of figure 3.6. There are no periodic orbits for (3.9) when $\lambda \leq 0$. So the appearance of $\mathcal{E}$ once $\lambda > 0$ indicates that a bifurcation occurs at $\lambda = 0$. We can see from equation (3.12) that states on $\mathcal{E}$ revolve about the origin with a constant angular velocity. The appearance of an attracting periodic orbit when an attracting fixed point turns into a repelling fixed point is another characteristic feature of the Hopf bifurcation.

We now begin the process of showing that $\mathcal{E}$ is an attracting set. A straightforward way to do this would be to work in polar coordinates. However, we will take a longer route here, for expository purposes. We will first show that $\mathcal{E}$ is a relative equilibrium, and then show it is attracting. This gives us an opportunity to work with relative equilibria in a simple example and to preview some of the techniques we will use later. There are two conditions that an invariant set needs to satisfy in order to be a relative equilibrium :

1. The dynamical system is equivariant with respect to a transformation group that acts on the dynamical system's state space.
2. The invariant set needs to be an "equivalence class" of the transformation group that acts on the state space.[18]

To show that $\mathcal{E}$ is a relative equilibrium, we first establish condition (2) by describing the equivalence classes of **SO**(2). This will make clear that $\mathcal{E}$ is one

---

18. The "equivalence classes" of a transformation group are usually called "orbits". We use the term "equivalence classes" here to avoid confusion with the orbits of dynamical systems.

of these equivalence classes. To establish condition (1), we use a theorem that in this particular case, amounts to showing that F in equation (3.9) commutes with $\mathbf{SO}(2)$.

In general, a transformation group can be used to define an equivalence relation on the space it acts on. Suppose G is some transformation group acting on some topological space S. We will take a state $s_1 \in S$ to be related to a state $s_2 \in S$ if and only if there is a transformation $g \in G$ such that $s_2 = g(s_1)$. We explain in section 5.5 how, in general, this gives us an equivalence relation on S.

In the case of $\mathbf{SO}(2)$ acting on $\mathbf{R}^2$ the points in the plane are related by rotations in $\mathbf{SO}(2)$. To show that this is an equivalence relation, we show that it is reflexive, symmetric, and transitive.

- Reflexive: Any $(x, y)^T \in \mathbf{R}^2$ is related to itself by the identity map on $\mathbf{R}^2$, which is a member of $\mathbf{SO}(2)$ corresponding to a rotation about the origin by $0°$.

- Symmetric: If $(x_1, y_1)^T$ is related to $(x_2, y_2)^T$ by the rotation $R_\Theta \in \mathbf{SO}(2)$, then $(x_2, y_2)^T$ is related to $(x_1, y_1)^T$ by the rotation by the opposite angle, $R_{(-\Theta)} \in \mathbf{SO}(2)$.

- Transitivity: If $(x_1, y_1)^T$ is related to $(x_2, y_2)^T$ by $R_{\Theta_1} \in \mathbf{SO}(2)$ and $(x_2, y_2)^T$ is related to $(x_3, y_3)^T$ by $R_{\Theta_2} \in \mathbf{SO}(2)$ then $(x_1, y_1)^T$ related to $(x_3, y_3)^T$ by the composition of these two rotations $R_{\Theta_1} \circ R_{\Theta_2} = R_{(\Theta_1 + \Theta_2)} \in \mathbf{SO}(2)$.

The equivalence classes of an equivalence relation forms a partition. In this case, the state space $\mathbf{R}^2$ is partitioned by the action of $\mathbf{SO}(2)$ into circles centered at the origin and the origin itself. This is shown in the bottom right panel of figure 3.6. By comparing the equivalence classes in the bottom right panel with the phase phase portrait in the top middle panel, we can see that there is one invariant set that is also an equivalence class. This is the periodic orbit $\mathcal{E}$. So $\mathcal{E}$ satisfies condition (2) for being a relative equilibrium.

We now proceed to show that $\mathcal{E}$ satisfies condition (1) for being a relative equilibrium, which is to say that the dynamical system given by differential equation (3.9) is equivariant with respect to $\mathbf{SO}(2)$. To do this, we make use of a result presented in (Field 1980), which we explain in greater detail in section 5.4 [it is summarized by equation (5.15)]. We present it here in a simpler form:

$$DR_\Theta \circ F = F \circ R_\Theta.$$

where F is the function in differential equation (3.9). Since each rotation $R_\Theta \in \mathbf{SO}(2)$ is already a linear transformation, its total derivative is itself. So in this case we just need to show that:

$$R_\Theta \circ F = F \circ R_\Theta$$

That is, the function F has to commute with every $R_\Theta \in \mathbf{SO}(2)$. For the left-hand side we get:

$$R_\Theta \left(F \begin{pmatrix} x \\ y \end{pmatrix} \right) = R_\Theta \left(\left(\left(\lambda - x^2 - y^2 \right) R_0 + R_{(\pi/2)} \right) \begin{pmatrix} x \\ y \end{pmatrix} \right)$$

$$= R_\Theta \left(\left(\lambda - x^2 - y^2 \right) R_0 \begin{pmatrix} x \\ y \end{pmatrix} \right) + R_\Theta \, R_{(\pi/2)} \begin{pmatrix} x \\ y \end{pmatrix}.$$

The quantity $\left(\lambda - x^2 - y^2 \right)$ is unaffected by rotations about the origin, and we can reverse the order in which we perform rotations about the origin, so:

$$R_\Theta \left(\left(\lambda - x^2 - y^2 \right) R_0 \begin{pmatrix} x \\ y \end{pmatrix} \right) + R_\Theta \, R_{(\pi/2)} \begin{pmatrix} x \\ y \end{pmatrix}$$

$$= \left(\lambda - x^2 - y^2 \right) R_\Theta \, R_0 \begin{pmatrix} x \\ y \end{pmatrix} + R_\Theta \, R_{(\pi/2)} \begin{pmatrix} x \\ y \end{pmatrix}$$

$$= \left(\lambda - x^2 - y^2 \right) R_0 \, R_\Theta \begin{pmatrix} x \\ y \end{pmatrix} + R_{(\pi/2)} \, R_\Theta \begin{pmatrix} x \\ y \end{pmatrix}$$

$$= \left(\left(\lambda - x^2 - y^2 \right) R_0 + R_{(\pi/2)} \right) R_\Theta \begin{pmatrix} x \\ y \end{pmatrix}$$

$$= F \left(R_\Theta \begin{pmatrix} x \\ y \end{pmatrix} \right).$$

This holds for all $(x, y)^T \in \mathbf{R}^2$ and all $R_\Theta \in \mathbf{SO}(2)$. So the dynamical systems given by differential equation (3.9) are equivariant with respect to $\mathbf{SO}(2)$.[19] Therefore $\mathcal{E}$ satisfies both conditions for being a relative equilibrium.

Having established that $\mathcal{E}$ is a relative equilibrium, we use linearization to show that it is attracting. To do this, we introduce a rotating frame to counteract the rotation that occurs in the original frame of reference, which we call the "rest frame." To express the differential equation in the rotating frame, we use the coordinates $(x', y')^T$:

$$\begin{pmatrix} x' \\ y' \end{pmatrix} = R_{(-t)} \begin{pmatrix} x \\ y \end{pmatrix}. \tag{3.13}$$

To get the differential equation that $(x', y')^T$ satisfies, we take the derivative with respect to time of both sides of equation (3.13). By the product rule we

19. This example is discussed briefly in Matsui (2001) and in Golubitsky and Stewart (2003).

get:

$$\begin{pmatrix} \ddot{x}' \\ \ddot{y}' \end{pmatrix} = R_{(-t)} \begin{pmatrix} \ddot{x} \\ \ddot{y} \end{pmatrix} + \dot{R}_{(-t)} \begin{pmatrix} x \\ y \end{pmatrix}.$$

Since $\dot{R}_{(-t)} = -R_{(-t)} R_{(\pi/2)}$, we can rewrite the right-hand side as:

$$R_{(-t)} \begin{pmatrix} \ddot{x} \\ \ddot{y} \end{pmatrix} + \dot{R}_{(-t)} \begin{pmatrix} x \\ y \end{pmatrix} = R_{(-t)} \begin{pmatrix} \ddot{x} \\ \ddot{y} \end{pmatrix} - R_{(-t)} R_{(\pi/2)} \begin{pmatrix} x \\ y \end{pmatrix}$$

$$= R_{(-t)} \left(\begin{pmatrix} \ddot{x} \\ \ddot{y} \end{pmatrix} - R_{(\pi/2)} \begin{pmatrix} x \\ y \end{pmatrix} \right).$$

Substituting for $(\dot{x}, \dot{y})^T$ from equation (3.9) and using equation (3.13) gives:

$$R_{(-t)} \left(\begin{pmatrix} \ddot{x} \\ \ddot{y} \end{pmatrix} - R_{(\pi/2)} \begin{pmatrix} x \\ y \end{pmatrix} \right)$$

$$= R_{(-t)} \left(\left((\lambda - x^2 - y^2) R_0 + R_{(\pi/2)} \right) \begin{pmatrix} x \\ y \end{pmatrix} - R_{(\pi/2)} \begin{pmatrix} x \\ y \end{pmatrix} \right)$$

$$= R_{(-t)} \left((\lambda - x^2 - y^2) R_0 \begin{pmatrix} x \\ y \end{pmatrix} + R_{(\pi/2)} \begin{pmatrix} x \\ y \end{pmatrix} - R_{(\pi/2)} \begin{pmatrix} x \\ y \end{pmatrix} \right)$$

$$= R_{(-t)} \left((\lambda - x^2 - y^2) R_0 \begin{pmatrix} x \\ y \end{pmatrix} \right)$$

$$= (\lambda - x^2 - y^2) R_{(-t)} \begin{pmatrix} x \\ y \end{pmatrix}$$

$$= (\lambda - x^2 - y^2) \begin{pmatrix} x' \\ y' \end{pmatrix}.$$

Since we are only rotating about the origin, $x'^2 + y'^2 = x^2 + y^2$. So the differential equation in the rotating frame is:

$$\begin{pmatrix} \dot{x}' \\ \dot{y}' \end{pmatrix} = (\lambda - x'^2 - y'^2) \begin{pmatrix} x' \\ y' \end{pmatrix}. \tag{3.14}$$

As can be seen from equation (3.14), the velocity of $(x', y')^T$ is a scalar multiple of $(x', y')^T$. Each state moves radially away from or toward the origin, and every line that passes through the origin is an invariant set. This is illustrated in the phase portrait in the top right panel of figure 3.6.

Note that the dynamical system given by differential equation (3.14) is not equivalent to the dynamical system given by differential equation (3.9). It is the case that at each moment in time the change of coordinates in equation (3.13) is a homeomorphism. However we do not get just one homeomorphism from equation (3.13). As time varies, the homeomorphism also varies and this alters the dynamics given by equation (3.14). However, we will see that equation (3.14) can still help us analyze equation (3.9).

We now proceed to the linearization of differential equation (3.14). The fixed points are found by solving the algebraic equation:

$$\begin{pmatrix} 0 \\ 0 \end{pmatrix} = \left(\lambda - x'^2 - y'^2 \right) \begin{pmatrix} x' \\ y' \end{pmatrix}$$

for $(x', y')^T$. So either $(x', y')^T = (0, 0)^T$ or $\lambda - x'^2 - y'^2 = 0$. The fixed point $(x', y')^T = (0, 0)^T$ is the same as the fixed point $(x, y)^T = (0, 0)^T$ in the rest frame. That leaves $\lambda - x'^2 - y'^2 = 0$, which is the equation for a circle of radius $\sqrt{\lambda}$ about the origin. So all the points on this circle are fixed points. Since $\mathcal{E}$ is defined in equation (3.12) to be the circle with radius $\sqrt{\lambda}$ about the origin, and since this periodic orbit is mapped to itself in the rotating frame, it follows that $\mathcal{E}$ has become a circle of fixed points. A periodic orbit in the rest frame has been turned into a circle of fixed points in the rotating frame, a "fixed circle" (see the top right of figure 3.6). The only fixed points in the rotating frame are the origin and the points in $\mathcal{E}$.

We are now set up to linearize at the relative equilibrium $\mathcal{E}$, the circle of fixed points. We can apply the technique to any of these fixed points. The total derivative of the right-hand side of equation (3.14) is:

$$\begin{pmatrix} \dfrac{\partial}{\partial x'} \left(\lambda - x'^2 - y'^2 \right) x' & \dfrac{\partial}{\partial y'} \left(\lambda - x'^2 - y'^2 \right) x' \\ \dfrac{\partial}{\partial x'} \left(\lambda - x'^2 - y'^2 \right) y' & \dfrac{\partial}{\partial y'} \left(\lambda - x'^2 - y'^2 \right) y' \end{pmatrix} =$$

$$\begin{pmatrix} \lambda - 3x'^2 - y'^2 & -2x'y' \\ -2x'y' & \lambda - x'^2 - 3y'^2 \end{pmatrix}.$$

A state $(x', y')^T$ is in $\mathcal{E}$ if and only if $x'^2 + y'^2 = \lambda$. This implies

$$\lambda - 3x'^2 - y'^2 = -2x'^2 \qquad \text{and} \qquad \lambda - x'^2 - 3y'^2 = -2y'^2.$$

So the total derivative evaluated at any state, $(x', y')^T \in \mathcal{E}$ simplifies to:

$$-2 \begin{pmatrix} x'^2 & x'y' \\ x'y' & y'^2 \end{pmatrix}.$$

We can check that for any $(x', y')^T \in \mathcal{E}$ this matrix has eigenvalues -2λ and 0:

$$-2 \begin{pmatrix} x'^2 & x'y' \\ x'y' & y'^2 \end{pmatrix} \begin{pmatrix} x' \\ y' \end{pmatrix} = -2 \begin{pmatrix} (x'^2 + y'^2) x' \\ (x'^2 + y'^2) y' \end{pmatrix} = -2\lambda \begin{pmatrix} x' \\ y' \end{pmatrix}$$

$$-2 \begin{pmatrix} x'^2 & x'y' \\ x'y' & y'^2 \end{pmatrix} \begin{pmatrix} -y' \\ x' \end{pmatrix} = -2 \begin{pmatrix} -x'^2 y' + x'^2 y' \\ -x'y'^2 + x'y'^2 \end{pmatrix} = 0 \begin{pmatrix} -y' \\ x' \end{pmatrix}.$$

The eigenvalues are the same for every point in the fixed circle, but the eigenvectors are not. Because one of the eigenvalues is 0, each fixed point in $\mathcal{E}$ is nonhyperbolic. This is to be expected because none of the fixed points are isolated. Since λ must be positive for $\mathcal{E}$ to exist, the nonzero eigenvalue must be negative.

We can use these facts about the linearization of the fixed points in $\mathcal{E}$ to make conclusions about the original system, first in the rotating frame and then in the rest frame. For each of these fixed points, an eigenvector for -2λ is $(x', y')^T$, which is orthogonal to $\mathcal{E}$ at that point. That is the stable eigenspace for that fixed point. For each fixed point, an eigenvector for 0 is $(-y', x')^T$, which is tangent to $\mathcal{E}$ at that point. This is the center eigenspace for that fixed point. So for each fixed point on the circle, there is an orthogonal stable eigenspace, and a tangent center eigenspace.

The stable eigenspaces in the linearization correspond to stable manifolds in the original system (in the rotating frame), and the center eigenspaces corresponds to center manifolds. All of this can be seen in the top right panel of figure 3.6. The stable manifold for each $(x', y')^T \in \mathcal{E}$ is the open half-line that radiates from the origin and that passes through $(x', y')^T$. Each fixed point in the relative equilibrium is an attracting fixed point within its own stable manifold. The union of the stable manifolds for all of the points in $\mathcal{E}$ is the whole plane minus the origin. This is the stable manifold for $\mathcal{E}$ as a whole. A center manifold for each $(x', y')^T$ is the relative equilibrium $\mathcal{E}$ itself and $\mathcal{E}$ is the only center manifold that is common to all $(x', y')^T \in \mathcal{E}$.

While each of the fixed points in $\mathcal{E}$ is nonhyperbolic, the relative equilibrium $\mathcal{E}$ as a whole is an example of a "normally hyperbolic invariant set". The word "normal" here is an old-fashioned term for orthogonal. We think of $\mathcal{E}$ as a whole as being hyperbolic in the directions orthogonal to it. Every point on the fixed circle is attracting within the half-line passing through it in the normal direction. It is only nonhyperbolic in the tangential directions to the fixed circle. Normally hyperbolic invariant sets are discussed further in section 5.6.

That $\mathcal{E}$ is an attracting set follows from the linearization at any one of its fixed points in the rotating frame. The negative eigenvalue -2λ for the orthogonal direction to $\mathcal{E}$ tells us that states sufficiently close to $\mathcal{E}$ converge to it in the orthogonal direction as time increases without bound. In other words, in the rotating frame there is an open neighborhood of $\mathcal{E}$ in which each state converges to whichever state in $\mathcal{E}$ is closest to it. So $\mathcal{E}$ is an attracting set in the rotating frame.

The motion of the states in the rest frame can be obtained from the motion of the states in the rotating frame. Reversing equation (3.13) gives:

$$\begin{pmatrix} x(t) \\ y(t) \end{pmatrix} = R_t \begin{pmatrix} x'(t) \\ y'(t) \end{pmatrix}. \tag{3.15}$$

At each moment in time, the distance of $(x(t), y(t))^T$ from $\mathcal{E}$ is the same as the distance of $(x'(t), y'(t))^T$ from $\mathcal{E}$. In the rotating frame, the distance of $(x'(t), y'(t))^T$ from $\mathcal{E}$ converges to 0 so the distance of $(x(t), y(t))^T$ from $\mathcal{E}$ converges to 0. In the rest frame, the distance of any state sufficiently close to $\mathcal{E}$ goes to 0 even though it does not converge to any point in $\mathcal{E}$. This means $\mathcal{E}$ is also an attracting set in the rest frame. In the rest frame, $\mathcal{E}$ contains a single orbit so that orbit is dense inside it. Therefore $\mathcal{E}$ is an attractor. It is, more specifically, an attracting limit cycle.

For (supercritical) Hopf bifurcations in general, an attracting fixed point with a pair of complex conjugate eigenvalues becomes nonattracting as the eigenvalues cross the imaginary axis and an attracting limit cycle emerges from the fixed point. The size of the limit cycle tends to grow as the parameter moves away from its bifurcation value. In general, the topology of the limit cycle is that of a circle, but (unlike in this example) it is usually not a geometrically perfect circle after the bifurcation.

Unlike example (3.9), the relative equilibria of the Braitenberg system contain an infinite number or orbits. When they are attracting, they are attracting sets not attractors. The basic idea is that these invariant sets correspond to collections of congruent configurations, and congruent configurations can occur in many different locations and directions in the plane. The structure of the relative equilibria of the Braitenberg system is the subject of chapter 5. Hopf-like bifurcations of the relative equilibria are discussed in chapters 6 and 7.

4 The Open Dynamical Systems for a Pair of Braitenberg Vehicles

In this chapter, we provide an overview of open dynamical systems and formally specify the open dynamical systems for a pair of Braitenberg vehicles.

4.1 Overview of Open Dynamical Systems

We have defined the term *open dynamical system* in previous works (Hotton and Yoshimi 2010, 2011), and we have shown how to analyze such systems in terms of their typical behaviors. We say these systems are "open" since they have an agent that is exposed to influences outside of itself. Open dynamical systems can be used to compare a system's intrinsic behavior with its behavior in an environmental context, or to compare its behavior in different environments.

In our implementation of Braitenberg vehicles, a single agent by itself engages in extremely simple behavior, moving in a straight line forever like a particle in a universe without forces. The internal state of the agent persists at just a single point when it is by itself, since its sensor activations persist unchanged. However, two Braitenberg vehicles interacting in a simple way can behave in a complicated fashion, which is manifested in the way they represent each other over time.

Open dynamical systems make use of three spaces and five functions. The three spaces are denoted by S_α, S_τ, and T. The *agent state space* S_α is the state space for an agent by itself. The *total state space* S_τ is the state space for an environment with one or more embedded agents.[1] The space T is the set of moments in time.

---

1. We will use the term "embedding" in the sense of cognitive science, not in a topological sense. In particular, the space S_α is not necessarily embedded topologically inside the space S_τ, although that is allowed.

Three of the five functions are dynamical systems:

$$\phi_\alpha : S_\alpha \times T \to S_\alpha$$

$$\phi_\tau : S_\tau \times T \to S_\tau$$

$$\varphi_\tau : S_\tau \times T \to S_\tau.$$

The *agent dynamical system* ϕ_α gives us the intrinsic dynamics of an agent by itself. In our implementation of Braitenberg vehicles, ϕ_α is the simplest possible dynamical system, the identity dynamical system, that maps all states to themselves. When left to themselves, an agent's neurons simply remain in their current state. The *total dynamical system* ϕ_τ gives us the overall dynamics. In our implementation of Braitenberg vehicles, ϕ_τ will give us the motion in a plane for a pair of interacting vehicles. These two dynamical systems are required to satisfy compatibility conditions that are expressed in terms of the dynamical system φ_τ, described shortly.

The remaining two of the five functions are

$$\pi : S_\tau \to S_\alpha$$

$$\sigma : S_\alpha \to S_\tau.$$

The map π associates states of the total system to states of an agent. We want each state of the total system to produce a unique state in each agent. This determines the function π (we will have two such functions, one for each agent). We assume that π is continuous and surjective: any sufficiently small change of the total state only results in a small change in the agents' state and for every agent state there is at least one total state that can produce it. We think of π as a type of projection from S_τ to S_α.

The function σ goes from the agent state space back to the total state space. We require σ to be continuous, and it must be compatible with the projection π: each state of an agent is mapped by σ to a unique state in the total system which is in turn mapped back to that agent state by π. In other words, the composition $\pi \circ \sigma$ must be the identity map on S_α. When the composition of two functions is the identity map, we say that they are left and right inverses of each other. The fact that π is a left inverse of σ or that σ is a right inverse of π implies that σ is one to one. Therefore σ is a continuous injection. The function σ can satisfy further conditions such as being an immersion or even a topological embedding, but these properties are not required for σ in the definition of an open dynamical system.

There are constraints on what the total dynamical system can be, given the agent dynamical system. That is, there must be a coherent way to relate the agent dynamical system ϕ_α to the total dynamical system ϕ_τ. These constraints are described by the dynamical system φ_τ, which roughly speaking describes a dynamical system on the total state space S_τ that is compatible with the agent dynamics. Specifically, we require the set $\sigma(S_\alpha)$ to be invariant under the dynamical system φ_τ and the restriction of φ_τ to $\sigma(S_\alpha)$ to be equivalent[2] to the dynamical system ϕ_α on S_α. Finally, we require that it be possible to continuously deform the dynamical system φ_τ to the dynamical system ϕ_τ.

These conditions relate the intrinsic dynamics of an agent by itself to the dynamics of the same agent in an environment. When these conditions are met, then S_α, S_τ, T, ϕ_α, ϕ_τ, φ_τ, π, and σ together constitute an open dynamical system.

The concept of an open dynamical system can shed light on the analysis of embedded agents as we show in section 6.7. However, they also present some complications not present in a closed dynamical system, so we introduce some mathematical tools to facilitate their analysis.

We call the projection (via π) of an orbit of the total system ϕ_τ to the agent space S_α a *path* . A set of paths displayed in an agent space is an *open phase portrait* (some examples are are shown in figure 6.8). The topology of paths for an open dynamical system are much more varied than the topology of orbits for a dynamical system. Unlike the orbits of a dynamical system, distinct paths of an open dynamical system can cross each other. For instance, a path can be a figure eight curve that crosses itself once or it can be a curve that crosses itself an infinite number of times. It's even possible for the same path to be traversed in opposite directions at different times.

It is common in dynamical systems theory to focus on exceptional orbits, often without calling attention to their exceptional character. For instance, the isolated fixed points of a dynamical system are often the targets of analysis. Isolated fixed points are exceptional orbits. There are no other fixed points in a small enough neighborhood of an isolated fixed point, and what happens qualitatively in that neighborhood can often be determined by linearizing the dynamical system at that fixed point. Thus, we can learn about the typical orbits of a dynamical system from the exceptional orbits.

We can partition the state space of a dynamical system into exceptional orbits and the regions between them, which contain generic orbits. Open dynamical systems cannot be analyzed in quite the same way, because their paths cross each other and so there is no straightforward partition of the agent state space into exceptional orbits and the regions between them. So here we briefly introduce some concepts we have developed for analyzing the agent

---

2. More precisely stated, "equivalent" here means "topologically conjugate" for discrete-time dynamical systems and "topologically equivalent" for continuous-time dynamical systems.

state space of an open dynamical system. For further details on these concepts, see (Hotton and Yoshimi 2010, 2011). We say a path is a *generic path* if an orbit in its preimage under π is contained by an open invariant set all of whose orbits have the same topology. Otherwise we say it is an *exceptional path*. Note that generic and exceptional paths can overlap each other in an agent state space.

When a path in an agent space is a single point, we call it a fixed point (section 3.1). As with classical dynamical systems, the fixed points of an open dynamical system tend to be exceptional. We distinguish two kinds of fixed points for an open dynamical system. A *fully fixed point* is one whose preimage under π is an invariant set of the total dynamical system ϕ_τ. Whenever the total system is in a state that projects to a fully fixed point, it stays in states that project to that point. Thus paths in the open phase portrait never cross a fully fixed point. A fully fixed point is like a fixed point of a dynamical system: paths can converge to it, but they cannot pass through it. A *partially fixed point* is a fixed point whose preimage is not an invariant set but that does contain at least one orbit of ϕ_τ. In some cases, the system will converge to or remain in a partially fixed point, but in other cases, the system will simply pass through that point. Paths in an open phase portrait can cross a partially fixed point.

When an agent state begins at a partially fixed point, it may or may not remain in place in the agent state space, depending on whether or not the initial condition of the total dynamical system is in an orbit contained by the preimage of the partially fixed point. We will see that the only fixed points in the agent state spaces for the Braitenberg vehicles are partially fixed points: they are agent states that remain fixed for some but not all of the configurations of the vehicles that produce them.

A fully or partially fixed point is a type of *attracting path* if its preimage contains an attracting set. If the state of the total dynamical system is sufficiently close to an attracting set in the preimage of an attracting fixed point, then the state in the agent state space converges toward the fixed point.

4.2 The Braitenberg Vehicles as Open Dynamical Systems

For the Braitenberg vehicles, the total state space, S_τ, will be the configuration space for the two vehicles in a Euclidean plane. The configuration is fully characterized without redundancies by the positions, $(x_n, y_n)^T$, and headings, θ_n, of the two vehicles, $n = 1, 2$. We put these variables together into a hextuple:

$$s = (x_1, y_1, \theta_1, x_2, y_2, \theta_2)^T.$$

The form of the total state space can be concisely expressed as $S_\tau \cong (\mathbf{R}^2 \times \mathbf{S}^1)^2$. We let the time space T be the set of real numbers $\mathbf{R}$.

Since we have two agents, we actually have two open dynamical systems, one for each vehicle. They share the same total state space, but they each have their own internal states and their own agent state spaces , which we denote by S_{α_n} for $n = 1, 2$. Thus, for each of the two agents, $n = 1, 2$, we have a projection, an injection, and a dynamical system:

$$\pi_n: S_T \to S_{\alpha_n},$$

$$\sigma_n: S_{\alpha_n} \to S_T,$$

$$\phi_{\alpha_n}: S_{\alpha_n} \times \mathbf{R} \to S_{\alpha_n}.$$

The state of each agent corresponds to the activation levels of its two sensory neurons, so that the members of S_{α_n} can be written as the ordered pair $(a_{(\ell,n)}, a_{(r,n)})^T$. The activation levels go from 0 to 1, so S_{α_n} is a subset of $[0, 1]^2$. It might seem at first that agent n's state space S_{α_n} should be *all* of $[0, 1]^2$. However, not all of these states are possible because there are pairs of numbers in $[0, 1]^2$ that cannot be produced by any possible configuration of vehicles. For example, a sensor activation can only be 1 when the other agent is located on that sensor. Since the other agent cannot be located on both sensors at the same time, the state $(1, 1)^T \in [0, 1]^2$ is not possible for the agent. Thus, S_{α_n} is a proper subset of $[0, 1]^2$; specifically, it is the pentagon shown in the right panel of figure 4.1 (the open circles ○ in that figure correspond to states where the other agent is on top of a sensor). In section 4.3, we determine the precise form of this pentagon [see equation (4.5)].

Agent states in an open dynamical system can often be interpreted as representations (Hotton and Yoshimi 2011). For the Braitenberg system, states in the agent space S_{α_n} can be interpreted as representations of the position of the other agent ($\neg n$). The question of what a representation is has a long history in cognitive science and philosophy and continues to be a topic of discussion and debate (Markman and Dietrich 2000; Barack and Krakauer 2021). As we will see, the states of S_{α_n} reliably indicate the presence and location of the other agent, in the sense that given an agent state we can make inferences about whether the other agent is present, and if it is, where it is located. Reliable indication is a general feature of most theories of representation.[3]

3. Other components are often added to the concept of a representation. For example, it should be possible to activate an agent state absent its normal cause (representations are "detachable"; they are not "stimulus bound"), which allows for cases of mis-representation, planning in the absence of the object, and so on. For the Braitenberg system, we can easily imagine the sensors being stimulated by an external force absent a normal stimulus. It is generally assumed that representations are used by the agent to guide its behavior, which is the case for the Braitenberg system. An additional assumption, not made by the Braitenberg model, is that computational processes transform representations into other representations, to support adaptive behaviors. These ideas are compatible with extensions of our model, but we emphasize that philosophical issues about representation

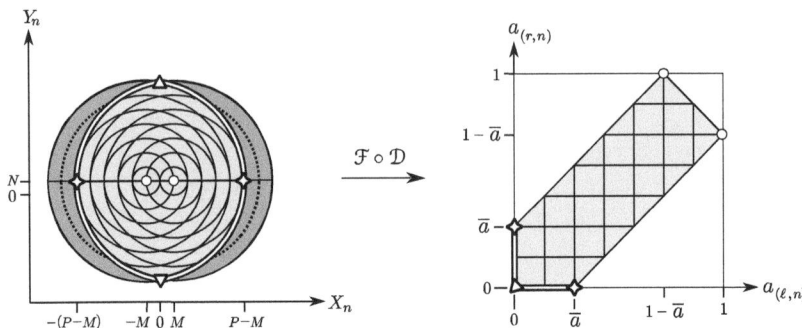

Figure 4.1
(left) The fields of view in the body frame for agent n. (right) The agent state space S_{α_n} for agent n. The functions $\mathcal{F}$ and $\mathcal{D}$ are defined in section 4.3. The value $\bar{a}$ corresponds to the maximum value one sensor can take when the other sensor is inactive. This activation occurs when the other agents is located at the four-pointed stars in the left panel. It is also defined in section 4.3.

A framework for interpreting these representations is given in figure 4.1. The left panel shows the fields of view for agent n. Recall that the central lune in agent n's body frame is where the fields of view of its sensors overlap, that is, $D_{(\ell,n)} \cap D_{(r,n)}$. The central lune is shaded light gray in the left panel of the figure, and it produces states drawn in the same shade in the interior of the pentagonal agent space, shown in the right panel of the figure. These agent states represent the position of the other agent as being in the front half of the central lune.

The left lune in agent n's body frame is the portion of the left sensor's field of view that is outside of the right sensor's field of view, that is, $D_{(\ell,n)} \setminus D_{(r,n)}$. The left lune is shaded dark gray in the left panel of the figure. In the agential representation, these states have collapsed to the common boundary of the left and central lunes, shown as a white circular arc in the left panel of the figure. They produce states on the bottom edge of S_{α_n}, shown as a white line segment in the right panel of the figure.

The right lune in agent n's body frame is the portion of the right sensor's field of view that is outside of the left sensor's field of view, that is, $D_{(r,n)} \setminus D_{(\ell,n)}$. The right lune is shaded dark gray in the left panel of the figure. In the agential representation, these states have collapsed to the common boundary of the right and central lunes, shown as a white circular arc in the left panel of the figure.

are logically independent of the utility of the open dynamical systems framework (one could be skeptical about representations and still make use of this framework).

They produce states on the left edge of S_{α_n}, shown as a white line segment in the right panel of the figure.

The vertical line segments in the interior of S_{α_n} correspond to the other agent being located at a fixed positive distance from the left sensor (i.e., $a_{(\ell,n)}$ is a positive constant). They correspond to the other agent being located on concentric circles around the left sensor. The horizontal line segments in the interior of S_{α_n} correspond to the other agent being located at a fixed positive distance from the right sensor (i.e., $a_{(r,n)}$ is a positive constant). They correspond to concentric circles around the right sensor. The intersections of the vertical and horizontal lines correspond to a specific location of the other agent relative to the two sensors. These representations are ambiguous, since there are two positions of the other agent that could produce each of these agent states: one in front of the agent, and one behind the agent. We focus on representations in the front, which is consistent with how we define the continuous injections σ_n.[4]

The cases where the other agent is only in view of one sensor and at a distance of P from the other sensor corresponds to the left and right circular arcs, shown in white in the left panel of the figure. Note that different positions along these arcs are the same maximal distance from one sensor, but at different distances from the other sensor. Only one sensor is active in these cases. These arcs map to the left and bottom edges of the agent space, which are shown in white in the right panel of the figure. Again, they correspond to the other agent being out of view of one sensor and at different distances from the other sensor. The bottom edge corresponds to the other agent being out of view of the right sensor ($a_{(r,n)} = 0$), and the left edge corresponds to the other agent being out of view of the left sensor ($a_{(\ell,n)} = 0$). The vertex $(0, 0)^T$ of the pentagonal agent space (marked by the small triangle in the bottom left of S_{α_n} in the right panel of the figure) corresponds the other agent being entirely out of view of both sensors.

The remaining three edges of S_{α_n} are the black diagonal lines. They correspond to the other agent being located in the physical plane on the line that passes through the sensors (the horizontal line running through the center of the left panel of figure 4.1). The two agent states that correspond to the other agent being located at a sensor are marked by open circles ∘ in figure 4.1. These are vertices of the pentagonal agent state space.

Paths that emerge from the left edge of S_{α_n} correspond to the other agent coming into the agent's field of view from the left. Paths that emerge from the

---

4. In a more sophisticated agent, other factors could disambiguate the representation, for example, a third sensor, an interpretation of motion, or even the supposition that nothing else can occupy the space inside an agent's body.

bottom edge of S_{α_n} correspond to the other agent coming into view from the right. Paths in the interior of S_{α_n} correspond to representations of the other agent's position within the agent's body frame.

We will study two types of attracting paths in S_{α_n}. For certain sensor weights, there is an attracting partially fixed point in S_{α_n} that corresponds to a representation of the other agent that remains at a fixed location in the agent's body frame. For other sensor weights, there is an attracting path in the form of a simple closed curve that corresponds to representations of the other agent moving periodically in the agent's body frame.

The intrinsic agent dynamics, ϕ_{α_n}, is the identity dynamical system (defined in section 3.4) on the state space S_{α_n}. In other words, when the agents are considered independently of an environment, they simply remain in whatever state they begin in. If an agent's two neurons are set to some initial state, they stay in that state indefinitely. Thus, the Braitenberg system is dominated by interactions between the two agents. In previous work (Hotton and Yoshimi 2010, 2011), we have considered agents that engage in more internal processing, but our focus here is primarily on the interaction of the agents with each other.

Even though interactions dominate the dynamics of this system, what happens in the agent spaces in isolation is still revealing. We can interpret the state of an agent as a representation of the other agent, on the basis of which it turns left or right, depending on its weights. This observation can be used to relate *representational dynamics* in the agent spaces with the dynamics of the total system, in ways that are relevant to current debates in cognitive science (Hotton and Yoshimi 2011; Yoshimi 2012).

It is notable that even with the trivial intrinsic dynamics for the agents, the open representational dynamics we observe are complicated. This is consistent with Braitenberg's original purpose in introducing his vehicles. In the following chapters, we will see a wealth of representational processes, including attracting and nonattracting partially fixed points and closed attracting paths as well as bifurcations between them as the quadruple of weights for the two vehicles are varied.

One advantage of this framework is that it allows us to be precise about the degree to which representations are determinate or not. The state $(0, 0)^T$ implies that there is no other agent in view, and is thus indeterminate with respect to where the other agent is (agent $\neg n$ could be anywhere out of view of agent n and produce the agent state $(0, 0)^T \in S_{\alpha_n}$). States on the bottom and left edges of the agent space are compatible with the other agent being anywhere in the left and right lunes, respectively (states where the agent is only in view of one sensor). States in the interior of the pentagonal agent space for agent n are within view of both of agent n's sensors and are consistent with the presence of the other agent in exactly two possible locations (one in front of agent n and one behind it). This indeterminacy could be removed by the addition of a third sensor, as discussed in footnote 4. So, the system could easily be enriched to reduce representational indeterminacy. On the other hand, the system could also be pared down, simplifying the system but increasing its indeterminacy. For example, the set of possible differences between the sensors could have been treated as a one-dimensional agent space, in which case agent states for agent n would represent the relative angular position of the other agent, $\neg n$. This representational scheme is indeterminate in the sense of being consistent with the other agent being any distance from the agent. Such a representation is sufficient to produce behaviors we describe, such as pursuit (cf. the discussion of tiger beetles in chapter 8). However, an agent endowed with such simple representations would not be able to perform more complicated behaviors in response to the relative distance of a pursuer, for example. Thus, different representational schemes are possible, and are more or less useful, depending on the needs of an agent.

4.3   The Projections π_n

We obtain the two projections, π_n, by composing three functions. The first function takes a state of the total system S_τ and uses equation (2.4) to say where the other agent is in agent n's body frame. The second function uses equation (2.5) to say what the distances are between agent n's sensors and the other agent. The third function uses the scaling function from equation (2.6) in equation (2.7) to specify the activation levels of agent n's nodes from these distances. The composition of these functions maps a hextuple $(x_1, y_1, \theta_1, x_2, y_2, \theta_2)^T$ in S_τ to the ordered pair:

$$\begin{pmatrix} a_{(\ell,n)} \\ a_{(r,n)} \end{pmatrix} = \begin{pmatrix} f\left(d_{(\ell,n)}\left(R_{(\pi/2-\theta_n)} \begin{pmatrix} x_{(\neg n)} - x_n \\ y_{(\neg n)} - y_n \end{pmatrix} \right) \right) \\ f\left(d_{(r,n)}\left(R_{(\pi/2-\theta_n)} \begin{pmatrix} x_{(\neg n)} - x_n \\ y_{(\neg n)} - y_n \end{pmatrix} \right) \right) \end{pmatrix} \qquad (4.1)$$

in the agent space S_{α_n}, for $n = 1, 2$.

We need the projections π_n to be surjective, that is, onto the agent state space, so we first determine the image of this composition of functions. To do so, it will help to introduce the symbols $\mathcal{C}_n$, $\mathcal{D}$, and $\mathcal{F}$ for the three functions being composed.

After obtaining the agent state spaces, we work backwards through these three functions (in the following section) to obtain continuous right inverses for them, which we denote by $\mathcal{C}_n^{-1}$, $\mathcal{D}^{-1}$, and $\mathcal{F}^{-1}$. We obtain continuous right inverses to π_n by composing these three inverse functions. This gives us the functions σ_n, which we use to continuously inject the agent state spaces S_{α_n} into the total state space S_τ.

We denote the coordinate transformation that maps a hextuple from the total state space S_τ to a point $(X_n, Y_n)^T$ in the body frame of agent n by $\mathcal{C}_n : S_\tau \to \mathbf{R}^2$. By equation (2.4), this function is given by the formula:

$$\mathcal{C}_n \begin{pmatrix} x_1 \\ y_1 \\ \theta_1 \\ x_2 \\ y_2 \\ \theta_2 \end{pmatrix} = R_{(\pi/2 - \theta_n)} \begin{pmatrix} x_{(\neg n)} - x_n \\ y_{(\neg n)} - y_n \end{pmatrix} \tag{4.2}$$

for $n = 1, 2$. This function is already surjective. The location of the other agent in an agent's body frame, $(X_n, Y_n)^T$, can be anywhere since the pair of agents can be located anywhere in the rest frame.

A preview for the next part of the argument is shown in figure 4.2. As in figure 4.1, the left panel shows the fields of view for the agent's two sensors, but the right panel shows the image of $\mathcal{D}$ which is the semi-infinite strip $\mathcal{H}$. This is the set of all possible ordered pairs of distances from the agent's two sensors to the other agent. The light gray pentagon in the right panel of figure 4.2 is the set of pairs of distances for the agent when the other agent is in the central lune. Its top edge corresponds to the other agent being located on the white circular arc that forms the boundary of the left and central lunes, shown in the left panel. Its right edge corresponds to the other agent being located on the white circular arc that forms the boundary of the right and central lunes, shown in the left panel.

Distances, by definition, are non-negative real numbers, that is, members of $\mathbf{R}_{\geq 0}$. We denote the function that takes the location of the other agent in agent n's body frame and returns the pair of distances to agent n's sensors by $\widetilde{\mathcal{D}} : \mathbf{R}^2 \to \mathbf{R}^2_{\geq 0}$. From equation (2.5), this function is given by the formula:

$$\widetilde{\mathcal{D}} \begin{pmatrix} X_n \\ Y_n \end{pmatrix} = \left(d_{(\ell, n)} \begin{pmatrix} X_n \\ Y_n \end{pmatrix}, \, d_{(r, n)} \begin{pmatrix} X_n \\ Y_n \end{pmatrix} \right)^T .$$

This function is not surjective. For instance, if the other agent were located at one of agent n's sensors, then it must be at a distance of $2M$ from the other sensor since $2M$ is the distance between the sensors. We now determine the image of $\widetilde{\mathcal{D}}$ and use it to define the function $\mathcal{D}$ which agrees with $\widetilde{\mathcal{D}}$ but whose range is the image of $\widetilde{\mathcal{D}}$.

The constraint on the range of $\widetilde{\mathcal{D}}$ is geometric in nature. An agent's two sensors and the center of the other agent are the vertices of a triangle in the

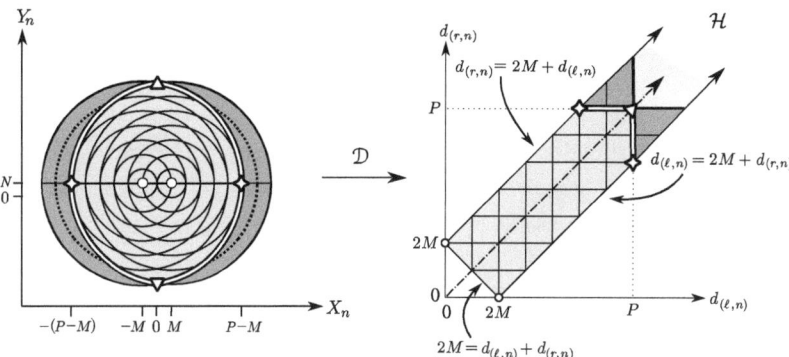

Figure 4.2
(left) The fields of view for agent n. (right) The semi-infinite strip $\mathcal{H}$ defined by the three inequalities in (4.3). Its boundaries are the three line segments indicated by their equations. In this case $P = 7M$ and the solid horizontal and vertical lines inside $\mathcal{H}$ are spaced apart at a distance of M. $\mathcal{H}$ is partitioned into four subsets according to whether $d_{(\ell,n)}$ or $d_{(r,n)}$ are greater than or less than P. The function $\mathcal{D}$ maps the central lune to the pentagon in $\mathcal{H}$ shaded light gray. It maps the left lune to the dark gray triangle above the pentagon, it maps the right lune to the dark gray triangle to the right of the pentagon, and it maps everything outside of the lunes to the light gray region above and to the right of the pentagon where $d_{(\ell,n)}$ and $d_{(r,n)}$ are greater than P.

plane. The triangle inequality theorem states that no side of a triangle can be longer than the sum of the lengths of the other two sides. The lengths of the sides of the triangle formed by an agent's sensors and the other agent are $d_{(\ell,n)}$, $d_{(r,n)}$, and $2M$. They must satisfy the three inequalities:

$$d_{(\ell,n)} \leq 2M + d_{(r,n)}$$

$$d_{(r,n)} \leq 2M + d_{(\ell,n)} \tag{4.3}$$

$$2M \leq d_{(\ell,n)} + d_{(r,n)}.$$

Equality happens for two of these inequalities when the other agent is located at one of agent n's sensors. Equality happens for just one of these inequalities when the other agent is not at one of agent n's sensors but is collinear with them.

The converse to the triangle inequality theorem states that if three nonnegative numbers, such as $d_{(\ell,n)}$, $d_{(r,n)}$, and $2M$ satisfy these three inequalities, then there is a triangle whose sides have these lengths. Thus, the image of $\widetilde{\mathcal{D}}$ is the subset of ordered pairs $(d_{(\ell,n)}, d_{(r,n)})^T \in \mathbf{R}^2_{\geq 0}$ that satisfy inequalities (4.3). We denote this subset by:

$$\mathcal{H} = \{(d_{(\ell,n)}, d_{(r,n)})^T \mid d_{(\ell,n)} \leq 2M + d_{(r,n)}, \ d_{(r,n)} \leq 2M + d_{(\ell,n)}, \ 2M \leq d_{(\ell,n)} + d_{(r,n)} \}.$$

This is the semi-infinite strip shown in the right panel of figure 4.2 and the left panel of figure 4.3. For each point in this strip, there is a triangle formed in the agent frame with edges of length $2M$, $d_{(\ell,n)}$, and $d_{(r,n)}$. For any point outside this strip, no such triangle can be formed. The vertex $(0, 2M)^T$ of $\mathcal{H}$ corresponds to the other agent being located at agent n's left sensor, and the vertex $(2M, 0)^T$ of $\mathcal{H}$ corresponds to the other agent being located at agent n's right sensor. If the other agent is directly between the two sensors, then $(d_{(\ell,n)}, d_{(r,n)})^T$ is directly between these two vertices of $\mathcal{H}$.

The strip $\mathcal{H}$ is positioned symmetrically about the identity line, $d_{(\ell,n)} = d_{(r,n)}$. The horizontal and vertical distance from any point in $\mathcal{H}$ to the identity line is at most $2M$. So the pair of distances can vary from being equal to each other [this occurs when $(d_{(\ell,n)}, d_{(r,n)})^T$ is on the identity line], to differing by as much as $2M$ from each other [this occurs when $(d_{(\ell,n)}, d_{(r,n)})^T$ is on the upper or lower bounding half-lines].

We set the function $\mathcal{D} \colon \mathbf{R}^2 \to \mathcal{H}$ equal to $\widetilde{\mathcal{D}}$ for all members of their common domain, but we make the technical distinction that the range of $\mathcal{D}$ is $\mathcal{H}$ so that it will be surjective.

The third function, $\widetilde{\mathcal{F}} \colon \mathcal{H} \to [0, 1]^2$, for the composition of π_n, takes the pair of distances to the sensors returned by $\mathcal{D}$ and maps them to a pair of sensor activations using the function from equation (2.7). It is given by the formula:

$$\widetilde{\mathcal{F}} \begin{pmatrix} d_{(\ell,n)} \\ d_{(r,n)} \end{pmatrix} = \begin{pmatrix} f\left(d_{(\ell,n)}\right) \\ f\left(d_{(r,n)}\right) \end{pmatrix}.$$

This function is not surjective. If we had made the domain of $\widetilde{\mathcal{F}}$ be all of $\mathbf{R}^2_{\geq 0}$ then it would be onto $[0, 1]^2$. However, to compose $\widetilde{\mathcal{F}}$ with $\mathcal{D}$ we need the domain of $\widetilde{\mathcal{F}}$ to be limited to the image of $\mathcal{D}$, which is $\mathcal{H}$. This forces the image of $\widetilde{\mathcal{F}}$ to be a proper subset of $[0, 1]^2$. So we need to determine the image of $\widetilde{\mathcal{F}}$ to obtain the surjective function $\mathcal{F}$.

The function $\widetilde{\mathcal{F}}$ is piecewise linear since f is piecewise linear. The region $\mathbf{R}^2_{\geq 0}$ is partitioned into four subsets according to whether $d_{(\ell,n)}$ or $d_{(r,n)}$ are greater than or less than the sensor range P. These four subsets are:

$$[0, P]^2, \qquad [0, P] \times (P, \infty), \qquad (P, \infty) \times [0, P], \qquad \text{and} \qquad (P, \infty)^2.$$

Each of these subsets intersects the semi-infinite strip $\mathcal{H}$ as shown in the right panel of figure 4.2 and the left panel of figure 4.3.

The set $\mathcal{H} \cap [0, P]^2$ is a solid convex pentagon whose interior is shaded light gray in the left panel of figure 4.3. Its set of vertices is:

$$\left\{ \begin{pmatrix} 0 \\ 2M \end{pmatrix}, \begin{pmatrix} 2M \\ 0 \end{pmatrix}, \begin{pmatrix} P-2M \\ P \end{pmatrix}, \begin{pmatrix} P \\ P \end{pmatrix}, \begin{pmatrix} P \\ P-2M \end{pmatrix} \right\}.$$

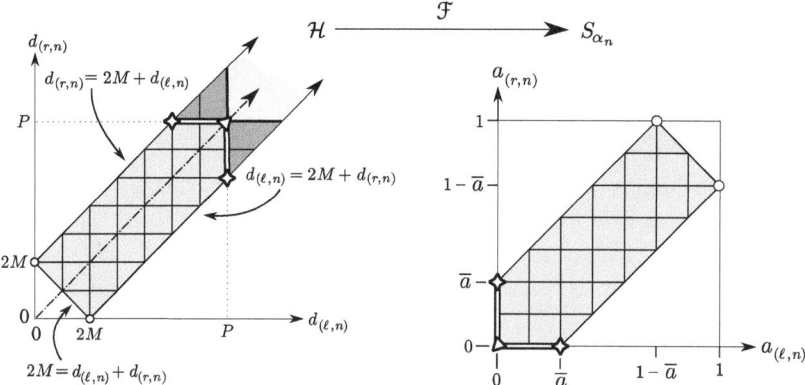

Figure 4.3
(left) The semi-infinite strip $\mathcal{H}$. (right) The pentagonal agent state space S_{α_n}. The image of the function $\mathcal{F}$ is S_{α_n}. The map $\mathcal{F}$ gives us a one to one correspondence between the light gray pentagon in $\mathcal{H}$ and the agent space S_{α_n}. It collapses the dark gray triangle above the pentagon in $\mathcal{H}$ to the bottom edge of S_{α_n}. It collapses the dark-colored triangle to the right of the pentagon in $\mathcal{H}$ to the left edge of S_{α_n}. And it collapses the light gray region above and to the right of the pentagon in $\mathcal{H}$ to the bottom left vertex of S_{α_n}, which is marked with a small triangle.

For $(d_{(\ell,n)}, d_{(r,n)})^T \in \mathcal{H} \cap [0,P]^2$ the formula for $\widetilde{\mathcal{F}}$ reduces to:

$$\widetilde{\mathcal{F}} \begin{pmatrix} d_{(\ell,n)} \\ d_{(r,n)} \end{pmatrix} = \begin{pmatrix} 1 - d_{(\ell,n)}/P \\ 1 - d_{(r,n)}/P \end{pmatrix} = \begin{pmatrix} 1 \\ 1 \end{pmatrix} - \frac{1}{P} \begin{pmatrix} d_{(\ell,n)} \\ d_{(r,n)} \end{pmatrix}. \tag{4.4}$$

Dividing $(d_{(\ell,n)}, d_{(r,n)})^T$ by P is a contraction towards the origin. Multiplying by -1 is a half-turn about the origin. Adding $(1,1)^T$ is a translation that slides the origin to $(1,1)^T$. So the restriction of $\mathcal{F}$ to $\mathcal{H} \cap [0,P]^2$ has the form of a Euclidean similarity transformation. The image of $\mathcal{H} \cap [0,P]^2$ under $\widetilde{\mathcal{F}}$ is a geometrically similar pentagon in the right panel of figure 4.3. Its set of vertices is:

$$V = \left\{ \begin{pmatrix} 1 \\ 1-\overline{a} \end{pmatrix}, \begin{pmatrix} 1-\overline{a} \\ 1 \end{pmatrix}, \begin{pmatrix} 0 \\ \overline{a} \end{pmatrix}, \begin{pmatrix} 0 \\ 0 \end{pmatrix}, \begin{pmatrix} \overline{a} \\ 0 \end{pmatrix} \right\}.$$

where $\overline{a} = 2M/P$ is the maximum activation either sensor can have when the other sensor's activation is 0.

The region $\mathcal{H} \cap ([0,P] \times (P,\infty))$ is the dark shaded triangle above the pentagon in the left panel of figure 4.3. Its image under $\widetilde{\mathcal{F}}$ is the white line segment, $[0,\overline{a}] \times \{0\}$ in the right panel of figure 4.3. The region $\mathcal{H} \cap ((P,\infty) \times [0,P])$ is the dark shaded triangle to the right of the pentagon in the left panel of figure 4.3. Its image under $\widetilde{\mathcal{F}}$ is the white line segment, $\{0\} \times [0,\overline{a}]$ in the right panel of figure 4.3. The unbounded region $\mathcal{H} \cap (P,\infty)^2$ is shaded light gray in the left panel of figure 4.3. Its image is the point $(0, 0)^T$ marked by the small triangle in the right panel of figure 4.3. The image of all of $\mathcal{H}$ under $\widetilde{\mathcal{F}}$ is the solid closed pentagon shown in the right panel of figure 4.3.

So for each pair of activations, $(a_{(\ell,n)}, a_{(r,n)})^T$, in the pentagon, $\widetilde{\mathcal{F}}(\mathcal{H})$, there is at least one pair of distances, $(d_{(\ell,n)}, d_{(r,n)})^T$, between the sensors of one agent and the position of the other agent that can generate those activations. Otherwise, there is no pair of distances that can generate those activations. Therefore, this pentagon must be the agent state space. Concisely stated:

$$S_{\alpha_n} = \text{Convexhull}(V) \tag{4.5}$$

for $n = 1, 2$.

We set the function $\mathcal{F}: \mathcal{H} \to S_{\alpha_n}$ equal to $\widetilde{\mathcal{F}}$ for all members of their common domain, but we limit the range of $\mathcal{F}$ to S_{α_n} so that it will be surjective.

To summarize, we have now defined three continuous functions: $\mathcal{C}_n$, $\mathcal{D}$, and $\mathcal{F}$. The function $\mathcal{C}_n$ maps each state of the total system to a position in the body frame for agent n. The function $\mathcal{C}_n$ is surjective. For every position in the body frame for agent n there is a state of the total system that gets mapped by $\mathcal{C}_n$ to the given position in agent n's body frame. The function $\mathcal{D}$ maps a position in the body frame for agent n to an ordered pair of distances from agent n's sensors to the other agent. The set of all ordered pairs of numbers that can be the distances from the sensors to the other agent forms a semi-infinite planar strip called $\mathcal{H}$. The image of the function $\mathcal{D}$ is all of $\mathcal{H}$. The function $\mathcal{F}$ maps each ordered pair of distances to an ordered pair of activations for agent n's sensors. The set of all possible ordered pairs of activations forms a solid pentagon. This is the agent state space S_{α_n}. The function $\mathcal{F}$ is onto S_{α_n}.

The functions π_n as given by the formula in equation (4.1) maps total states to agent states. We can now express π_n as a composition of three continuous surjective functions:

$$\pi_n = \mathcal{F} \circ \mathcal{D} \circ \mathcal{C}_n \tag{4.6}$$

for $n = 1, 2$. The composition of continuous functions is a continuous function and the composition of surjective functions is a surjective function. Therefore the functions π_n are continuous surjections and they satisfy the definition for the projections of an open dynamical system.

4.4 The Continuous Injections σ_n

Now that we have the continuous surjective functions π_n, we need to find continuous right inverses for them. We can do this by finding continuous right inverses to each of $\mathcal{F}$, $\mathcal{D}$, and $\mathcal{C}_n$ and composing the inverses together in the reverse order:

$$\pi_n : S_\tau \xrightarrow{\;\mathcal{C}_n\;} \mathbf{R}^2 \xrightarrow{\;\mathcal{D}\;} \mathcal{H} \xrightarrow{\;\mathcal{F}\;} S_{\alpha_n}$$

$$S_\tau \xleftarrow{\;\mathcal{C}_n^{-1}\;} \mathbf{R}^2 \xleftarrow{\;\mathcal{D}^{-1}\;} \mathcal{H} \xleftarrow{\;\mathcal{F}^{-1}\;} S_{\alpha_n} : \sigma_n.$$

To show that we can find σ_n by finding these three continuous right inverses, suppose they exist. Then by the associativity of the composition of functions:

$$(\mathcal{F} \circ \mathcal{D} \circ \mathcal{C}_n) \circ (\mathcal{C}_n^{-1} \circ \mathcal{D}^{-1} \circ \mathcal{F}^{-1}) = \mathcal{F} \circ (\mathcal{D} \circ (\mathcal{C}_n \circ \mathcal{C}_n^{-1}) \circ \mathcal{D}^{-1}) \circ \mathcal{F}^{-1}.$$

We can replace $(\mathcal{F} \circ \mathcal{D} \circ \mathcal{C}_n)$ on the left-hand side by π_n. If there is a right inverse, $\mathcal{C}_n^{-1}$, to $\mathcal{C}_n$ then the composition $(\mathcal{C}_n \circ \mathcal{C}_n^{-1})$ is the identity map on the space of positions for the other agent in agent n's body frame. The composition of any function with the identity map just gives the function back. So we get:

$$\pi_n \circ (\mathcal{C}_n^{-1} \circ \mathcal{D}^{-1} \circ \mathcal{F}^{-1}) = \mathcal{F} \circ (\mathcal{D} \circ \mathcal{D}^{-1}) \circ \mathcal{F}^{-1}.$$

By the same reasoning if there is a right inverse, $\mathcal{D}^{-1}$, to $\mathcal{D}$ then $(\mathcal{D} \circ \mathcal{D}^{-1})$ is the identity map on $\mathcal{H}$ and we get:

$$\pi_n \circ (\mathcal{C}_n^{-1} \circ \mathcal{D}^{-1} \circ \mathcal{F}^{-1}) = \mathcal{F} \circ \mathcal{F}^{-1}.$$

Again if there is a right inverse, $\mathcal{F}^{-1}$, to $\mathcal{F}$ then the right-hand side of this equation is the identity map on S_{α_n}. This implies that $(\mathcal{C}_n^{-1} \circ \mathcal{D}^{-1} \circ \mathcal{F}^{-1})$ is a right inverse to π_n. Furthermore, if $\mathcal{C}_n^{-1}$, $\mathcal{D}^{-1}$, and $\mathcal{F}^{-1}$ are continuous, then $(\mathcal{C}_n^{-1} \circ \mathcal{D}^{-1} \circ \mathcal{F}^{-1})$ is a continuous right inverse for π_n. So we can set

$$\sigma_n = \mathcal{C}_n^{-1} \circ \mathcal{D}^{-1} \circ \mathcal{F}^{-1} \tag{4.7}$$

for $n = 1, 2$ to get our continuous right inverses for π_n. The functions σ_n defined this way are one to one. They are a continuous injections of the agent state spaces S_{α_n} in the total state space S_τ.

So to find σ_n we find the continuous right inverses $\mathcal{F}^{-1}$, $\mathcal{D}^{-1}$, and $\mathcal{C}_n^{-1}$. As we will see, there is only one choice for $\mathcal{F}^{-1}$, two choices for $\mathcal{D}^{-1}$ (depending on whether the other agent is in front of or behind agent n), and infinitely many choices for $\mathcal{C}_n^{-1}$ because congruent configurations of agents can be located anywhere in the rest frame.

We will start by finding a continuous right inverse for $\mathcal{F}$ and work our way in. A continuous right inverse for $\mathcal{F}$ will tell us how far the other agent has to be from agent n's sensors to produce a given pair of sensor activations $(a_{(\ell,n)}, a_{(r,n)})^T$. Each point in the half-open subset $S_{\alpha_n} \cap (0, 1]^2$ has a unique preimage in $\mathcal{H}$ under the function $\mathcal{F}$. So, on this subset, the right inverse to $\mathcal{F}$ must be given by the inverse to the Euclidean similarity in equation (4.4),

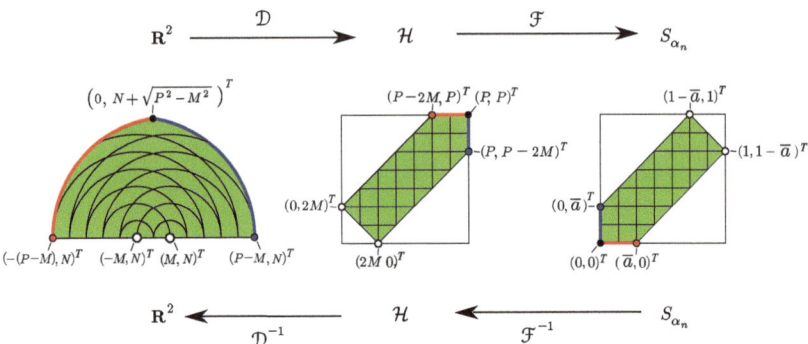

Figure 4.4
(left) The front half of the central lune is shaded lime green. (middle) The space $\mathcal{H} \cap [0,P]^2$, the interior of which is also shaded lime green. (right) The agent state space, S_{α_n}, is the solid pentagon, the interior of which is shaded lime green. The image of S_{α_n} under the similarity transformation $\mathcal{F}^{-1}$ is $\mathcal{H} \cap [0,P]^2$. The image of $\mathcal{H} \cap [0,P]^2$ under the function $\mathcal{D}^{-1}$ is the closure of the front half of the central lune. The function $\left(\mathcal{D}^{-1} \circ \mathcal{F}^{-1}\right)$ takes vertical lines in S_{α_n} to circular arcs about the left sensor at $(-M,N)^T$ and horizontal lines in S_{α_n} to circular arcs about the right sensor at $(M,N)^T$.

which is also a Euclidean similarity:

$$\begin{pmatrix} d_{(\ell,n)} \\ d_{(r,n)} \end{pmatrix} = \mathcal{F}^{-1} \begin{pmatrix} a_{(\ell,n)} \\ a_{(r,n)} \end{pmatrix} = P \begin{pmatrix} 1-a_{(\ell,n)} \\ 1-a_{(r,n)} \end{pmatrix}. \qquad (4.8)$$

This Euclidean similarity is a continuous injection, and it has a unique continuous extension to the closure of $S_{\alpha_n} \cap (0,1]^2$, which happens to be all of S_{α_n}. Thus equation (4.8) defines $\mathcal{F}^{-1}$ for all $(a_{(\ell,n)}, a_{(r,n)})^T$ in S_{α_n}. This is our first right inverse. It is the only continuous right inverse there is for $\mathcal{F}$.

We now find a continuous right inverse for $\mathcal{D}$, which tells us where the other agent must be in an agent's body frame, that is, coordinates X_n and Y_n needed to produce a given pair of distances $d_{(\ell,n)}$ and $d_{(r,n)}$. There are in general two choices for these coordinates, and we pick the one for which Y_n is positive. We start with the solution to equation (2.5) for $(X_n, Y_n)^T$ in terms of $(d_{(\ell,n)}, d_{(r,n)})^T$. This is:

$$\begin{pmatrix} X_n \\ Y_n \end{pmatrix} = \begin{pmatrix} 0 \\ N \end{pmatrix} + \frac{1}{4M} \begin{pmatrix} d_{(\ell,n)}^2 - d_{(r,n)}^2 \\ \pm\sqrt{U} \end{pmatrix}. \qquad (4.9)$$

where

$$U = -\left(4M^2 - (d_{(\ell,n)} - d_{(r,n)})^2\right)\left(4M^2 - (d_{(\ell,n)} + d_{(r,n)})^2\right).$$

We need to contend with the fact that real valued solutions to equation (2.5) do not exist for some values of $(d_{(\ell,n)}, d_{(r,n)})^T$. The value of U is a fourth-degree polynomial of $d_{(\ell,n)}$ and $d_{(r,n)}$. We can re-write the inequalities in (4.3) as:

$$2M + (d_{(\ell,n)} - d_{(r,n)}) \geq 0$$

$$2M - (d_{(\ell,n)} - d_{(r,n)}) \geq 0$$

$$2M - (d_{(\ell,n)} + d_{(r,n)}) \leq 0.$$

Since $d_{(\ell,n)}$, $d_{(r,n)}$, $2M$ are always nonnegative their sum is nonnegative

$$2M + (d_{(\ell,n)} + d_{(r,n)}) \geq 0.$$

The function U is the negative of the product of the left-hand side of these four inequalities. So U is non-negative for all $(d_{(\ell,n)}, d_{(r,n)})^T$ in $\mathcal{H}$. Thus $\sqrt{U}$ is a continuous function for all $(d_{(\ell,n)}, d_{(r,n)})^T$ in $\mathcal{H}$. The remaining arithmetic operations in equation (4.9) are continuous and therefore the point $(X_n, Y_n)^T$ depends continuously on $(d_{(\ell,n)}, d_{(r,n)})^T$.

Unlike with the continuous injection $\mathcal{F}^{-1}$, the continuous injection $\mathcal{D}^{-1}$ is not uniquely determined. If $U = 0$ then $Y_n = N$ and the other agent's location is collinear with the sensors. Otherwise, we have two choices for Y_n, which depend on the choice of sign for the radical in equation (4.9). The positive choice places the other agent in front of the agent's sensors. The negative choice places it behind the agent's sensors. We choose so that the σ_n we get will place the other agent in front of the sensors. Our choice for $\mathcal{D}^{-1}$ is the continuous injection:

$$\begin{pmatrix} X_n \\ Y_n \end{pmatrix} = \mathcal{D}^{-1}\begin{pmatrix} d_{(\ell,n)} \\ d_{(r,n)} \end{pmatrix} = \begin{pmatrix} 0 \\ N \end{pmatrix} + \frac{1}{4M}\begin{pmatrix} d_{(\ell,n)}^2 - d_{(r,n)}^2 \\ \sqrt{U} \end{pmatrix}. \qquad (4.10)$$

The composite function $(\mathcal{D}^{-1} \circ \mathcal{F}^{-1})$ is a continuous injection. It relates agent n's sensor activations to a location $(X_n, Y_n)^T$ in its body frame where the other agent can produce those activations.

As we have seen, the coordinate grid in S_{α_n} gets mapped to circular arcs centered about the two sensors. A state in S_{α_n} lies at the intersection of a vertical and horizontal line segment. These line segments are typically mapped to a pair of circular arcs that have a unique intersection point in front of the agent. This is illustrated in figure 4.4. Each vertical line segment in S_{α_n} corresponds to the other agent being located at a fixed distance from the agent's left sensor. Thus $(\mathcal{D}^{-1} \circ \mathcal{F}^{-1})$ maps the vertical line segments to arcs of circles centered about the left sensor. Similarly for horizontal line segments in S_{α_n} and the right sensor. We can summarize the effect of $(\mathcal{D}^{-1} \circ \mathcal{F}^{-1})$ as follows. If either of the activation levels is 1 then the other agent must be located at the corresponding sensor. Otherwise the two activation levels determine a pair of semicircles centered about the sensors, and the other agent is located at the semicircles' intersection point.

We now proceed to find continuous right inverses for each $\mathcal{C}_n$. They give us a hextuple $s = (x_1, y_1, \theta_1, x_2, y_2, \theta_2)^T \in S_\tau$ that produces the body frame coordinates $(X_n, Y_n)^T$. There are infinitely many choices for $\mathcal{C}_n^{-1}$, and we choose one that is convenient. To go from the body frame to the rest frame, we can let the agent be located at the origin of the rest frame, and let it be headed in the vertical direction. This way the rest frame position of the other agent is the same as the body frame position of the other agent. An agent's state is unaffected by the heading of the other agent, so it does not matter what heading we give the other agent. For convenience, we also give it a vertical heading. This results in slightly different continuous injections for the two agents :

$$\mathcal{C}_1^{-1}(X_1, Y_1)^T = \left(0, 0, \frac{\pi}{2}, X_1, Y_1, \frac{\pi}{2}\right)^T$$

$$\mathcal{C}_2^{-1}(X_2, Y_2)^T = \left(X_2, Y_2, \frac{\pi}{2}, 0, 0, \frac{\pi}{2}\right)^T.$$

The total state obtained by σ_n from the agent state can be explicitly written out as:

$$\sigma_1 \begin{pmatrix} a_{(\ell,1)} \\ a_{(r,1)} \end{pmatrix} = \left(0, 0, \frac{\pi}{2}, (\mathcal{D}^{-1} \circ \mathcal{F}^{-1}) \begin{pmatrix} a_{(\ell,1)} \\ a_{(r,1)} \end{pmatrix}, \frac{\pi}{2}\right)^T$$

(4.11)

$$\sigma_2 \begin{pmatrix} a_{(\ell,2)} \\ a_{(r,2)} \end{pmatrix} = \left((\mathcal{D}^{-1} \circ \mathcal{F}^{-1}) \begin{pmatrix} a_{(\ell,2)} \\ a_{(r,2)} \end{pmatrix}, \frac{\pi}{2}, 0, 0, \frac{\pi}{2}\right)^T.$$

where $(\mathcal{D}^{-1} \circ \mathcal{F}^{-1})$ can be evaluated from equations (4.8) and (4.10).

We now have our projections $\pi_n: S_\tau \to S_{\alpha_n}$ and continuous injections $\sigma_n: S_{\alpha_n} \to S_\tau$ so that $\pi_n \circ \sigma_n$ is the identity map on S_{α_n}. Note that the reverse composition, $\sigma_n \circ \pi_n$, does not give us the identity map on S_τ. By itself, the condition that $\pi_n \circ \sigma_n$ be the identity map imposes little constraint on how the agents can move through the plane. Further constraints do involve the functions π_n and σ_n, but they also involve the dynamical systems ϕ_{α_n}, ϕ_τ, and φ_{τ_n}.

In the next two sections, we specify ϕ_τ and φ_{τ_n} to complete the definition of an open dynamical system.

4.5   The Total Dynamical System ϕ_τ

The dynamical system for ϕ_τ is obtained as the solution to a differential equation which we derive now. The magnitude of the nth Braitenberg vehicle's velocity is fixed at v_n. The direction of its velocity is the same as its heading, that is, θ_n. So for $n = 1, 2$

$$\begin{pmatrix} \dot{x}_n \\ \dot{y}_n \end{pmatrix} = v_n \begin{pmatrix} \cos(\theta_n) \\ \sin(\theta_n) \end{pmatrix}.$$ (4.12)

The rate of change in the heading of a Braitenberg vehicle at any moment in time is the weighted difference between the activation level of its two sensors. This is expressed by the turning function in equation (2.8). Using equation (4.2) we re-express this as a function of the state variables:

$$\dot{\theta}_n = \bar{\omega}_n \left(\mathcal{C}_n(x_1, y_1, \theta_1, x_2, y_2, \theta_2)^T \right)$$

$$= \omega_n (x_1, y_1, \theta_1, x_2, y_2, \theta_2)^T.$$ (4.13)

Note that the function $\bar{\omega}_n$ is the turning function for the body frame, and ω_n is the turning function for the total space.

We combine equations (4.12) and (4.13) into the equation of motion for the two Braitenberg vehicles:

$$\begin{pmatrix} \dot{x}_1 \\ \dot{y}_1 \\ \dot{\theta}_1 \\ \dot{x}_2 \\ \dot{y}_2 \\ \dot{\theta}_2 \end{pmatrix} = F \begin{pmatrix} x_1 \\ y_1 \\ \theta_1 \\ x_2 \\ y_2 \\ \theta_2 \end{pmatrix} = \begin{pmatrix} v_1 \cos(\theta_1) \\ v_1 \sin(\theta_1) \\ \omega_1(x_1, y_1, \theta_1, x_2, y_2, \theta_2)^T \\ v_2 \cos(\theta_2) \\ v_2 \sin(\theta_2) \\ \omega_2(x_1, y_1, \theta_1, x_2, y_2, \theta_2)^T \end{pmatrix}.$$ (4.14)

The function F is defined to be equal to the right side of (4.14). The total dynamical system, $\phi_\tau : S_\tau \times \mathbf{R} \to S_\tau$, of our open dynamical systems is the general solution[5] to differential equation (4.14)

$$\frac{d}{dt} \phi_\tau(s, t) = F(\phi(s, t))$$

for any initial condition

$$s = (x_1, y_1, \theta_1, x_2, y_2, \theta_2)^T \in S_\tau$$

and all $t \in \mathbf{R}$. We present particular solutions in equations (6.2) and (7.6). For the most part though, we rely on equivariant dynamical systems theory (discussed in chapter 5) and numerical integration to study the total dynamical system ϕ_τ.

---

5. It is not difficult to check that the function F is bounded and differentiable on almost all of S_τ. This is compatible with F satisfying the existence and uniqueness theorem for differential equations (e.g., see Perko [2001]; Meiss [2007]). Numerical integrations show no sign of any particular solution going to infinity in finite time.

4.6   The Continuous Deformation of φ_τ to ϕ_τ

We require that, when the agent space S_{α_n} of an open dynamical system is placed in the total space using σ_n, there be a way to relate the intrinsic dynamics of ϕ_{α_n} (which in this case is the identity dynamical system) to the total dynamics ϕ_τ. That is, we require some way of saying how the intrinsic dynamics of the agents are compatible with the total dynamics when the agents interact. We do this by using another dynamical system φ_{τ_n}.

We require the subset $\sigma_n(S_{\alpha_n})$ of S_τ be an invariant set for the dynamical system φ_{τ_n} and the restriction of φ_{τ_n} to $\sigma_n(S_{\alpha_n})$ be equivalent to the dynamical system ϕ_{α_n} on S_{α_n}. Finally, we require that it be possible to continuously deform φ_{τ_n} to ϕ_τ.

For the Braitenberg vehicles, these conditions can be easily obtained. We let φ_{τ_n} be the identity dynamical system on S_τ. This makes $\sigma_n(S_{\alpha_n})$ an invariant set and the identity dynamical system restricted to $\sigma_n(S_{\alpha_n})$ is equivalent to the identity dynamical system on S_{α_n}.

Since φ_{τ_n} is the identity dynamical system, it is fairly easy to continuously deform ϕ_τ to φ_{τ_n}. Let $\beta \in [0, 1]$. Since $\phi_\tau(s, t)$ is the solution to differential equation (4.14), the dynamical system $\phi_\tau(s, \beta t)$ is a solution to the differential equation:

$$\dot{s} = \beta F(s).$$

For $\beta = 1$, we have $\phi_\tau(s, \beta t) = \phi_\tau(s, t)$. For $\beta \in (0, 1)$, the dynamical system $\phi_\tau(s, \beta t)$ is equivalent to $\phi_\tau(s, t)$. The only difference is that the orbits are traversed more slowly. For $\beta = 0$, motion stops altogether and the solution to the differential equation is the identity dynamical system φ_{τ_n}. So we can continuously deform ϕ_τ to φ_{τ_n}.

5   The Relative Equilibria of the Total Dynamical System

By definition, our abstract Braitenberg vehicles only turn in response to each other; there are no other forces that act on them. The dynamics depends only on the relative position and heading of the two vehicles. Since the speed of the Braitenberg vehicles have fixed positive values, it is not possible for the total dynamical system to have fixed points, but the vehicles can engage in something resembling fixed point behavior. Under the right circumstances, the position and headings of the Braitenberg vehicles relative to one another can remain fixed over time. Fixed points are often referred to as "equilibria." When Braitenberg vehicles are motionless relative to each other, the dynamical system has an invariant set called a "relative equilibrium."[1]

The concept of a relative equilibrium has its origins in Newtonian mechanics (Newton 1846). Newton put forward the idea that there is an absolute space within which bodies move. To explain this, he considered a circularly symmetric vessel hung from a twisted cord and filled with water. The water was thought of as a continuous medium. Initially the vessel is held fixed and the water in it is motionless. When the cord is allowed to untwist, the vessel starts to turn about its axis of symmetry, and as it turns, it drags the water along with it. Sometime after the vessel has reached a nearly constant angular velocity, the water in it will be nearly at rest with respect to the vessel. Ideally no portion of the water moves relative to any other portion. We now call this state of the water a "hydrostatic equilibrium."

The water's exposed surface is flat when it is at rest. Otherwise, when the water is in a hydrostatic equilibrium, the surface is lower at the rotation axis than along the vessel's sides. Because of the many ways in which the water could move about, it may seem that determining its precise shape when it is in a hydrostatic equilibrium is a difficult problem. However the problem can be simplified by making use of the fact that the water in a hydrostatic equilibrium

---

1. Demonstrations of continuity will be omitted throughout this chapter.

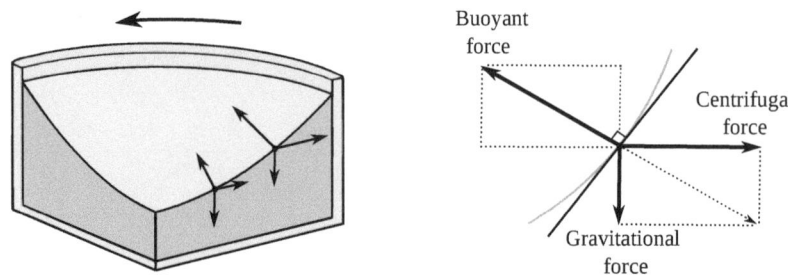

Figure 5.1
(left) A quarter slice of water in a hydrostatic equilibrium. At each point on the surface, the gravitational, centrifugal, and buoyant forces sum to zero. (right) At each point, the surface is orthogonal to the buoyant force.

is motionless in a rotating frame.[2] This turns it into a problem in statics, and the shape of the upper surface can be deduced from the balance of forces acting on it.

In the rotating frame, there are two external forces acting at each point on the water's surface: the force of gravity that pushes the water down, and the centrifugal force that pushes the water away from the rotation axis and towards the sides of the vessel. In order for the surface to remain motionless, the two external forces must counterbalance the buoyant force that derives from the pressure exerted by the water against its surface. Therefore the buoyant force is opposite to the sum of the gravitational and centrifugal forces (see figure 5.1). Using calculus, it can be shown that as a result of these interacting forces, the surface of the water is a circular paraboloid.

It is also the case that despite the many ways in which Braitenberg vehicles can move, they often become virtually motionless with respect to each other. This simplifies their analysis. Even without a general solution to the governing equations (4.14), finding relative equilibria for Braitenberg vehicles can be reduced to a geometrical problem. This is done in sections 6.2 and 7.2. This is analogous to how, despite the many ways water can move, it has a fairly simple shape when it is in a hydrostatic equilibrium.

A relative equilibrium is a subset of states of a dynamical system that are collectively "steady." A hydrostatic equilibrium is an example of a relative equilibrium. The water appears to be at a standstill in the rotating frame. Similarly, Braitenberg vehicles appear to be at a standstill when viewed in an appropriate frame when the system is in a relative equilibrium.

---

2. Ignoring the microscopic motion of the water molecules.

Relative equilibria are conceived of in terms of a mathematical structure known as a "transformation group," which are the allowed ways we can transform a system (e.g., the water, or the vehicles). For the water in the spinning vessel, the transformation group is the set of rotations about the fixed axis established by the cord in three-dimensional Euclidean space. If we could instantaneously shift the spinning vessel even further around its spin axis without any other alterations, its subsequent behavior would be the same as before. The water would continue to appear to be at a standstill in the rotating frame.

For the Braitenberg vehicles, the transformation group is the set of all rotations and translations of the Euclidean plane. These transformations are known as the "proper congruences" of a Euclidean plane. The transformation group as a whole is denoted by $\mathbf{SE}(2)$. If we were to instantaneously apply a proper congruence to a pair of Braitenberg vehicles, they would appear to each other to be in the same relative position, and their subsequent behavior would remain essentially the same. They would continue to move in such a way as to be relatively motionless with respect to each other.

In this section, we present the modern mathematical definition of a relative equilibrium and apply it to our six-dimensional dynamical system. The abstract concept of a relative equilibrium links two ideas: the concept of an invariant set for a dynamical system and the concept of equivalence class for a transformation group.

As we saw in chapter 3, an invariant set of a dynamical system is a subset of its state space with the following property: if the system begins in that subset and we run the dynamics, it will stay in that subset for all time. Invariant sets can be attracting, repelling, or neither depending on what happens in a small neighborhood of the invariant set. An attracting invariant set is called an *attracting set* and a repelling invariant set is called a *repelling set*.

The attracting sets of a dynamical system determine the long-term behavior of the system. For initial conditions far from the attracting set, there can be a period of transient behavior that is unlike the behavior near the attracting set. Figure 5.2 shows transient behavior by the Braitenberg system before it nears a relative equilibrium in which the agents revolve around a common center. During the transient period, the agents trace out loops of increasing size as the distance between the agents grows to the radius of revolution. After that, the agents follow a circular path for all time.

The attracting sets of dynamical systems are often single orbits, e.g., a fixed point or a periodic orbit, but in other cases the attracting sets can be collections of orbits, e.g., a quasiperiodic or chaotic attracting set. For our six-dimensional system, the attracting sets are collections of orbits. For example, two agents can revolve around a common center for all time, but any point in the plane

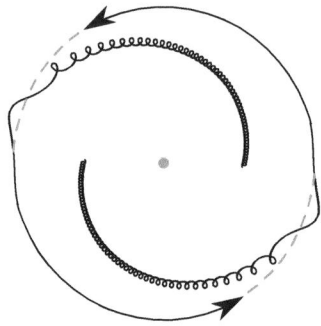

Figure 5.2
Two vehicles engaging in transient behavior. Their physical paths are colored black. In this example, the agents begin in a configuration that is symmetrical under a half-turn whose center is marked by the gray dot. The configuration remains symmetrical under the half-turn while the agents trace out loops of increasing size. After reaching the attracting set, they cease to trace out loops and simply revolve around their common center.

can be the center of revolution. Revolving about a common center corresponds to a single orbit of the dynamical system. The collection of all orbits for the revolving agents constitutes the attracting set. In this case, a single orbit is not an attracting set because, for instance, if we perturb the system by translating one agent slightly, the system need not return to revolving around the same center but instead it can revolve around a nearby point. What is attracting in this example is not a single orbit but an entire collection of orbits that constitute an attracting set of ϕ_τ.

An attracting set is an attractor if it contains an orbit that is dense in it, that is, an orbit that passes arbitrarily close to every point in the invariant set.[3] Fixed points and periodic orbits can be attractors since they only contain one orbit. Quasiperiodic and chaotic attracting sets are often attractors even though they contain more than one orbit. The attracting sets for our Braitenberg vehicles are not attractors. Given any orbit, we can find another orbit in the same invariant set that is arbitrarily far apart from it. Thus, no orbit in the attracting set can pass close to every point in the attracting set.

The six-dimensional system also has an intrinsic symmetry: if we change the positions and headings of the two agents in the rest frame without changing the positions and headings of the two agents in their body frames, they will behave in the same way. We can describe this mathematically in terms of the transformation group **SE**(2) acting on the entire six-dimensional total space.[4]

---

3. This is closely related to the concept of ergodicity.
4. It might seem that a better way to capture the symmetry would be to consider a three-dimensional space describing the relative positions and angle of the two agents. This space removes three degrees of freedom, which are from a certain perspective redundant. In this reduced-dimensional space, what we call relative equilibria become fixed points and relative periodic orbits become periodic orbits. However, we do not follow this strategy because (1) such an analysis makes it difficult to study the remarkable physical behaviors observed in this system, which are comparable to motions of real-world agents, e.g., ascidian larvae, as we'll see; (2) we do in fact see something like this reduced system in the agent spaces (seeing how the agents make use of

We define a group action so that the rotations and translations in $SE(2)$ can be applied to S_τ to rotate or translate configurations of two vehicles. Applying a member of $SE(2)$ to a configuration of vehicles—a point in S_τ—produces an "equivalent" configuration. If we take all the points in S_τ that can be reached from some initial point using $SE(2)$—all the ways of rotating and translating the two agents rigidly—we get an equivalence class in S_τ, a set of points all of which are equivalent in that the relative positions and headings of the agents are the same. We also call these "proper congruence classes." . We can in fact partition all of S_τ into these equivalence classes in which the agents have congruent configurations. Each of these equivalence classes is a copy of $SE(2)$, so that we can partition S_τ into a collection of copies of $SE(2)$.

For a dynamical system equivariant with respect to a group action, a relative equilibrium can occur when an invariant set of the dynamical system is also an equivalence class of the group action. For example, when the agents move around each other in a circle, they exhibit a stable type of behavior in an invariant set that is also an equivalence class of the group action. The whole set of these rotating behaviors is an attracting set of the dynamical system, and it is also the set obtained by applying arbitrary rotations and translations in the symmetry group $SE(2)$ to any configuration in this set. Thus we see how dynamical systems and group actions come together in the definition of a relative equilibrium.

This chapter is organized as follows. Section 5.1 explains the modern concept of a congruence and a proper congruence for the Euclidean plane. The set of all proper congruences of the plane (translations, rotations, and the identity map) forms the group $SE(2)$. Section 5.2 discusses the basic facts we will need to know about transformation groups, and $SE(2)$ in particular. Section 5.3 shows how the rotations and translations in $SE(2)$ act on S_τ to transform configurations of Braitenberg vehicles in the six-dimensional space S_τ. Section 5.4 applies a modern mathematical definition of equivariance to pairs of Braitenberg vehicles. We show that ϕ_τ is equivariant with respect to $SE(2)$. Section 5.5 introduces a modern mathematical definition of relative equilibrium, formalizing the idea that when we have an equivariance between a dynamical system and a group action, a relative equilibrium is an invariant set of the dynamical system, which is also an equivalence class of the group action. The two main types of relative equilibria for the Braitenberg vehicles are distinguished using one-parameter subgroups. We call these the "revolving type" and "translating type" relative equilibria. Section 5.6 reviews the history of equivariant

---

this reduced system is itself of some philosophical interest); and (3) analyzing the symmetry using group actions places this analysis alongside other discussions of symmetry in modern physics.

dynamical systems theory and introduces techniques (used in the rest of the book) for using a moving frame to convert an orbit of a relative equilibrium to a collection of fixed points and then linearizing about one of these fixed points. In section 5.7, relative periodic orbits are introduced, which are sets of equivalence classes the system moves through periodically. These correspond to meandering behaviors in the physical plane. We distinguish between revolving and translating types of these meandering behaviors. In sections 6.2 and 7.2, we identify revolving type and translating type relative equilibria along with relative periodic orbits for the system.

5.1   The Concept of Congruence in Modern Geometry

Geometry has had the concept of "congruence" ($\iota\sigma o \varsigma$) since antiquity (Euclid 1956). The basic idea is that congruent figures have the same size and shape. Congruence was specifically defined for some types of geometric figures. For instance, two circles were said to be congruent if and only if their diameters had the same length.

In addition to definitions of congruence for various types of geometric figures, there are theorems that can simplify the task of deciding whether two figures are congruent. For example, two rhombi are congruent if and only if the pair of diagonals of one rhombus have the same lengths as the pair of diagonals of the other rhombus. The lengths of a rhombus's diagonals determines its other geometric properties such as the lengths of its sides and the angles at its vertices.

Felix Klein proposed, in his Erlangen program (Klein 1893), that geometry could be founded on a more general conception of congruence. This reconception of geometry had its origins in the study of the motion of rigid bodies, such as spinning tops and oscillating pendulums. After a sufficiently large body of knowledge about rigid body motions had developed, it was possible to use this knowledge to create a different type of foundation for geometry. Rigid motions could then be understood as paths in a group of congruences. Each point in the group of congruences corresponds to a congruence that maps an initial configuration to a subsequent configuration.

In this section, we give a brief introduction to a modern classification of the congruences in a Euclidean plane.[5] The classification is summarized in table 5.1. This introduction will rely on physical intuition about the continuous motion of objects through space or within a plane.

---

5. A more thorough elementary introduction can be found in Prenowitz and Swain (1966) and a more advanced introduction can be found in Carne (2012).

	Number of fixed points		
	0	**1**	**> 1**
proper	translation	rotation	identity map
improper	glide-reflection		reflection

Table 5.1
A classification of the congruences of a Euclidean plane into five types. There are no improper congruences that fix exactly one point.

By assumption, each individual Braitenberg vehicle moves as a rigid body in the plane. The two of them can undergo the same rigid motion together. When they do the rigid motion they perform together is one of two types: the uniform linear motions and the uniform circular motions. These are defined in equations (5.1) and (5.2).

A *congruence* is a distance preserving function from a Euclidean plane to itself.[6] We will denote an arbitrary congruence by $g \colon \mathbf{R}^2 \to \mathbf{R}^2$. Not surprisingly, Euclidean congruences can preserve other geometric properties such as angles and areas. This is analogous to how the lengths of a rhombus' diagonals determine its other geometric properties. In modern geometry two figures are said to be *congruent* if there is a congruence that maps one figure to the other. Congruences establish a one to one correspondence between the points of two congruent figures as well as between their other geometric properties such as angles.

The congruences of the Euclidean plane can be classified into five types as shown in table 5.1. This classification is based on the number of points in the plane the congruence leaves fixed and how the congruence can be generated by a rigid motion. A rigid motion generates a congruence if the starting configuration of the rigid motion is mapped by the congruence to the final configuration of the rigid motion.

A *proper* congruence is one that can be generated by a rigid motion that occurs entirely within the plane. An improper congruence can only be generated by a rigid motion that has to "pick the rigid body up" and take it outside of the plane before moving it to its final destination back in the plane.

---

6. A distance preserving function is generally called an "isometry" but it is useful to call an isometry of Euclidean space to itself a congruence because of the fundamental role such functions play. In non-Euclidean geometry a congruence need not be a distance preserving map. One of the goals of Klein's Erlangen program was to unify non-metric geometries, such as projective geometry, with metric geometries, such as Euclidean geometry (Klein 1893; Stillwell 2005). It is also not uncommon for people to call a "congruence" a "rigid motion" but we will use the phrase "rigid motion" to describe the continuous motion of a rigid body.

We now review each of the five types of congruences going clockwise through table 5.1. A translation adds a fixed nonzero vector $(x_g, y_g)^T$ to each point in the plane. A translation has the form:

$$g \begin{pmatrix} x \\ y \end{pmatrix} = \begin{pmatrix} x_g \\ y_g \end{pmatrix} + \begin{pmatrix} x \\ y \end{pmatrix}.$$

No point in the plane is fixed by a translation.

Translations can be generated by sliding the plane in a fixed direction at a fixed speed. This is called "uniform linear motion." A uniform linear motion can be expressed as adding scalar multiples of $(x_g, y_g)^T$ to $(x, y)^T$, that is,

$$\begin{pmatrix} x \\ y \end{pmatrix} \mapsto t \begin{pmatrix} x_g \\ y_g \end{pmatrix} + \begin{pmatrix} x \\ y \end{pmatrix}. \tag{5.1}$$

As t goes from 0 to 1 this motion takes each point, $(x, y)^T$, to the point, $(x_g, y_g)^T + (x, y)^T$, at a constant speed along a line segment within the plane.

A rotation around an arbitrary point $(\bar{x}, \bar{y})^T$ adds a fixed angle θ_g to the direction each point is from $(\bar{x}, \bar{y})^T$. The point $(\bar{x}, \bar{y})^T$ is called the center of the rotation. A rotation can be implemented by first translating the plane by $-(\bar{x}, \bar{y})^T$. This sends the point $(\bar{x}, \bar{y})^T$ to the origin. This is followed by rotating the plane around the origin by the angle θ_g. Finally we translate back by $(\bar{x}, \bar{y})^T$. A rotation has the form:

$$g \begin{pmatrix} x \\ y \end{pmatrix} = \begin{pmatrix} \bar{x} \\ \bar{y} \end{pmatrix} + R_{\theta_g} \begin{pmatrix} x - \bar{x} \\ y - \bar{y} \end{pmatrix}.$$

The only point in the plane fixed by a rotation is its center.

Rotations can be generated by spinning the plane at a constant angular speed around the center of rotation. This is called "uniform circular motion". A uniform circular motion can be expressed as adding multiples of θ_g to the direction each point, $(x, y)^T$, is from the center $(\bar{x}, \bar{y})^T$, that is,

$$\begin{pmatrix} x \\ y \end{pmatrix} \mapsto \begin{pmatrix} \bar{x} \\ \bar{y} \end{pmatrix} + R_{(t\theta_g)} \begin{pmatrix} x - \bar{x} \\ y - \bar{y} \end{pmatrix}. \tag{5.2}$$

As t goes from 0 to 1 this motion takes each point $(x, y)^T$ to $(\bar{x}, \bar{y})^T + R_{\theta_g}(x - \bar{x}, y - \bar{y})^T$ at a constant angular speed along a circular arc within the plane.

The identity map is a congruence that fixes every point in $\mathbf{R}^2$. It counts as a proper congruence since we do not have to move a body out of the plane to move it to itself.

The improper congruences will also play a role in our analysis of Braitenberg vehicles so we briefly discuss them here. A reflection fixes a line in the plane called the axis of reflection. It sends every other point in the plane to the opposite side of the axis so that the perpendicular bisector of the point and its image is the axis of the reflection.

A reflection can be generated by spinning the plane around the reflection axis but this takes the plane outside of itself. A reflection cannot be generated by a motion that occurs entirely within the plane so it is an improper congruence.

A glide-reflection is the result of composing a reflection with a translation in a direction that is not orthogonal to the reflection axis. An improper congruence composed with a proper congruence is an improper congruence, so a glide reflection is an improper congruence. A glide reflection fixes no points.

Proper congruences have the general form:

$$g\begin{pmatrix} x \\ y \end{pmatrix} = \begin{pmatrix} x_g \\ y_g \end{pmatrix} + R_{\theta_g} \begin{pmatrix} x \\ y \end{pmatrix}. \tag{5.3}$$

If $\theta_g = 0$ and $(x_g, y_g)^T = (0,0)^T$ then g is the identity map. If $\theta_g = 0$ and $(x_g, y_g)^T \neq (0,0)^T$ then g is a translation by $(x_g, y_g)^T$. If $\theta_g \neq 0$ then g is a rotation by angle θ_g around the center:

$$\begin{pmatrix} \bar{x} \\ \bar{y} \end{pmatrix} = (I - R_{\theta_g})^{-1} \begin{pmatrix} x_g \\ y_g \end{pmatrix}.$$

Each proper congruence g is uniquely determined by specifying values for the three numbers x_g, y_g, and θ_g in equation (5.3). We put these three numbers in an ordered triple:

$$(x_g, y_g, \theta_g)^T \leftrightarrow ((x_g, y_g)^T, \theta_g)^T$$

$$\mathbf{R} \times \mathbf{R} \times S^1 \cong \mathbf{R}^2 \times S^1$$

There is a one to one correspondence between the proper congruences of the Euclidean plane and the set of ordered triples $(x_g, y_g, \theta_g)^T$. Since this bijection and its inverse are continuous, the group of all proper congruences, $\mathbf{SE}(2)$, has the topology of $\mathbf{R}^2 \times S^1$, that is, the plane times a circle. It also has the same topology as an open solid torus.

5.2   The Congruence Group SE(2)

The set of all congruences of a Euclidean plane forms a transformation group (Klein 1893; Bredon 1972). The congruences of a Euclidean plane are transformations that preserve distances and other geometric properties. Not all transformations are congruences, for example a shear in the Euclidean plane is a transformation that is not a Euclidean congruence.

Recall from section 3.3 that a set of transformations is a transformation group if the composition of any two members of the set is in the set and the inverse of any member of the set is in the set. The composition of two congruences is a congruence since distances are preserved in applying one congruence after the other. It is straightforward to find the inverse of a congruence. The identity map and the reflections are their own inverses. A translation adds a fixed vector to each point in the plane. Its inverse adds the opposite of that fixed vector to each point in the plane. The inverse of a rotation is a rotation about the same center by the opposite angle. The inverse of a glide-reflection is another glide-reflection which is the result of composing the same reflection with a translation in the opposite direction. Therefore the set of Euclidean congruences is a transformation group.

We call a transformation group made up of congruences a *congruence group*. From a modern standpoint, a congruence group together with a compatible topological space defines a geometry. The set of all congruences of the Euclidean plane is a congruence group called the *Euclidean group* and it is commonly denoted by $E(2)$. The set of proper congruences, the congruences we are primarily interested in, forms a subgroup of $E(2)$. It is called the *special Euclidean group*. As we have stated it is commonly denoted by $SE(2)$ and it has the topology of an open solid torus.

Transformations groups have *subgroups*, that is, subsets of the transformation group that also satisfy the conditions for being a transformation group on their own. It is convenient to regard the identity map both as a type of translation (i.e., the trivial translation by no distance) and as a type of rotation (i.e., the trivial rotation by $0°$). This allows us to say that the set of all translations and the set of all rotations about the origin each form subgroups of $SE(2)$. Each translation is given by the addition of some fixed vector (possibly $(0,0)^T$) in $\mathbf{R}^2$ to all of the points in $\mathbf{R}^2$. The subgroup of all translations has the topology of $\mathbf{R}^2$. The set of all rotations that fix the origin forms another subgroup. Each rotation about the origin is specified by the angle (possibly $0°$) by which the plane is turned. This subgroup has the topology of $\mathbf{S}^1$. The topology of $SE(2)$ is the product topology of these two subgroups: translations in $\mathbf{R}^2$ and rotations in $\mathbf{S}^1$.

These are not the only subgroups of $SE(2)$. Within $SE(2)$ there are infinitely many subgroups corresponding to translations in particular directions and rotations about different centers. An important class of subgroups of $SE(2)$ are the "one-parameter subgroups", which describe continuous motions. A *one-parameter subgroup* is a continuous map:

$$h\colon \mathbf{R} \to SE(2)$$

such that

$$h(t_1 + t_2) = h(t_1) \circ h(t_2) \tag{5.4}$$

for all t_1, $t_2 \in \mathbf{R}$.

One-parameter subgroups are useful in the study of motion where it is convenient to think of the domain of the function as time.

Every one-parameter subgroup maps 0 to the identity map in $\mathbf{SE}(2)$. To see this, observe from the definition of h that $h(0)$ equals itself when it is composed with itself:

$$h(0) = h(0+0) = h(0) \circ h(0)$$

Equation (5.3) implies that the only proper congruence that equals itself when composed with itself is the identity map. Therefore the congruence $h(0)$ must be the identity map on $\mathbf{R}^2$. So any one-parameter subgroup is a curve whose image is inside of the transformation group $\mathbf{SE}(2)$ and which passes through the identity map of $\mathbf{SE}(2)$ at $t = 0$.

Two distinct one-parameter subgroups can have the same image in $\mathbf{SE}(2)$. What distinguishes two such one-parameter subgroups is the speed at which the image in $\mathbf{SE}(2)$ is traversed.

There are three types of one-parameter subgroups for $\mathbf{SE}(2)$. They turn out to correspond to the three simple types of rigid motions that generate the three types of proper congruences. One type of one-parameter subgroup corresponds to the absence of motion, that is, $h(t)$ is the identity map of the plane for all $t \in \mathbf{R}$. We shall not concern ourselves with this one-parameter subgroup. The uniform linear motions in equation (5.1) and the uniform circular motions in equation (5.2) are the other two types of one-parameter subgroups of $\mathbf{SE}(2)$:

$$h : t \mapsto \left(\begin{pmatrix} x \\ y \end{pmatrix} \mapsto t \begin{pmatrix} x_g \\ y_g \end{pmatrix} + \begin{pmatrix} x \\ y \end{pmatrix} \right)$$

$$h : t \mapsto \left(\begin{pmatrix} x \\ y \end{pmatrix} \mapsto \begin{pmatrix} \bar{x} \\ \bar{y} \end{pmatrix} + R_{(t\theta_g)} \begin{pmatrix} x - \bar{x} \\ y - \bar{y} \end{pmatrix} \right)$$

These are the types of rigid motions the Braitenberg vehicles engage in when they undergo the same rigid motion together. When the two vehicles undergo the same rigid motion together then they are of course motionless with respect to each other.

5.3 The Action of SE(2) on S_τ

When Braitenberg vehicles undergo the same rigid motion together their configuration at each moment in time remains properly congruent to their initial configuration. In this section we explain how proper congruences in $\mathbf{SE}(2)$ can act on the six-dimensional configuration space for two Braitenberg vehicles, S_τ. We will see that this group action is *effective*, which means that no two proper congruences have the same effect on the vehicle's configuration.

Each proper congruence in **SE**(2) acts differently on S_τ from any other proper congruence in **SE**(2). On this basis we also show that the space S_τ can be partitioned into copies of **SE**(2). The equivalence classes of this partition can be invariant sets of the total dynamical system and it is these invariant sets that are the relative equilibria of ϕ_τ.

Suppose we have two pairs of Braitenberg vehicles whose configurations are denoted by

$$s = (x_1, y_1, \theta_1, x_2, y_2, \theta_2)^T$$
$$s' = (x'_1, y'_1, \theta'_1, x'_2, y'_2, \theta'_2)^T$$

If the configurations are properly congruent then there is a proper congruence, g, mapping one configuration to the other. The positions and headings of the second pair of Braitenberg vehicles s' can be written in terms of the positions and headings of the first pair s in terms of g:

$$\begin{pmatrix} x'_1 \\ y'_1 \end{pmatrix} = g \begin{pmatrix} x_1 \\ y_1 \end{pmatrix} \qquad \begin{pmatrix} x'_2 \\ y'_2 \end{pmatrix} = g \begin{pmatrix} x_2 \\ y_2 \end{pmatrix}$$
$$\theta'_1 = \theta_1 + \theta_g \qquad\qquad \theta'_2 = \theta_2 + \theta_g$$

The corresponding transformation of the configuration space S_τ maps the point which corresponds to the first configuration to the point which corresponds to the second configurations. This can be written out as:

$$s' = G_g(s) = \begin{pmatrix} g \begin{pmatrix} x_1 \\ y_1 \end{pmatrix} \\ \theta_1 + \theta_g \\ g \begin{pmatrix} x_2 \\ y_2 \end{pmatrix} \\ \theta_2 + \theta_g \end{pmatrix} \tag{5.5}$$

It useful to regard the symbol "G" by itself as standing for a function. This function maps a proper congruence, g, to a transformation of the configuration space S_τ which we denote by G_g. In other words G is a function from one function space, **SE**(2), to another function space which is a group of transformations of S_τ to itself. It is convenient to put the argument to the function G in its subscript. Thus G_g is the transformation of S_τ that results when we apply the function G to $g \in$ **SE**(2). The argument to the transformation G_g is written in the usual way for arguments to functions. Giving a configuration $s \in S_\tau$ as the argument to the transformation G_g is expressed as $G_g(s)$.

An important property of the function G is that it is a "homomorphism", which maps a composition of transformations to a composition of transformations. We leave it to the reader to check that for $g_1, g_2 \in \mathbf{SE}(2)$,

$$G_{g_1 \circ g_2} = G_{g_1} \circ G_{g_2}. \tag{5.6}$$

This is like condition (5.4). In general, a function from one transformation group to another that satisfies condition (5.6) is a *homomorphism*. We also say that G defines a *group action* of $\mathbf{SE}(2)$ on S_τ.

We proceed to show that the homomorphism G is one to one, that is, for any distinct $g_1, g_2 \in \mathbf{SE}(2)$ the transformations G_{g_1}, G_{g_2} of S_τ are distinct. We show this in three steps. First, note that for any $g \in \mathbf{SE}(2)$ the function $g \circ g^{-1}$ is the identity map on the plane. It does not change the vehicles' configuration. So $G_{(g \circ g^{-1})}$ must be the identity map on S_τ. By the same reasoning, $G_{(g^{-1} \circ g)}$ is the identity map on S_τ. By the homomorphism condition (5.6),

$$G_{(g \circ g^{-1})} = G_g \circ G_{(g^{-1})}$$
$$G_{(g^{-1} \circ g)} = G_{(g^{-1})} \circ G_g.$$

Since composing G_g with $G_{(g^{-1})}$ in either order gives the identity map on S_τ, the transformation $G_{(g^{-1})}$ must be the inverse of G_g:

$$G_g^{-1} = G_{(g^{-1})}. \tag{5.7}$$

Next, note that for any $g \in \mathbf{SE}(2)$, aside from the identity map, no configuration in S_τ is fixed by G_g. Translations change the positions of the vehicles, while rotations change their headings. So no configuration is fixed by a proper congruence other than the identity map. Conversely, if $G_g(s) = s$ for some configuration $s \in S_\tau$, then g must be the identity map on the plane.

Finally, suppose for some $g_1, g_2 \in \mathbf{SE}(2)$ there is a configuration $s \in S_\tau$ such that the transformations G_{g_1} and G_{g_2} of S_τ have the same effect on s:

$$G_{g_1}(s) = G_{g_2}(s). \tag{5.8}$$

Applying the inverse transformation, $G_{g_2}^{-1}$, to both sides of equation (5.8) gives:

$$(G_{g_2}^{-1} \circ G_{g_1})(s) = s.$$

Substituting from equation (5.7) gives:

$$(G_{g_2^{-1}} \circ G_{g_1})(s) = s.$$

By the homomorphism condition (5.6):

$$G_{(g_2^{-1} \circ g_1)}(s) = s.$$

So $g_2^{-1} \circ g_1$ must be the identity map on the plane. Proper congruences are invertible functions, so they have unique inverses and therefore $g_1 = g_2$.

In summary, if G_{g_1} and G_{g_2} have the same effect on any state $s \in S_\tau$, then g_1 and g_2 must be the same proper congruence. No two distinct proper congruences of the plane are mapped to the same transformation of S_τ by the homomorphism G. The homomorphism G is one to one.

A homomorphism that is one to one is called an *isomorphism*.[7] So G is an isomorphism. An isomorphism between transformation groups establishes a one-to-one correspondence between its domain and image that is compatible with the composition of transformations. This basically means, the domain and image of an isomorphism are equivalent as transformation groups. The image of the isomorphism G is a copy of **SE**(2) that acts on S_τ. We can also say that G defines an effective action of **SE**(2) on S_τ.

The action of **SE**(2) on S_τ will be used to define an equivalence relation on S_τ. We explain in section 5.5 that each relative equilibrium for our two Braitenberg vehicles is an equivalence class of this equivalence relation.[8]

We define a relation on S_τ by asserting that a pair of points $s, s' \in S_\tau$ satisfy the relation if and only if there is a $g \in$ **SE**(2) such that $G_g(s) = s'$. It follows from the fact that **SE**(2) is a transformation group that this is an equivalence relation on S_τ. Equivalence relations are reflexive, symmetric, and transitive. The reflexivity property follows from the fact that the identity map is in **SE**(2). The symmetry property follows from the fact that the inverse of every member of **SE**(2) is in **SE**(2). The transitivity property follows from the fact that the composition of any two members of **SE**(2) is in **SE**(2). So if for $s, s' \in S_\tau$, there is a $g \in$ **SE**(2) such that $G_g(s) = s'$ then s is equivalent to s' in the sense that they are properly congruent configurations.

We call the equivalence classes of this equivalence relation *proper congruence classes* since equivalent points in S_τ correspond to configurations of vehicles that are properly congruent. Since there are configurations that are not congruent to each other, there is more than one proper congruence class in S_τ. This highlights a distinction between how **SE**(2) acts on the Euclidean plane and how it acts on S_τ. For any given point in a plane, there are infinitely many proper congruences in **SE**(2) that can map it to any other given point. However, there are points in S_τ that cannot be mapped to each other by any proper congruence in **SE**(2).

We now show that S_τ can be partitioned into copies of **SE**(2). We first define a function from the transformation group **SE**(2) to each proper congruence class in S_τ. We begin by choosing an arbitrary point in the state space $s \in S_\tau$

7. Not to be confused with an isometry.
8. These equivalence classes are usually called "group orbits," not to be confused with dynamical system orbits.

and using it to define the function, $\alpha_s \colon \mathbf{SE}(2) \to S_\tau$. For each $g \in \mathbf{SE}(2)$, we set[9]:

$$\alpha_s(g) = G_g(s). \tag{5.9}$$

We now show that α_s is a bijection from $\mathbf{SE}(2)$ to the proper congruence class containing s. In other words, when we apply the function α_s to the entire group of proper congruences, $\mathbf{SE}(2)$, we get a copy of $\mathbf{SE}(2)$ in S_τ.

By definition, there is a $g \in \mathbf{SE}(2)$ such that $s' = G_g(s)$ if and only if s' is in the same proper congruence class as s. So by equation (5.9), $s' = \alpha_s(g)$ if and only if s' is in the same proper congruence class as s. Therefore, the function α_s is onto the proper congruence class containing s.

To show that α_s is one to one, suppose that $\alpha_s(g_1) = \alpha_s(g_2)$ for some $g_1, g_2 \in \mathbf{SE}(2)$. Then equation (5.8) holds because of equation (5.9). Equation (5.8) implies $g_1 = g_2$. Therefore the function α_s is one to one. Since the function α_s is one to one and onto it must be a bijection from $\mathbf{SE}(2)$ to the proper congruence class containing s.

Since $s \in S_\tau$ was arbitrary, this argument can be applied to any proper congruence class in S_τ. They are all copies of $\mathbf{SE}(2)$. The action of $\mathbf{SE}(2)$ on S_τ creates a partition of S_τ into proper congruence classes, each of which is a copy of $\mathbf{SE}(2)$.

5.4 The Equivariance of ϕ_τ with SE(2)

We can now say precisely what it means for the dynamical system ϕ_τ to be equivariant with respect to the transformation group $\mathbf{SE}(2)$. Recall that intuitively speaking this means that if we translate or rotate both vehicles the same way, then they will continue to behave in essentially the same way in their new location. The way the Braitenberg vehicles move only on depends on how the vehicles are situated relative to each other. Where they happen to be in the plane does not matter.

For instance, suppose that time is momentarily "paused" for two pursuers that are revolving around each other and that the same congruence is applied to both vehicles. When time is "turned back on," the vehicles will continue to revolve around each other like before.

Formally, we say that ϕ_τ is *equivariant* with respect to $\mathbf{SE}(2)$ if for all $s \in S_\tau$, $t \in \mathbf{R}$, and $g \in \mathbf{SE}(2)$

$$\phi_\tau(G_g(s), t) = G_g(\phi_\tau(s, t)). \tag{5.10}$$

---

9. Compare with Bredon (1972), 40.

In other words, if we apply the transformation G_g to the initial configuration s and determine the resulting configuration under the dynamical system at time t, we get the same configuration as if we apply the dynamical system to the initial configuration s to determine the resulting configuration at time t and then apply the transformation G_g to that configuration.

To show that the dynamical system ϕ_τ is equivariant with respect to $\mathbf{SE}(2)$, let s be some fixed member of S_τ, g be some fixed member of $\mathbf{SE}(2)$, and set $s' = G_g(s)$. The functions of time $t \mapsto \phi_\tau(s, t)$ and $t \mapsto \phi_\tau(s', t)$ are curves in S_τ. At time $t = 0$, the curve for $\phi_\tau(s, t)$ passes through the point s and the curve for $\phi_\tau(s', t)$ passes through s'.

We apply the transformation G_g to the entire image of the curve for $\phi_\tau(s, t)$ to get a another curve that passes through s' at time $t = 0$. By the chain rule and the fact that ϕ_τ satisfies the equation of motion (4.14), the tangent vectors to $G_g(\phi_\tau(s, t))$ are given by multiplying the total derivative DG_g times the function F from equation (4.14):

$$\frac{d}{dt} G_g(\phi_\tau(s, t)) = DG_g \left(\frac{d}{dt} \phi_\tau(s, t) \right)$$

$$= DG_g \left(F(\phi_\tau(s, t)) \right). \tag{5.11}$$

The computation of DG_g from equation (5.5) is facilitated by writing the orbits of s and s' as time-dependent hextuples:

$$(x_1(t), y_1(t), \theta_1(t), x_2(t), y_2(t), \theta_2(t))^T = \phi_\tau(s, t)$$
$$\left(x_1'(t), y_1'(t), \theta_1'(t), x_2'(t), y_2'(t), \theta_2'(t) \right)^T = \phi_\tau(s', t)$$

The matrix DG_g has a block diagonal form:

$$DG_g = \begin{pmatrix} R_{\theta_g} & & & \\ & 1 & & \\ & & R_{\theta_g} & \\ & & & 1 \end{pmatrix}$$

Notice that all the partial derivatives in DG_g are constant so DG_g is independent of time. From the trigonometric addition rules for cosine and sine:

$$R_{\theta_g} \begin{pmatrix} v_n \cos(\theta_n(t)) \\ v_n \sin(\theta_n(t)) \end{pmatrix} = \begin{pmatrix} v_n \cos(\theta_n(t) + \theta_g) \\ v_n \sin(\theta_n(t) + \theta_g) \end{pmatrix}$$

for $n = 1, 2$. Therefore $DG_g(F(\phi_\tau(s, t))) =$

$$
\begin{pmatrix}
v_1 \cos(\theta_1(t) + \theta_g) \\
v_1 \sin(\theta_1(t) + \theta_g) \\
\omega_1(x_1(t), y_1(t), \theta_1(t), x_2(t), y_2(t), \theta_2(t))^T \\
v_2 \cos(\theta_2(t) + \theta_g) \\
v_2 \sin(\theta_2(t) + \theta_g) \\
\omega_2(x_1(t), y_1(t), \theta_1(t), x_2(t), y_2(t), \theta_2(t))^T
\end{pmatrix}.
\tag{5.12}
$$

We now show this is equal to $F(G_g(\phi_\tau(s, t)))$. By equation (5.5), the transformation G_g maps $\theta_n(t)$ to $\theta'_n(t) = \theta_n(t) + \theta_g$ for $n = 1, 2$. So when we apply F to $G_g(\phi_\tau(s, t))$, we get the first, second, fourth, and fifth components in (5.12).

For the third and sixth components of (5.12), we note that because s and s' are congruent configurations the conversion of their positions in the rest frame to the body frames, that is, the image of C_n, is the same regardless of which of these configurations (s or s') we use:

$$
C_n(s') = C_n(G_g(s)) = R_{(\pi/2 - (\theta_n(t) + \theta_g))} \left(g \begin{pmatrix} x_{(\neg n)}(t) \\ y_{(\neg n)}(t) \end{pmatrix} - g \begin{pmatrix} x_n(t) \\ y_n(t) \end{pmatrix} \right)
$$

$$
= R_{(\pi/2 - (\theta_n(t) + \theta_g))} \left(R_{\theta_g} \begin{pmatrix} x_{(\neg n)}(t) \\ y_{(\neg n)}(t) \end{pmatrix} - R_{\theta_g} \begin{pmatrix} x_n(t) \\ y_n(t) \end{pmatrix} \right)
$$

$$
= R_{(\pi/2 - (\theta_n(t) + \theta_g))} R_{\theta_g} \left(\begin{pmatrix} x_{(\neg n)}(t) \\ y_{(\neg n)}(t) \end{pmatrix} - \begin{pmatrix} x_n(t) \\ y_n(t) \end{pmatrix} \right)
\tag{5.13}
$$

$$
= R_{(\pi/2 - \theta_n(t))} \left(\begin{pmatrix} x_{(\neg n)}(t) \\ y_{(\neg n)}(t) \end{pmatrix} - \begin{pmatrix} x_n(t) \\ y_n(t) \end{pmatrix} \right) = C_n(s)
$$

for $n = 1, 2$. So by equation (4.13),

$$
\omega_n(s') = \bar{\omega}_n(C_n(s')) = \bar{\omega}_n(C_n(s)) = \omega_n(s)
\tag{5.14}
$$

for $n = 1, 2$. Consequently, the third and sixth components of $DG_g(F(\phi_\tau(s, t)))$ are the same as in (5.12). Therefore:

$$
DG_g(F(\phi_\tau(s, t))) = F(G_g(\phi_\tau(s, t))).
\tag{5.15}
$$

Substituting this into equation (5.11) gives:

$$
\frac{d}{dt} G_g(\phi_\tau(s, t)) = F(G_g(\phi_\tau(s, t))).
$$

This shows that $\phi_\tau(s', t)$ and $G_g(\phi_\tau(s, t))$ both satisfy differential equation (4.14). Since

$$
\phi_\tau(s', 0) = s'
$$

$$
G_g(\phi_\tau(s, 0)) = G_g(s) = s'.
$$

these two solutions begin with the same initial condition s'. It then follows from the uniqueness of solutions to initial value problems for ordinary differential equations that these two solutions are the same:

$$\phi_\tau(s', t) = \phi_\tau(G_g(s), t) = G_g(\phi_\tau(s, t))$$

for all $t \in \mathbf{R}$. Since $s \in S_\tau$ and $g \in \mathbf{SE}(2)$ were arbitrary, condition (5.10) is satisfied. Therefore the dynamical system ϕ_τ is equivariant with respect to $\mathbf{SE}(2)$.[10]

5.5   Translating and Revolving Type Relative Equilibria

We can now proceed to give a formal definition for the relative equilibria of our abstract Braitenberg vehicles and to classify the types of relative equilibria that occur for them.

Traditional definitions of relative equilibria have often been phrased in terms of individual orbits of a dynamical system. We follow the more recent definition for a relative equilibrium by Krupa (1990) and applied to spiral tip meander in Golubitsky, LeBlanc, and Melbourne (1997), although some of our terminology is different. By this definition, a relative equilibrium is a union of orbits that satisfy two conditions. For our system, the first condition requires that ϕ_τ be equivariant with respect to $\mathbf{SE}(2)$. We showed this condition holds in section 5.4. Second, we require that a proper congruence class in S_τ be invariant under the dynamical system ϕ_τ. That is, an equivalence class of the group action is an invariant set of the dynamical system. If these two conditions are satisfied, the proper congruence class is a *relative equilibrium* of ϕ_τ. We show that particular proper congruence classes are relative equilibria for ϕ_τ in sections 6.2 and 7.2.

Some properties of relative equilibria are as follows. Suppose the proper congruence class containing some initial state s is a relative equilibrium. The orbit of s must be fully contained in the relative equilibrium. Since the map α_s defined in equation (5.9) is a bijection from $\mathbf{SE}(2)$ to the proper congruence class containing s, we can invert it. We map the orbit of s into $\mathbf{SE}(2)$. We will show that the image of the orbit is a one-parameter subgroup of $\mathbf{SE}(2)$. Anticipating this result, we set:

$$h(t) = \alpha_s^{-1}(\phi_\tau(s, t)). \tag{5.16}$$

---

10. This argument has its origin in Field (1980).

To be a one-parameter subgroup, the map h must satisfy equation (5.4). This will be shown by making use of the dynamical system property :

$$\phi_\tau(\phi_\tau(s, t_2), t_1) = \phi_\tau(s, t_2 + t_1)$$
$$= \phi_\tau(s, t_1 + t_2) \tag{5.17}$$

(see Hotton and Yoshimi [2010, 2011]; Robinson [1995]; Hasselblatt and Katok [2002]). The dynamical system ϕ_τ can be expressed in terms of the map h defined in equation (5.16):

$$\phi_\tau(s, t) = \alpha_s(h(t)) = G_{h(t)}(s). \tag{5.18}$$

Using equation (5.18) twice with $t = t_2$ and $t = t_1$ we get:

$$\phi_\tau(\phi_\tau(s, t_2), t_1) = \phi_\tau(G_{h(t_2)}(s), t_1)$$
$$= G_{h(t_1)}(G_{h(t_2)}(s))$$
$$= (G_{h(t_1)} \circ G_{h(t_2)})(s).$$

Since G is an isomorphism, we can use equation (5.6) to get:

$$\phi_\tau(\phi_\tau(s, t_2), t_1) = G_{(h(t_1) \circ h(t_2))}(s). \tag{5.19}$$

We also get from equation (5.18):

$$\phi_\tau(s, t_1 + t_2) = G_{h(t_1 + t_2)}(s). \tag{5.20}$$

Therefore by equations (5.17), (5.19), and (5.20) we get

$$G_{h(t_1 + t_2)}(s) = G_{(h(t_1) \circ h(t_2))}(s).$$

Recall that equation (5.8) implied $g_1 = g_2$. By the same reasoning, with this equation we get:

$$h(t_1 + t_2) = h(t_1) \circ h(t_2).$$

Thus h as defined in equation (5.16) is a homomorphism $\mathbf{R} \to \mathbf{SE}(2)$, that is, a one-parameter subgroup.

Recall from section 5.2 that there are three types of one-parameter subgroups for $\mathbf{SE}(2)$, that is, the absence of motion, uniform linear motion, and uniform circular motion. The absence of motion case does not occur for the Braitenberg vehicles because they have fixed nonzero speeds. So whenever the initial condition for ϕ_τ is in a relative equilibrium, the Braitenberg vehicles will either undergo uniform linear motion or uniform circular motion.

It turns out that there are no relative equilibria in which some of the orbits correspond to uniform linear motion and some to uniform circular motion. All of the orbits in a relative equilibrium either correspond to uniform linear

motion or all of them correspond to uniform circular motion. We can characterize the relative equilibrium as a whole by whether its orbits correspond to
uniform linear motion or to uniform circular motion.

To show this, suppose we choose some other initial condition $s' \in S_T$ from
the relative equilibrium containing s but not from the orbit of s. We define the
map $\alpha_{s'} : \mathbf{SE}(2) \to S_T$ as:

$$\alpha_{s'}(g) = G_g(s')$$

and set

$$h'(t) = \alpha_{s'}^{-1}(\phi_T(s', t))$$

as in equation (5.16). By the same reasoning as with h, it follows that h' is a
one-parameter subgroup and that

$$\phi_T(s', t) = G_{h'(t)}(s'). \tag{5.21}$$

On the other hand, there must be some $g \in \mathbf{SE}(2)$ such that $s' = G_g(s)$ and so:

$$\phi_T(s', t) = \phi_T(G_g(s), t).$$

Since ϕ_T is equivariant with respect to $\mathbf{SE}(2)$

$$\phi_T(s', t) = G_g(\phi_T(s, t)).$$

From equation (5.18),

$$\phi_T(s', t) = G_g(G_{h(t)}(s)).$$

By inverting $s' = G_g(s)$ and substituting for s, we get:

$$\phi(s', t) = G_g(G_{h(t)}(G_g^{-1}(s')))$$
$$= (G_g \circ G_{h(t)} \circ G_g^{-1})(s').$$

Since G is a homomorphism,

$$\phi(s', t) = G_{(g \circ h(t) \circ g^{-1})}(s').$$

Combining this with equation (5.21) gives:

$$G_{h'(t)}(s') = G_{(g \circ h(t) \circ g^{-1})}(s').$$

Recall that equation (5.8) implied $g_1 = g_2$. By the same reasoning we get:

$$h'(t) = g \circ h(t) \circ g^{-1}.$$

We say that the one-parameter subgroups h and h' are *conjugate* to each other
in $\mathbf{SE}(2)$.

If two one-parameter subgroups are conjugate to each other, then either they
both correspond to uniform linear motion or they both correspond to uniform
circular motion. This can be seen by considering the points in the plane fixed
by the one-parameter subgroups. Recall from table 5.1 that there are two kinds

of nontrivial proper congruences: translations which do not fix any points in the plane and rotations which fix exactly one point. So the uniform linear motions, which are generated by translations, do not fix any points, while the uniform circular motions, which are generated by rotations, fix exactly one point called the center.

Suppose the one-parameter subgroup h corresponds to uniform circular motion with center $(\bar{x}, \bar{y})^T$. Then h fixes $(\bar{x}, \bar{y})^T$ for all $t \in \mathbf{R}$

$$(h(t)) \begin{pmatrix} \bar{x} \\ \bar{y} \end{pmatrix} = \begin{pmatrix} \bar{x} \\ \bar{y} \end{pmatrix} + R_{(\theta_g t)} \begin{pmatrix} \bar{x} - \bar{x} \\ \bar{x} - \bar{y} \end{pmatrix} = \begin{pmatrix} \bar{x} \\ \bar{y} \end{pmatrix}.$$

And h' fixes $g(\bar{x}, \bar{y})^T$ for all $t \in \mathbf{R}$.

$$\left(h'(t) \right) \left(g \begin{pmatrix} \bar{x} \\ \bar{y} \end{pmatrix} \right) = (g \circ h(t) \circ g^{-1}) \left(g \begin{pmatrix} \bar{x} \\ \bar{y} \end{pmatrix} \right)$$

$$= (g \circ h(t)) \begin{pmatrix} \bar{x} \\ \bar{y} \end{pmatrix}$$

$$= g \begin{pmatrix} \bar{x} \\ \bar{y} \end{pmatrix}.$$

Thus h' corresponds to uniform circular motion with center $g(\bar{x}, \bar{y})^T$.

By the same reasoning, if h' corresponds to uniform circular motion with center $(\bar{x}, \bar{y})^T$ then h corresponds to uniform circular motion with center $g^{-1}(\bar{x}, \bar{y})^T$.

Therefore, either both h and h' fix some point in the plane or both of them do not. Conjugate one-parameter subgroups either both correspond to uniform circular motion or both correspond to uniform linear motion.

Since s and s' are arbitrary points in a relative equilibrium, we can conclude that every orbit in a relative equilibrium corresponds to uniform circular motion or every orbit in a relative equilibrium corresponds to uniform linear motion. This allows us to distinguish the relative equilibria of ϕ_τ into two types.

We call a relative equilibrium whose orbits correspond to uniform circular motion a *revolving type* relative equilibrium. The uniform circular motion applies to both vehicles so they both revolve about a common center with the same angular velocity. The center is not the same for every orbit in a revolving type relative equilibrium. Wherever the center happens to be the vehicles revolve about it with the same angular velocity.

We call a relative equilibrium whose orbits correspond to uniform linear motion a *translating type* relative equilibrium. The uniform linear motion applies to both vehicles so they both travel in the same direction with the same linear velocity. The direction is not the same for every orbit in a translating type relative equilibrium. Whatever direction the vehicles happen to be traveling in they move with the same linear velocity.

The concept of relative equilibria is fairly abstract but for the Braitenberg vehicles it reduces to a fairly simple and intuitive idea. The simplest types of motion exhibited by the agents is either uniform circular motion or uniform linear motion. These are the only ways the agents can remain motionless relative to each other over time. In chapter 6, we will find four kinds of revolving type relative equilibria for the Braitenberg vehicles, which we will denote by $\mathcal{E}_{(\ell,\text{in})}$, $\mathcal{E}_{(\ell,\text{out})}$, $\mathcal{E}_{(r,\text{in})}$, and $\mathcal{E}_{(r,\text{out})}$. In chapter 7, we will find several kinds of translating type relative equilibria, one of which will be particularly important. We will denote it by $\mathcal{E}_T$.

5.6   Invariant Manifolds

Historically the dynamical systems studied in science were usually derived from differential equations. Early in the development of calculus, the concept of a derivative was defined for functions from the real numbers to itself. The concept was generalized to functions between vector spaces, and later it was generalized to functions between manifolds, that is, spaces that are locally like vector spaces. This led to differential equations on manifolds from which dynamical systems could be derived (e.g., Smale [1967]; Pugh and Shub [1970]). The total dynamical system for the Braitenberg vehicles, ϕ_τ, is an example of such a system. The state space, $S_\tau \cong (\mathbf{R}^2 \times \mathbf{S}^1)^2$, is not a vector space, but it is a manifold since it is a Cartesian product of the manifolds $\mathbf{R}$ and $\mathbf{S}^1$.

In the mid-twentieth century, dynamical systems theorists (e.g., Sacker [1964]; Hirsch, Pugh, and Shub [1970]) began to consider invariant sets with topologies beyond that of fixed points and periodic orbits, particularly invariant sets that are submanifolds of the dynamical system's state space, that is, invariant manifolds of dynamical systems. The relative equilibria for the Braitenberg vehicles are examples of invariant manifolds. They are copies of the manifold $\mathbf{SE}(2) \cong \mathbf{R}^2 \times \mathbf{S}^1$ embedded in S_τ.

A pair of vector spaces can be associated to each point of an invariant manifold. One of them is the tangent space to the invariant manifold at the given point and the other is the orthogonal complement to the tangent space at the given point, often called the normal space to the invariant manifold at that point. See section 3.6 and in particular the top right panel of figure 3.6 for a simple example.

Since $\mathbf{SE}(2)$ is three-dimensional, the tangent spaces to the relative equilibria are three-dimensional. And since S_τ has three more dimensions than the relative equilibria, the normal spaces to the relative equilibria are three-dimensional.

A well-behaved type of invariant manifold that emerged from research around the 1960s is known as a normally hyperbolic invariant set (e.g., Fenichel [1972]; Hirsch, Pugh, and Shub [1977]; Mane [1978]; Wiggins [1994]; Robinson [1995]). We will apply techniques that developed in this area for studying the tangent and normal spaces of relative equilibria in the Braitenberg system. Roughly speaking, at each point of a normally hyperbolic invariant manifold, the normal space is a direct sum of two subspaces such that the dynamical system is locally expanding on one subspace and locally contracting on the other.

The expanding subspace can be zero-dimensional, in which case all states sufficiently close to the invariant manifold contract towards it under the dynamical system, that is, the normally hyperbolic invariant manifold would be an attracting invariant manifold. Or the contracting subspace can be zero-dimensional in which case all states sufficiently close to it move away from it under the dynamical system, that is, the normally hyperbolic invariant manifold would be a repelling invariant manifold.

More generally, the expanding and contracting subspaces of the normal space both have one or more dimensions, and the normally hyperbolic invariant manifold is neither attracting nor repelling. A bifurcation of a normally hyperbolic invariant manifold from attracting to non-attracting can be detected by the dimension of the expanding subspace of the normal space becoming positive.

Another research thread that occurred at about the same time involved dynamical systems with symmetry. It was eventually shown, roughly speaking, that in a neighborhood of a relative equilibrium the differential equation for the dynamical system can be decomposed into two differential equations, one for the tangent space and one for the normal space (Ruelle [1973]; Vanderbauwhede, Krupa, and Golubitsky [1989]; Golubitsky, Schaefer, and Stewart [1988]; Chossat and Golubitsky [1988]).

In the late twentieth century an important development occurred in research on the BZ reaction. Barkley reduced a reaction diffusion model for the BZ reaction to a dynamical system with a five-dimensional state space, (Barkley [1992, 1994]; Barkley and Kevrekidis [1994]). Like the Braitenberg system, his dynamical system is equivariant with respect to the group of proper Euclidean congruences **SE**(2), and it has relative equilibria which are copies of **SE**(2). The relative equilibria are exhibited in the reaction diffusion model for the BZ reaction as spiral waves that rotate as though they were rigid bodies.

When viewed in a rotating frame, the rotating spiral waves appear motionless, and the point in the state space which corresponds to the spiral wave is fixed. The resulting dynamical system can be linearized at this fixed point. Because of the equivariance with $SE(2)$, it does not matter which spiral wave we choose to make appear motionless. The linearization will yield the same set of eigenvalues.

Three of the eigenvalues correspond to the tangent space of the relative equilibrium. In Barkley's model one of these eigenvalues is zero and the other two are non-zero conjugate imaginary numbers. The remaining two eigenvalues correspond to the normal space of the relative equilibrium. When these are complex numbers they have the same real part. When the real part is negative, the dynamical system is locally contracting on the normal space and the relative equilibrium is an attracting set. When the real part is positive the dynamical system is locally expanding on the normal space and the relative equilibrium is a repelling set.

As the real part of the eigenvalues for the normal space go from negative to positive the dynamical system undergoes a Hopf-like bifurcation, as is discussed further in section 8.3. The relative equilibrium goes from attracting to repelling while a new attracting set emerges with a secondary oscillation. This is exhibited in the reaction diffusion model by the meandering of the tip of the spiral wave as the wave revolves.

In the following sections we adapt Barkley's method to our analysis of the Braitenberg vehicles. The relative equilibria in our system are exhibited by the Braitenberg vehicles revolving around each other with the same angular velocity or translating together with the same linear velocity. The agents appear motionless in a suitable moving frame. The dynamical system can be linearized at a fixed point. Again, because of the equivariance no matter which configuration in the relative equilibrium we linearize at we get the same set of eigenvalues.

For the Braitenberg vehicles the state space is six-dimensional (Barkley's dynamical system is five-dimensional) but the relative equilibria are still copies of the three-dimensional manifold $SE(2)$. Three of the eigenvalues correspond to the tangent space of the relative equilibrium. One of them is zero and the other two are non-zero imaginary conjugates. The extra dimension of the state space gives us a third eigenvalue for the normal space but this eigenvalue turns out to be the sum of the other two, which we call the "salient" eigenvalues. When the salient eigenvalues have negative real part, all three eigenvalues have negative real part. When the salient eigenvalues have positive real part, all three eigenvalues have positive real part.

As the real part of the salient eigenvalues goes from negative to positive, the third normal eigenvalue passes through zero. The dynamical system undergoes a Hopf-like bifurcation that is particularly degenerate. Not only are all six of the eigenvalues on the imaginary axis but the eigenvalue 0 has an algebraic multiplicity of 2.

Numerical integration shows that this Hopf-like bifurcation resembles the Hopf-like bifurcation in Barkley's dynamical system. The relative equilibrium goes from attracting to repelling while a new attracting set emerges with a secondary oscillation. This is exhibited in the Braitenberg vehicles by the two agents going from revolving around each other to meandering while the midpoint between them remains fixed.

The new attracting set which emerges from the Hopf-like bifurcation appears to be a relative periodic orbit which is a type of invariant set we discuss in the next section.

5.7   Relative Periodic Orbits

Another type of attracting set that occurs for the Braitenberg vehicles is called a "relative periodic orbit." Relative periodic orbits are slightly more complicated than relative equilibria. Instead of the vehicles being motionless relative to each other they move along closed curves relative to each other. The states of the agents also move along closed curves. Their physical paths resemble epicyclic motion (see the left panel of figure 6.9) in the rest frame.

Just as a relative equilibrium is not a single equilibrium point but a congruence class of S_τ, a relative periodic orbit is not a single orbit but a collection of congruence classes in S_τ. Roughly speaking, when the congruences classes of the group action are visited periodically by the dynamical system, their union forms a relative periodic orbit.

In many cases, for the Braitenberg vehicles, a relative periodic orbit arises from a bifurcation from a relative equilibrium. In chapter 6, we show how the revolving type relative equilibria $\mathcal{E}_{(\ell,\text{in})}$ and $\mathcal{E}_{(r,\text{in})}$ can undergo a bifurcation that produces a relative periodic orbit which we will denote by $\mathcal{P}_{(\ell,\text{in})}$ and $\mathcal{P}_{(r,\text{in})}$ respectively. We also describe numerical evidence for a saddle-node like bifurcation of relative periodic orbits, which we will denote by $\mathcal{P}_{(\ell,+)}$, $\mathcal{P}_{(\ell,-)}$, and $\mathcal{P}_{(r,+)}$, $\mathcal{P}_{(r,-)}$. In chapter 7, we show how the translating type relative equilibria $\mathcal{E}_T$ can undergo a bifurcation to produce a relative periodic orbit, which we will denote by $\mathcal{P}_T$. We also describe a numerical study of another type of relative periodic orbit, which we will denote by $\mathcal{P}_C$.

6 Revolving Type Relative Equilibria and Their Bifurcations

In this chapter, we study revolving type relative equilibria in which the vehicles move in circular paths around each other, as well as relative periodic orbits in which the vehicles meander around each other. These invariant sets occur when the sensor weights of the two agents are the same; that is, the four weights of the system are in W_{same} (see section 2.5). We denote the pair of sensor weights for each of the two agents with the row vector (w_ℓ, w_r), which corresponds to a linear combination of the basis vectors for W_{same} in equation (2.10):

$$(w_\ell, w_r) \cong w_\ell \, (1, 0, 1, 0)^T + w_r. \, (0, 1, 0, 1)^T$$

Recall that the subspace of W_{same} where all four weights are equal is called W_{eq}. It is the central vertical axis in figures 1.2 and 2.8, and it is the diagonal line in figures 6.1 and 6.10. Reflecting the plane W_{same} about the line W_{eq} amounts to switching the agents' two sensor weights. Above W_{eq} both agents move to the left, and below they turn to the right.

Within W_{same} there are saddle node-like bifurcations and Hopf like-bifurcations that divide it into regions that correspond to distinct behaviors (recall the discussion of dynamical regimes in section 3.4). The invariant sets associated with these regions are summarized in figure 6.1, which also shows the bifurcations that occur when traveling along the two paths in W_{same} shown as dotted lines. For the white regions in figure 6.1, no attracting sets have been observed numerically. In the blue regions, there is a pair of attracting and nonattracting relative equilibria in which both agents turn right along circular paths. The attracting and nonattracting pair of relative equilibria emerge via a saddle node-like bifurcation that occurs at the boundary between the white and blue regions (e.g., at the green stars in figure 6.1).

The circular paths for the attracting relative equilibria are smaller than the circular paths for the corresponding nonattracting relative equilibria. When these two circular paths are concentric, the circular path for the attracting

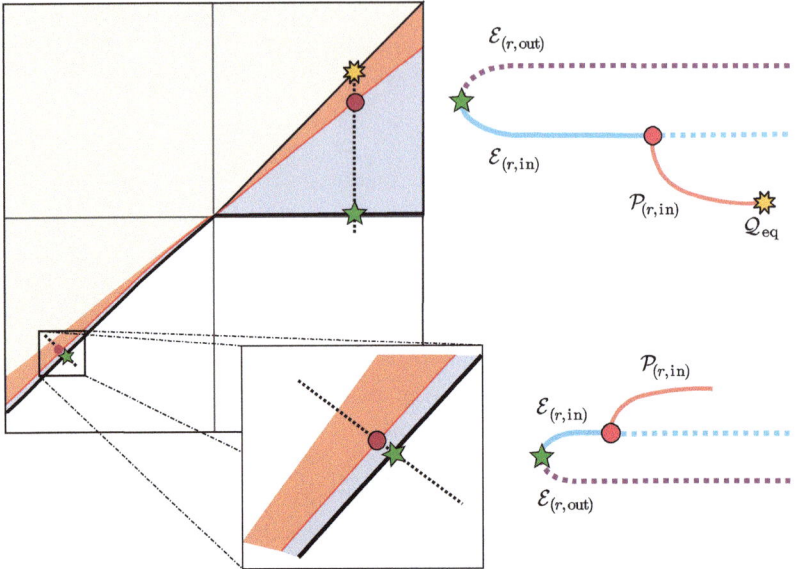

Figure 6.1

A summary of the bifurcations for the main relative equilibria and relative periodic orbits discussed in this chapter. (left) A square region of the weight space W_{same} centered at the origin. The dotted lines in W_{same} show paths traversed through the third quadrant (avoider-avoider) and the first quadrant (pursuer-pursuer). The bifurcations that occur along these two lines are similar. The green stars mark the location of saddle node-like bifurcations in which an attracting relative equilibrium $\mathcal{E}_{(r,\text{in})}$ appears along with a nonattracting relative equilibrium, $\mathcal{E}_{(r,\text{out})}$. The red dots mark the location of a Hopf-like bifurcation where $\mathcal{E}_{(r,\text{in})}$ becomes nonattracting and an attracting relative periodic orbit, $\mathcal{P}_{(r,\text{in})}$, appears. In the pursuer-pursuer quadrant, these relative periodic orbits generate meandering paths with loops that become increasingly elongated. At W_{eq}, the relative periodic orbit $\mathcal{P}_{(r,\text{in})}$ turns into the invariant set $\mathcal{Q}_{\text{eq}}$ in which the agents follow each other single-file. (right) Solid curves indicate attracting sets, and dotted curves indicate nonattracting sets. Other invariant sets and bifurcations that occur in W_{same} are discussed in the main text.

relative equilibrium is the one that is on the inside. We call the attracting relative equilibrium the "inner" relative equilibrium and the nonattracting relative equilibrium the "outer" relative equilibrium. We denote the right-turning inner relative equilibrium by $\mathcal{E}_{(r,\text{in})}$, the right-turning outer relative equilibrium by the right-turning outer $\mathcal{E}_{(r,\text{out})}$, the left-turning inner relative by $\mathcal{E}_{(l,\text{in})}$, and the left-turning outer relative equilibrium by $\mathcal{E}_{(l,\text{out})}$.

For the rose-colored regions in figure 6.1, the agents meander around each other along paths generated by a relative periodic orbit. These meandering paths can be partitioned into congruent "periodic arcs" often referred to as "loops" or "petals." In this region, there is an attracting right-turning relative periodic orbit that we call $\mathcal{P}_{(r,\text{in})}$. It emerges from $\mathcal{E}_{(r,\text{in})}$ via a Hopf-like bifurcation that occurs at the boundary between the blue- and rose-colored regions (e.g., the red dots in figure 6.1). By symmetry, the reflection of the rose-colored regions in figure 6.1 about W_{eq} is where there is an attracting left-turning relative periodic orbit, which we call $\mathcal{P}_{(\ell,\text{in})}$. The regions for right-turning and left-turning relative periodic orbits are shown together in the rose-colored regions of figure 6.10. The relative periodic orbit $\mathcal{P}_{(\ell,\text{in})}$ emerges from $\mathcal{E}_{(\ell,\text{in})}$ via a Hopf-like bifurcation that occurs along the boundary between the blue- and rose-colored regions in figure 6.10, which is the reflected image of the boundary of the blue- and rose-colored regions in figure 6.1.

The relative periodic orbits $\mathcal{P}_{(r,\text{in})}$ and $\mathcal{P}_{(\ell,\text{in})}$ generate meandering paths in the plane. As the weights are varied towards W_{eq}, within the pursuer-pursuer quadrant of W_{same}, the meandering paths become an alternating sequence of nearly straight arcs and sharp turns. When the weights reach W_{eq}, a bifurcation occurs in which the relative periodic orbit is transformed into another type of attracting set that is a union of relative equilibria. We denote this new invariant set by $\mathcal{Q}_{\text{eq}}$. For orbits in $\mathcal{Q}_{\text{eq}}$, the agents engage in single-file behavior and the physical paths become a single straight line.

There is also a trivial revolving type relative equilibrium, which does not appear to play a significant role in the dynamics. When the two agents are directly on top of each other and heading in the same direction, their sensors will be on top of each other and their sensor activations will be the same. Since the two agents have the same sensor weights, they will turn at the same rate. Since they also have the same forward velocity, they will move in unison and remain on top of each other with the same heading. When the left and right sensor weights are different, the agents will both engage in the same uniform circular motion. However, this relative equilibrium is not attracting. The agents will not return to this configuration after being perturbed slightly. We will not consider this trivial revolving type relative equilibrium any further.

In section 6.7, we consider these invariant sets and the bifurcations between them from the standpoint of the agents and their representations of each other. For weights in the white region, $\mathcal{O}_r$, in the top left panel of figure 6.2, and when the agents start out on each other's right side, they eventually travel out of each others' view never to return. Their internal state goes to an attracting partially fixed point in the agent space S_{α_n}, that is the point $(0,0)^T$ in the right panel of figure 4.1, which represents the absence of the other agent. For these weights, the agents can see each other at some locations, but as the agent goes out of view the representation decays to the "alone" state. For the non-white region, $\mathcal{R}_r$, in the top left panel of figure 6.2, there are other partially fixed points in the agent state space that correspond to an agent's representation of the other agent. When the agents revolve in the appropriate configuration, their position relative to each other remains fixed and hence their representations of each other are fixed. When the agents are meandering, there is an attracting periodic path in the agent space. It emerges in a Hopf-like bifurcation from the formerly attracting partially fixed point, and each agent observes the other agent to be moving along a simple closed curve. In other words, meandering paths in the physical plane correspond to closed curves in the agent space.

We provide a detailed analysis of the meandering paths and the geometry of their periodic arcs in section 6.8, and discuss several other invariant sets that occur in the system.

We assume in chapters 6 and 7 that both agents move at the same speed, so we drop the subscript from v throughout these chapters. We treat the physical parameters P, ψ, and v as given features of the Braitenberg vehicles (we sometimes alter these slightly for illustrative purposes in the figures, and when we do so we note the values we use for them). Also in this chapter (but not in chapter 7), we drop the subscript from the functions $\bar{\omega}$ since $\bar{\omega}_1 = \bar{\omega}_2$ when the two agents have the same sensor weights.

6.1 Curves for Saddle Node-Like Bifurcations in W_{same}

We denote the region of the weight space W_{same} where there are no non-trivial right-turning relative equilibria by $\mathcal{O}_r$ and the region where there are no non-trivial left-turning relative equilibria by $\mathcal{O}_\ell$. These are open regions, and the boundary of each region is a curve that is smooth everywhere except at a single point (marked by white boxes in figure 6.2). These boundary curves divide W_{same} into four regions according to the presence or absence of relative equilibria. We denote the complement of $\mathcal{O}_r$ in W_{same} by $\mathcal{R}_r$ and the complement of $\mathcal{O}_\ell$ by $\mathcal{R}_\ell$, as shown in the top middle panel of figure 6.2.

The region with no right- or left-turning relative equilibria is $\mathcal{O}_r \cap \mathcal{O}_\ell$. The region with right-turning relative equilibria and no left-turning relative equilibria is $\mathcal{R}_r \cap \mathcal{O}_\ell$. The region with left-turning relative equilibria and no right-turning relative equilibria is $\mathcal{O}_r \cap \mathcal{R}_\ell$. The region with both right- and left-turning relative equilibria is $\mathcal{R}_r \cap \mathcal{R}_\ell$. Each of these regions is unbounded and simply connected.

The boundary of $\mathcal{R}_r$ is denoted $\partial \mathcal{R}_r$, and similarly for $\partial \mathcal{R}_\ell$. In this case, the boundaries are curves. When the weights cross $\partial \mathcal{R}_r$, a saddle node-like bifurcation occurs in which a pair of right-turning relative equilibria appear, $\mathcal{E}_{(r,\text{in})}$ and $\mathcal{E}_{(r,\text{out})}$. This bifurcation curve is marked by the thick black curves and five-pointed stars in figure 6.1. Analogous results hold for $\partial \mathcal{R}_\ell$.

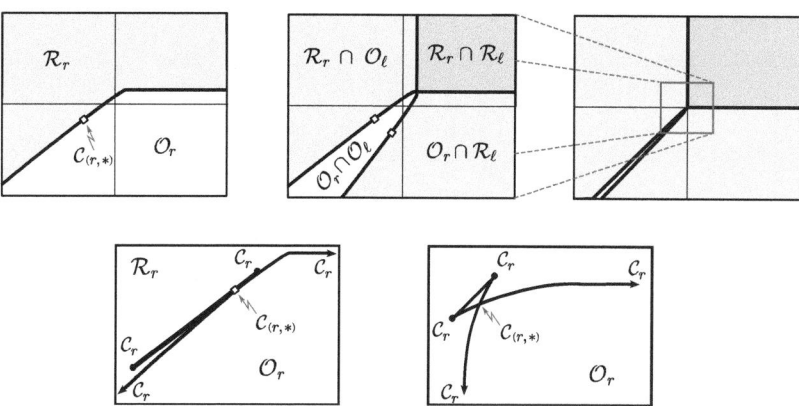

Figure 6.2

Regions where the left- and right-turning relative equilibria occur. The union of the shaded regions $\mathcal{R}_r$ and $\mathcal{R}_\ell$ contain the blue- and rose-colored regions in figure 6.10. (top left) As (w_ℓ, w_r) crosses the black curve $\partial\mathcal{R}_r$, a saddle node-like bifurcation occurs (except at the point $\mathcal{C}_{(r,*)}$ marked by a white box). (top middle) The two curves $\partial\mathcal{R}_r$, $\partial\mathcal{R}_\ell$ are colored black. They divide the parameter plane into four regions. (top right) A zoomed-out view of the parameter plane in which the region $\mathcal{O}_\ell \cap \mathcal{O}_r$ appears comparatively small, and the region $\mathcal{R}_r \cap \mathcal{R}_\ell$ appears to fill up the pursuer-pursuer quadrant, although it does not. (bottom left) A precise depiction of the curve $\mathcal{C}_r$ in black, $P = 8$, $\psi = \pi/4$, and $v = 1/100$. Its self-intersection at $\mathcal{C}_{(r,*)}$ is marked by a white box. The self-intersection produces an extremely narrow region inside $\mathcal{R}_r$. The only place where the tangent line to $\partial\mathcal{R}_r$ is not defined is at $\mathcal{C}_{(r,*)}$. It may appear as though there is another point in the upper right where a tangent line to $\partial\mathcal{R}_r$ is not defined, but this is just due to a sharp change in the curvature of $\partial\mathcal{R}_r$. (bottom right) A diffeomorphic transformation of $\mathcal{C}_r$ to make its topology in a neighborhood of $\mathcal{C}_{(r,*)}$ easier to see.

Recall from section 2.5 that the first quadrant of the weight space contains weights for pursuer-pursuer pairs, the third quadrant contains weights for avoider-avoider pairs, and the second and fourth quadrants contain "lateralized" pairs of vehicles that will pursue in one direction and avoid in the other. The region $\mathcal{R}_r \cap \mathcal{R}_\ell$ is contained in but does not coincide with the pursuer-pursuer quadrant: some pairs of pursuers only pursue each another transiently no matter where you place them. The regions $\mathcal{R}_r \cap \mathcal{O}_\ell$ and $\mathcal{O}_r \cap \mathcal{R}_\ell$ both overlap the avoider-avoider quadrant.

The white boxes in figure 6.2 mark a complication in the boundary curves $\partial\mathcal{R}_r$ and $\partial\mathcal{R}_\ell$ that we can largely ignore. The bottom panels of figure 6.2 show how $\partial\mathcal{R}_r$ extends to a self-intersecting curve we denote by $\mathcal{C}_r$. It produces a triangular "swallowtail" region inside $\mathcal{R}_r$. For weights inside the swallowtail, there is an extra pair of right-turning relative equilibria (which we do not name) in addition to $\mathcal{E}_{(r,\text{in})}$ and $\mathcal{E}_{(r,\text{out})}$. There is also a swallowtail region inside $\mathcal{R}_\ell$ for

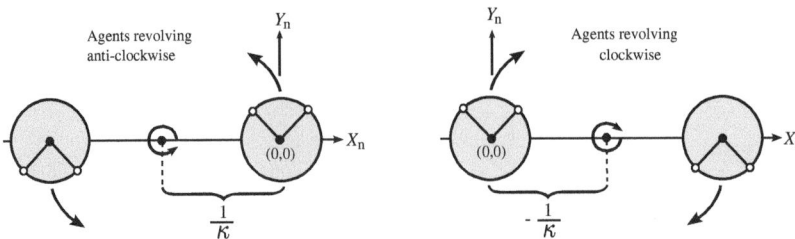

Figure 6.3
The position of the center of revolution in the body frame of the nth Braitenberg vehicle. (left) When the Braitenberg vehicles revolve anticlockwise, the center of revolution lies on the negative X_n-axis of the nth vehicle's body frame, while the curvature κ of the physical path is positive. (right) When the Braitenberg vehicles revolve clockwise, the center of revolution lies on the positive X_n-axis of the nth vehicle's body frame while κ is negative. In either case, the center of revolution is located at $(-1/\kappa, 0)^T$, and the other vehicle is located at $(-2/\kappa, 0)^T$.

which there is an extra pair of left-turning relative equilibria (also not named) in addition to $\mathcal{E}_{(\ell,in)}$ and $\mathcal{E}_{(\ell,out)}$. These swallowtail regions are so thin we defer discussion of them to appendix A.

There are relative equilibria associated with the blue regions in the avoider-avoider quadrant of figure 6.1 that are attracting. It might seem that avoiders would never revolve around one another, but the left and right sensor weights only have to be moderately different in order for them to do just that.

6.2 A Necessary and Sufficient Condition for the Existence of a Revolving Type Relative Equilibrium

We now describe conditions for the existence of revolving type relative equilibria. The velocity of a body revolving along a circular path must be tangent to the circular path. The heading of each Braitenberg vehicle points in the direction of its velocity so that when the vehicles are revolving around each other, each vehicle's heading is perpendicular to the diameter of its physical path (see figure 6.3).

Recall that each agent is located at $(0, 0)$ in its own body frame and that the other agent is located at (X_n, Y_n). When the agents are revolving around each other, the center of their revolution lies on the X_n-axis of each vehicle's body frame.

The radius of a circle is the absolute value of the reciprocal of the circle's curvature,[1] κ. When the Braitenberg vehicles revolve anticlockwise, the center

---

1. Curvature is explained in greater detail in section 6.8 on meandering paths.

of their revolution appears to the left of each vehicle and thus on the negative X_n-axis of its body frame. In this case, the curvature of their circular path is positive, their radius of revolution is $1/\kappa$, and the location of the other agent is $(-2/\kappa, 0)^T$. When the Braitenberg vehicles revolve clockwise, the center of their revolution is on the positive X_n-axis of each vehicle's body frame. In this case, the curvature of their circular path is negative, and so their radius of revolution is $-1/\kappa$. Notice that the the location of the other agent is still $(-2/\kappa, 0)^T$. No matter which way the vehicles are turning, the other agent is located at $(-2/\kappa, 0)^T$. Also note that in either case the diameter of revolution is $|X_n|$.

Since the Braitenberg vehicles' headings are always perpendicular to the radius of revolution, the angular velocity of revolution is the same as the angular velocity of the vehicles' headings, ω. The chain rule shows that when the vehicles are revolving, their angular velocity is equal to the product of the curvature of the circular path and the linear velocity v of the agents, both of which are constant as they revolve. That is for $n = 1, 2$:

$$\bar{\omega} = \frac{d\theta_n}{dt} = \frac{d\theta_n}{ds} \cdot \frac{ds}{dt} = \kappa v.$$

where s is the arc length along the path of revolution and $d\theta_n/ds$ is curvature[2]. The linear velocity v is a constant by assumption, and κ is constant when the agents are revolving.

We can express the rate of change in the vehicle's heading in terms of the diameter X_n:

$$X_n = -\frac{2}{\kappa} \quad \Longrightarrow \quad \kappa = -\frac{2}{X_n} \quad \Longrightarrow \quad \frac{d\theta_n}{dt} = -\frac{2v}{X_n}.$$

Since $d\theta_n/dt = \bar{\omega}(X_n, Y_n)^T$, this gives us a necessary condition for revolving type relative equilibria to exist:

$$\bar{\omega}(X_n, 0)^T = -\frac{2v}{X_n} \tag{6.1}$$

for $n = 1, 2$. This is our bifurcation equation.

Such behavior is possible for our total dynamical system ϕ_T, that is, equation (6.1) is also a sufficient condition for a revolving type relative equilibrium. This

---

2. The s here should not be confused with the s in equation (4.2).

can be seen by substituting the particular solution:

$$\phi_\tau \left((-X_*, 0, \pi, X_*, 0, \pi)^T/2\right) =$$

$$\begin{pmatrix} x_1(t) \\ y_1(t) \\ \theta_1(t) \\ x_2(t) \\ y_2(t) \\ \theta_2(t) \end{pmatrix} = \frac{1}{2} \begin{pmatrix} -X_* \sin(\Omega_* t + \pi/2) \\ X_* \cos(\Omega_* t + \pi/2) \\ 2(\Omega_* t + \pi/2) \\ -X_* \sin(\Omega_* t - \pi/2) \\ X_* \cos(\Omega_* t - \pi/2) \\ 2(\Omega_* t - \pi/2) \end{pmatrix} \tag{6.2}$$

into differential equation (4.14) for the total dynamical system, where X_* can be any solution for X_n in equation (6.1), and $\Omega_* = -2v/X_*$.

6.3 Approximate Solutions to the Bifurcation Equation

The bifurcation equation (6.1) can be used to compute the bifurcation curves shown in figure 6.2, which show where revolving type relative equilibria appear as the weights in W_{same} are varied. Equation (6.1) says that when the agents revolve around each other, their turning rate must be $-2v/X_n$. The graph of $-2v/X_n$ as a function of X_n is a hyperbola. To solve the bifurcation equation, we must find where this hyperbola intersects the graph of the turning function $\bar{\omega}_n(X_n, 0)^T$, as shown in the right panel of figure 6.4.

To solve equation (6.1) for X_n, we could try converting it into a polynomial. This requires us to eliminate the two radicals in equation (2.5). After the radicals are eliminated, we get an eighth-degree polynomial, and it is not feasible to find an algebraic expression for its roots. However, we have a good approximation available.

Recall from figure 2.7 that the function for the turning rate $\bar{\omega}_n(X_n, Y_n)^T$ is the weighted difference of two cones over the region $D_{(\ell,n)} \cup D_{(r,n)}$ in the body frame, that is, the weighted difference between the left and right sensor activations for agent n (see figure 2.7 and the left panel of figure 6.4). The line $Y_n = N$ in the physical plane passes through the Braitenberg vehicles' sensors and a vertical plane through the line $Y_n = N$ intersects each cone in two pairs of line segments. Consequently, the function $\bar{\omega}_n(X_n, N)^T$ is piecewise linear (see figure 2.6 and the right panel of figure 6.4). The line $Y_n = N$ is relatively close to the X_n-axis and when the sensor radius $P > 3M$ the graph of $\bar{\omega}_n(X_n, N)^T$ nearly conforms to the graph of $\bar{\omega}_n(X_n, 0)^T$.

Unlike equation (6.1), it is feasible to algebraically solve the equation

$$\bar{\omega}_n(X_n, N)^T = -\frac{2v}{X_n} \tag{6.3}$$

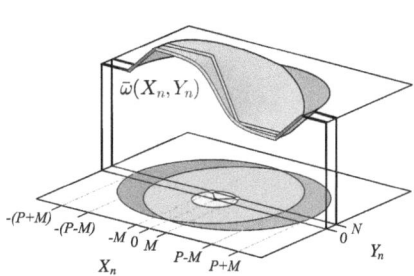 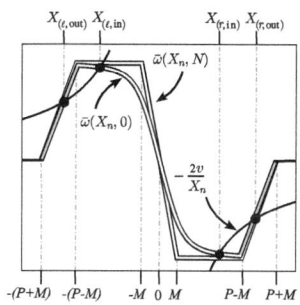

Figure 6.4
(left) The graph of the turning function $\bar{\omega}_n(X_n, Y_n)^T$, which shows how an agent turns based on where the other agent is. (right) Cross sections of $\bar{\omega}_n(X_n, Y_n)^T$ taken at $Y_n = 0$ and $Y_n = N$ along with the graph of $-2v/X_n$. The intersections between the graphs of $\bar{\omega}_n(X_n, 0)^T$ and $-2v/X_n$ are marked by the black dots. These intersection points can be approximated by the intersection points of the piecewise linear graph of $\bar{\omega}_n(X_n, N)^T$ with the graph of $-2v/X_n$.

since the intersection points are where line segments in the graph of $\bar{\omega}_n(X_n, N)^T$ meet the hyperbolic graph of $-2v/X_n$. We use the solutions to equation (6.3) to approximate the solutions to equation (6.1).

When $(w_\ell, w_r) = (0, 0)$, the agents don't turn at all, so the turning function always returns 0, and the graph of $\bar{\omega}_n(X_n, N)^T$ is the X_n-axis. As the sensor weights are varied from $(0, 0)$, the graph of $\bar{\omega}_n(X_n, N)^T$ becomes a piecewise linear curve as shown in figure 6.4.

In appendix A, we obtain approximations to the curves $\partial \mathcal{R}_r$, $\partial \mathcal{R}_\ell$ in W_{same} by computing those pairs of weights that correspond to the initial contacts that occur between the graphs of $\bar{\omega}_n(X_n, N)^T$ and $-2v/X_n$.

After an initial contact, there are typically two right-turning solutions to equation (6.1), denoted $X_{(r,\text{in})}$ and $X_{(r,\text{out})}$, which determine a pair of revolving type relative equilibria $\mathcal{E}_{(r,\text{in})}$ and $\mathcal{E}_{(r,\text{out})}$ (see figure 6.1). Similarly for the left-turning solutions. The smaller solutions denoted "in" correspond to an "inner" revolving path, and the larger solutions denoted "out" correspond to the "outer" revolving path. We use the symbol "$*$" as a wildcard encompassing the inner and outer cases in expressions like $\Omega_{(r,*)} = -2v/X_{(r,*)}$ and $\Omega_{(\ell,*)} = -2v/X_{(\ell,*)}$.

6.4 Converting to a Rotating Frame

As preparation for the linear stability analysis, we convert the equation of motion (4.14) to a rotating frame in which a subset of points of a relative equilibrium appear fixed (recall the discussion in section 3.6). In particular, we analyze the stability of revolving type relative equilibria for ϕ_τ by converting to a frame that rotates about the origin of the rest frame with the same angular velocity as a pair of revolving vehicles. In this rotating frame, there is a periodic orbit of ϕ_τ (which corresponds to a pair of vehicles revolving about the origin), all of whose points appear fixed.

We can linearize the system at any of the points that are fixed in the rotating frame. The eigenvalues end up being the same, because of the equivariance of the dynamical system ϕ_τ with respect to the congruence group $\mathbf{SE}(2)$. The equivariance also shows that this linearization gives us information about the stability of the whole relative equilibrium, beyond the points that are fixed in the rotating frame, since the points of a relative equilibrium are equivalent to each other under the action of $\mathbf{SE}(2)$.

To obtain the rotating frame, we counteract the revolving motion of a relative equilibrium using the one-parameter subgroup $t \mapsto R_{(-\Omega_{(r,*)}t)} \subset \mathbf{SE}(2)$:

$$\begin{pmatrix} x'_1 \\ y'_1 \\ \theta'_1 \\ x'_2 \\ y'_2 \\ \theta'_2 \end{pmatrix} = G_{R_{(-\Omega_{(r,*)}t)}} \begin{pmatrix} x_1 \\ y_1 \\ \theta_1 \\ x_2 \\ y_2 \\ \theta_2 \end{pmatrix} = \begin{pmatrix} R_{(-\Omega_{(r,*)}t)} \begin{pmatrix} x_1 \\ y_1 \end{pmatrix} \\ \theta_1 - \Omega_{(r,*)}t \\ R_{(-\Omega_{(r,*)}t)} \begin{pmatrix} x_2 \\ y_2 \end{pmatrix} \\ \theta_2 - \Omega_{(r,*)}t \end{pmatrix} . \tag{6.4}$$

Note that the prime symbol here denotes state variables in the rotating frame, which is an application of equation (5.5), where s' and s are states of the total system related by a proper congruence.

We can obtain the differential equation in the rotating frame from the differential equation in the rest frame (4.14) and equation (6.4). To obtain the first, second, fourth, and fifth components of the differential equation in the rotating frame, we apply the product rule to $(x'_n, y'_n) = R_{(-\Omega_{(r,*)}t)} (x_n, y_n)$ for $n = 1, 2$ and express the result in the rotating frame:

$$\begin{pmatrix} \dot{x}'_n \\ \dot{y}'_n \end{pmatrix} = R_{(-\Omega_{(r,*)}t)} \begin{pmatrix} \dot{x}_n \\ \dot{y}_n \end{pmatrix} + \dot{R}_{(-\Omega_{(r,*)}t)} \begin{pmatrix} x_n \\ y_n \end{pmatrix}$$

$$= R_{(-\Omega_{(r,*)}t)} \begin{pmatrix} v\cos(\theta_n) \\ v\sin(\theta_n) \end{pmatrix} - \Omega_{(r,*)} R_{(\pi/2-\Omega_{(r,*)}t)} \begin{pmatrix} x_n \\ y_n \end{pmatrix}$$

$$\begin{pmatrix} \dot{x}'_n \\ \dot{y}'_n \end{pmatrix} = R_{(-\Omega_{(r,*)}t)} \begin{pmatrix} v\cos(\theta'_n + \Omega_{(r,*)}t) \\ v\sin(\theta'_n + \Omega_{(r,*)}t) \end{pmatrix} - \Omega_{(r,*)} R_{(\pi/2-\Omega_{(r,*)}t)} R_{(\Omega_{(r,*)}t)} \begin{pmatrix} x'_n \\ y'_n \end{pmatrix}$$

$$= \begin{pmatrix} v\cos(\theta'_n) \\ v\sin(\theta'_n) \end{pmatrix} + \Omega_{(r,*)} \begin{pmatrix} y'_n \\ -x'_n \end{pmatrix}. \tag{6.5}$$

To obtain the third and sixth components of the differential equation in the rotating frame, we take the derivative of $\theta'_n = \theta_n - \Omega_{(r,*)}\,t$ in equation (6.4) and then substitute $\dot{\theta}_n = \omega_n(s(t))$ to get:

$$\dot{\theta}'_n = \dot{\theta}_n - \Omega_{(r,*)} = \omega_n(s(t)) - \Omega_{(r,*)}$$

for $n = 1, 2$. The $(X_n, Y_n)^T$ position of agent $\neg n$ relative to agent n is the same for congruent configurations of the vehicles. In the case here:

$$s'(t) = GR_{(-\Omega_{(r,*)}t)}(s(t)).$$

so we have:

$$\mathcal{C}_n(s'(t)) = \mathcal{C}_n(GR_{(-\Omega_{(r,*)}t)}(s(t))) = \mathcal{C}_n(s(t))$$

by similar reasoning as in equation (5.13). Thus by equation (4.13):

$$\omega_n(s'(t)) = \bar{\omega}_n\left(\mathcal{C}_n(s'(t))\right) = \bar{\omega}_n\left(\mathcal{C}_n(s(t))\right) = \omega_n(s(t)).$$

This result is intuitively plausible: if the configuration s' is congruent to the configuration s, then the turning function ω_n will produce the same value for both configurations. Thus:

$$\dot{\theta}'_n = \omega_n(s'(t)) - \Omega_{(r,*)} \tag{6.6}$$

for $n = 1, 2$.

Equations (6.5) and (6.6) yield the differential equation in the rotating frame:

$$\begin{pmatrix} \dot{x}'_1 \\ \dot{y}'_1 \\ \dot{\theta}'_1 \\ \dot{x}'_2 \\ \dot{y}'_2 \\ \dot{\theta}'_2 \end{pmatrix} = F' \begin{pmatrix} x'_1 \\ y'_1 \\ \theta'_1 \\ x'_2 \\ y'_2 \\ \theta'_2 \end{pmatrix} = \begin{pmatrix} F'_1(x'_1, y'_1, \theta'_1, x'_2, y'_2, \theta'_2)^T \\ F'_2(x'_1, y'_1, \theta'_1, x'_2, y'_2, \theta'_2)^T \\ F'_3(x'_1, y'_1, \theta'_1, x'_2, y'_2, \theta'_2)^T \\ F'_4(x'_1, y'_1, \theta'_1, x'_2, y'_2, \theta'_2)^T \\ F'_5(x'_1, y'_1, \theta'_1, x'_2, y'_2, \theta'_2)^T \\ F'_6(x'_1, y'_1, \theta'_1, x'_2, y'_2, \theta'_2)^T \end{pmatrix}$$

$$\begin{pmatrix} \dot{x}'_1 \\ \dot{y}'_1 \\ \dot{\theta}'_1 \\ \dot{x}'_2 \\ \dot{y}'_2 \\ \dot{\theta}'_2 \end{pmatrix} = \begin{pmatrix} v\cos(\theta'_1) + \Omega_{(r,*)}\,y'_1 \\ v\sin(\theta'_1) - \Omega_{(r,*)}\,x'_1 \\ \omega_1(x'_1, y'_1, \theta'_1, x'_2, y'_2, \theta'_2)^T - \Omega_{(r,*)} \\ v\cos(\theta'_2) + \Omega_{(r,*)}\,y'_2 \\ v\sin(\theta'_2) - \Omega_{(r,*)}\,x'_2 \\ \omega_2(x'_1, y'_1, \theta'_1, x'_2, y'_2, \theta'_2)^T - \Omega_{(r,*)} \end{pmatrix}. \tag{6.7}$$

6.5   Linear Stability Analysis

The rotating frame has been chosen so that the periodic orbit given by equation (6.2) becomes a set of fixed points. By symmetry, we can linearize about any one of these fixed points. We can use the eigenvalues of the system linearized at one of these fixed points to analyze the stability of the relative equilibrium as a whole.

For convenience, we linearize about the point that corresponds to the initial time $t = 0$ in equation (6.2), which is

$$s_{(r,*)} = \left(-X_{(r,*)},\ 0,\ \pi,\ X_{(r,*)},\ 0,\ -\pi\right)^T /2. \tag{6.8}$$

As a notational reminder $s_{(r,*)}$ stands for either of the two hextuples, $s_{(r,\text{in})}$ or $s_{(r,\text{out})}$. We also let $s_{(\ell,*)}$ stand for either $s_{(\ell,\text{in})}$ or $s_{(\ell,\text{out})}$. By symmetry, similar reasoning applies in the left-turning case as in the right-turning case, so we focus on the right-turning case. The same convention will apply to the subscripts of other quantities in this chapter, which makes it easy to transfer the reasoning from the right-turning to the left-turning case.

To simplify the subsequent formulas, we introduce symbols for three quantities:

$$\nu_{(r,*)} = -\frac{\partial}{\partial X_n}\, \bar{\omega}_n \left(X_{(r,*)},\ 0\right)^T$$

$$\mu_{(r,*)} = -\frac{\partial}{\partial Y_n}\, \bar{\omega}_n \left(X_{(r,*)},\ 0\right)^T \tag{6.9}$$

$$\mathcal{T}_{(r,*)} = \mu_{(r,*)}\, X_{(r,*)}.$$

We also introduce notation for the matrix of the derivative of F' evaluated at the fixed point $s_{(r,*)}$:

$$\mathbf{J}_{(r,*)} = DF'\left(s_{(r,*)}\right).$$

It is fairly easy to to compute the first, second, fourth, and fifth rows of $\mathbf{J}_{(r,*)}$ from equation (6.7). Most of the entries are 0. We compute the third and sixth rows in appendix B. The resulting matrix is:

$$
\mathbf{J}_{(r,*)} = \begin{pmatrix}
0 & \Omega_{(r,*)} & -v & 0 & 0 & 0 \\
-\Omega_{(r,*)} & 0 & 0 & 0 & 0 & 0 \\
\nu_{(r,*)} & \mu_{(r,*)} & \mathcal{T}_{(r,*)} & -\nu_{(r,*)} & -\mu_{(r,*)} & 0 \\
0 & 0 & 0 & 0 & \Omega_{(r,*)} & v \\
0 & 0 & 0 & -\Omega_{(r,*)} & 0 & 0 \\
\nu_{(r,*)} & \mu_{(r,*)} & 0 & -\nu_{(r,*)} & -\mu_{(r,*)} & \mathcal{T}_{(r,*)}
\end{pmatrix}. \tag{6.10}
$$

The entries of $\mathbf{J}_{(r,*)}$ and its eigenvalues are functions of the parameters for the total dynamical system ϕ_T. This allows us to study bifurcations of the total system as the sensor weights are varied. It is shown in appendix C that the set of eigenvalues for $\mathbf{J}_{(r,*)}$ is:

$$
\left\{ 0, \pm i\Omega_{(r,*)}, \mathcal{T}_{(r,*)}, \mathcal{S}_{(r,*)}^{\pm} \right\}, \tag{6.11}
$$

where

$$
\mathcal{S}_{(r,*)}^{\pm} = \frac{\mathcal{T}_{(r,*)} \pm \sqrt{\mathcal{T}_{(r,*)}^2 - 4\left(\Omega_{(r,*)}^2 + 2v\nu_{(r,*)}\right)}}{2}. \tag{6.12}
$$

We call these *salient eigenvalues* (the symbol "$\mathcal{S}$" is for "salient"). The eigenvalue 0 does not move, but the other eigenvalues trace out curves in the complex plane as the weights are varied. We will focus on the traces of the salient eigenvalues, to see where the bifurcations of the revolving type relative equilibrium occur.

Recall from section 5.6 that the eigenvectors of this linearization can be partitioned into two subsets, that is, the tangential and normal eigenvectors The first three eigenvalues in (6.11) are tangential, that is, their corresponding eigenvectors are tangent to the relative equilibrium at $s_{(r,*)}$. The last three eigenvalues are normal, that is, their corresponding eigenvectors are orthogonal to the relative equilibrium at $s_{(r,*)}$. The eigenvectors are given in appendix C.

There is a redundancy in both subsets of eigenvalues: one eigenvalue in each subset is the sum of the other two:

$$
0 = (i\Omega_{(r,*)}) + (-i\Omega_{(r,*)})
$$

$$
\mathcal{T}_{(r,*)} = \mathcal{S}_{(r,*)}^{+} + \mathcal{S}_{(r,*)}^{-}. \tag{6.13}
$$

It is also the case in the mathematical version of Braitenberg vehicles by (Rañó 2010) that one eigenvalue of the linearization is the sum of the other two.

The tangential eigenvalues describe how the system behaves inside the relative equilibrium. The traces of $\pm i\Omega_{(r,*)}$ never leave the imaginary axis, so the tangential eigenvalues are always on the imaginary axis. This is consistent with the fact that the orbits within a relative equilibrium neither attract nor repel one another. The relative equilibria are made up of many "parallel" orbits each of which is associated with a different revolution in the physical plane. The tangential eigenvalues will not play an important role in our bifurcation analysis.

The normal eigenvalues, which include the salient eigenvalues describe how the system behaves close to but outside of the relative equilibrium. If the real part of each of the normal eigenvalues is negative, then all states sufficiently close to the relative equilibrium tend towards it. If the real part of any of the normal eigenvalues is positive, then there are states arbitrarily close to the relative equilibrium that move away from it. Thus we can use the normal eigenvalues in our bifurcation analysis to determine whether a relative equilibrium is attracting or not.

The pair of salient eigenvalues, $\mathcal{S}^{\pm}_{(r,*)}$, can be real or complex conjugate numbers, while the other normal eigenvalue, $\mathcal{T}_{(r,*)}$, is always a real number. By equation (6.13), the eigenvalue $\mathcal{T}_{(r,*)}$ is negative whenever the real part of each of the salient eigenvalues is negative. Thus, we can determine the stability of the relative equilibrium $\mathcal{E}_{(r,*)}$ just from the sign of the real part of $\mathcal{S}^{\pm}_{(r,*)}$, which is why we call them the "salient eigenvalues."

There is a geometric interpretation for the salient eigenvalues, which is illustrated in figure 6.5. By equation (6.1), there is a relative equilibrium for each intersection point in the graphs of the functions $-2v/X_n$ and $\bar{\omega}_n(X_n, 0)^T$. The product of the salient eigenvalues for each relative equilibrium is proportional to the difference in the slopes of the two graphs at the intersection points.

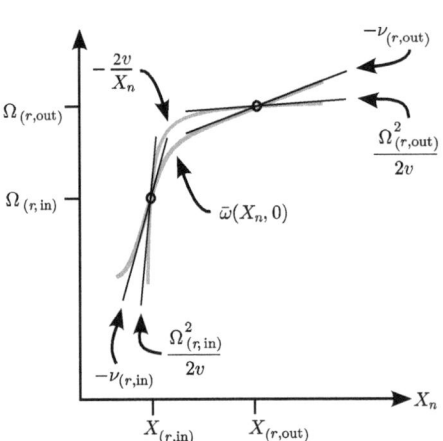

Figure 6.5

Graphical interpretation of $\Delta m_{(r,*)}$. The graphs of $-2v/X_n$ and $\bar{\omega}(X_n, 0)^T$ are shaded gray. They intersect at $X_n = X_{(r,\text{in})}$ and $X_n = X_{(r,\text{out})}$. The intersection points are marked with open circles. The tangent lines at the intersection points are depicted with thin black line segments, which are labeled with their slopes. The slopes are all positive in this case (recall that $v_{(r,*)}$ is defined to be the negative of the slope). The differences between these slopes at $X_{(r,\text{in})}$ and $X_{(r,\text{out})}$ correspond to $\Delta m_{(r,\text{in})}$ and $\Delta m_{(r,\text{out})}$ respectively.

By the power rule, the derivative of $-2v/X_n$ with respect to X_n evaluated at $X_{(r,*)}$ is $2v/X^2_{(r,*)}$. Since $\Omega_{(r,*)} = -2v/X_{(r,*)}$ we can express this slope as $\Omega^2_{(r,*)}/(2v)$. By definition, the derivative of $\bar{\omega}_n(X_n, 0)^T$ with respect to X_n evaluated at $X_{(r,*)}$ is $-\nu_{(r,*)}$.

We denote the difference between the slopes at $X_{(r,*)}$ by

$$\Delta m_{(r,*)} = \frac{\Omega^2_{(r,*)}}{2v} - (-\nu_{(r,*)}). \tag{6.14}$$

Rearranging this gives:

$$\Omega^2_{(r,*)} + 2v\nu_{(r,*)} = 2v\Delta m_{(r,*)}.$$

Substituting this into equation (6.12) gives us a fairly simple expression for the salient eigenvalues in terms of $\Delta m_{(r,*)}$, the difference in slopes at the intersection point:

$$\mathcal{S}^\pm_{(r,*)} = \frac{\mathcal{T}_{(r,*)} \pm \sqrt{\mathcal{T}^2_{(r,*)} - 8v\Delta m_{(r,*)}}}{2}. \tag{6.15}$$

We can compute the product of the salient eigenvalues from the equation above to get:

$$\mathcal{S}^+_{(r,*)}\mathcal{S}^-_{(r,*)} = 2v\Delta m_{(r,*)}.$$

Since the velocity v is always positive, the sign of the product of the salient eigenvalues is the same as the sign of $\Delta m_{(r,*)}$. So if $\Delta m_{(r,*)}$ is positive, then the real parts of the salient eigenvalues must have the same sign. If in addition they are both negative, the relative equilibrium is attracting. If $\Delta m_{(r,*)}$ is negative, then the real parts of the salient eigenvalues must have opposite signs, so one of them has positive real part, and so the relative equilibrium is not attracting. These facts will be useful in the bifurcation analysis.

6.6 Bifurcation Analysis

We now analyze the two main bifurcations that occur as the weights vary in W_{same}. The first bifurcation we consider is a saddle node-like bifurcation in which a pair of relative equilibria, $\mathcal{E}_{(r,\text{in})}$ and $\mathcal{E}_{(r,\text{out})}$, begin together. The second is a Hopf-like bifurcation in which the relative periodic orbit $\mathcal{P}_{(r,\text{in})}$ emerges from $\mathcal{E}_{(r,\text{in})}$. Both bifurcations can be understood in terms of the traces of the salient eigenvalues in the complex plane. Figures 6.6, 6.7, and 6.8 show typical examples of the traces made by the salient eigenvalues.

As explained in section 6.1, the saddle node-like bifurcation occurs as the weights cross the boundary $\partial\mathcal{R}_r$. When this happens, the graphs of the turning function and hyperbola form a new contact point, as explained in section 6.3. The initial contact can occur in two ways. First, the graphs can be tangent at

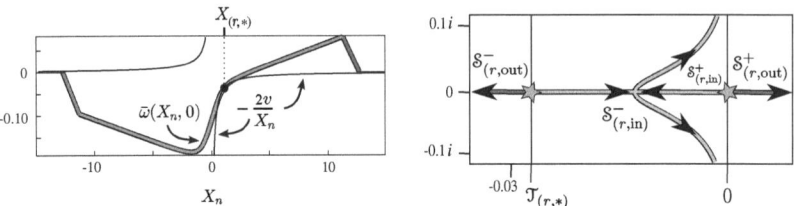

Figure 6.6
(left) Tangential initial contact between $\bar{\omega}(X_n, 0)^T$ and $-2v/X_n$ for weights in the avoider-avoider quadrant. (right) The salient eigenvalues $\mathcal{S}^{\pm}_{(r,*)}$ begin at the points $\mathcal{T}_{(r,*)}$ and 0 (marked by the green stars). The eigenvalue traces of $\mathcal{S}^{\pm}_{(r,\text{in})}$ (colored cyan) move inward and meet on the negative axes while the traces of $\mathcal{S}^{\pm}_{(r,\text{out})}$ (colored magenta) move outward along the real axis. Parameter values for the initial contact: $P = 12$, $\psi = \pi/4$, $v = 1/50$, and $(w_l, w_r) \approx (-0.8104, -0.6896)$.

the contact point. This happens for weights in the avoider-avoider quadrant as shown in figure 6.6. Otherwise, the graphs are not tangent at the contact point. This happens for weights in the pursuer-pursuer quadrant as shown in figure 6.7. We consider the tangential case first because the nontangential case can be understood in terms of the tangential case.

Wherever $-2v/X_n$ and $\bar{\omega}_n(X_n, 0)^T$ intersect tangentially, their slopes are the same at the intersection point so $\Delta m_{(r,*)} = 0$. The expression (6.15) for the salient eigenvalues reduces to:

$$\mathcal{S}^{\pm}_{(r,*)} = \frac{\mathcal{T}_{(r,*)} \pm |\mathcal{T}_{(r,*)}|}{2}.$$

The eigenvalue $\mathcal{T}_{(r,*)}$ is negative when the weights cross the bifurcation curve $\partial \mathcal{R}_r$. So when the initial contact is tangential the normal eigenvalues satisfy the conditions:

$$\mathcal{S}^{-}_{(r,*)} = \mathcal{T}_{(r,*)} < 0$$

$$\mathcal{S}^{+}_{(r,*)} = 0.$$

Thus, at this point, the relative equilibrium comes into existence and is marginally stable. The emergence of the salient eigenvalues with the values $\mathcal{T}_{(r,*)}$ and 0 is illustrated by the green stars in figure 6.6.

Typically, just after the initial contact, there are exactly two solutions to equation (6.1), $X_{(r,\text{in})}$ and $X_{(r,\text{out})}$, as shown in figure 6.5. We can determine the stability of the relative equilibria $\mathcal{E}_{(r,\text{in})}$ and $\mathcal{E}_{(r,\text{out})}$ from the difference in slopes of the two graphs at the intersection points, that is, from $\Delta m_{(r,\text{in})}$ and $\Delta m_{(r,\text{out})}$.

The hyperbola $-2v/X_n$ has a vertical asymptote at $X_n = 0$ while the graph of $\bar{\omega}_n(X_n, 0)^T$ is bounded from below. So for positive X_n sufficiently close to 0

the hyperbola is below the graph of $\bar{\omega}_n(X_n, 0)^T$. The value of $-2v/X_n$ increases rapidly as X_n increases. The first intersection point we reach is at $X_{(r,\text{in})}$. As X_n passes through $X_{(r,\text{in})}$ from left to right the hyperbola goes from being under the graph of $\bar{\omega}(X_n, 0)^T$ to being above it. So at this intersection point the slope of the tangent to the hyperbola is greater than the slope of the tangent to the graph of $\bar{\omega}_n(X_n, 0)^T$. Therefore $\Delta m_{(r,\text{in})}$ is positive, the product of the salient eigenvalues is positive, and the salient eigenvalues must have the same sign.

The only way the pair of salient eigenvalues can continuously go from $S^-_{(r,\text{in})}$ being negative and $S^+_{(r,\text{in})}$ being 0 to both of them having the same sign is for $S^+_{(r,\text{in})}$ to become negative. Numerical computation confirms this and shows, furthermore, that both of the salient eigenvalues, $S^\pm_{(r,\text{in})}$, move inward from the green stars as illustrated in figure 6.6. Since both salient eigenvalues are negative, the relative equilibrium $\mathcal{E}_{(r,\text{in})}$ is attracting after the saddle node-like bifurcation.

Before we proceed to the intersection point at $X_{(r,\text{out})}$, we consider what happens as the weights are varied further away from the bifurcation curve $\partial\mathcal{R}_r$. The value of $\Delta m_{(r,\text{in})}$ increases, which decreases the quantity under the radical in equation (6.15). As shown in figures 6.6 and 6.8, the pair of eigenvalues $S^\pm_{(r,\text{in})}$ continue to move towards each other on the negative real axis, and eventually meet and then move upwards and downwards. When $\Delta m_{(r,\text{in})}$ grows large enough so that $\mathcal{T}^2_{(r,*)} = 8v\Delta m_{(r,*)}$, then, according to equation (6.15), the eigenvalues $S^\pm_{(r,\text{in})}$ meet at:

$$S^+_{(r,\text{in})} = S^-_{(r,\text{in})} = \frac{1}{2}\mathcal{T}_{(r,\text{in})}.$$

As X_n increases, the value of $\Delta m_{(r,\text{in})}$ continues to increase, so that the quantity under the radical in equation (6.15) becomes negative and $S^\pm_{(r,\text{in})}$ become a pair of complex conjugate numbers whose real part is $\mathcal{T}_{(r,\text{in})}/2$. Recall that Hopf bifurcations are characterized by a pair of complex conjugate eigenvalues crossing the imaginary axis, which is also what happens with the salient eigenvalues.

We now consider the intersection point at $X_{(r,\text{out})}$. As X_n passes through $X_{(r,\text{out})}$ from left to right, the hyperbola $-2v/X_n$ goes from being above the graph of $\bar{\omega}(X_n, 0)^T$ to being below it. So at this intersection point the slope of the tangent to the hyperbola is less than the slope of the tangent to the graph of $\bar{\omega}_n(X_n, 0)^T$ (see figure 6.5). Consequently $\Delta m_{(r,\text{out})}$ is negative, the product of the salient eigenvalues is negative, and the salient eigenvalue $S^+_{(r,\text{out})}$ must be positive. Thus the relative equilibrium $\mathcal{E}_{(r,\text{out})}$ is nonattracting after the saddle

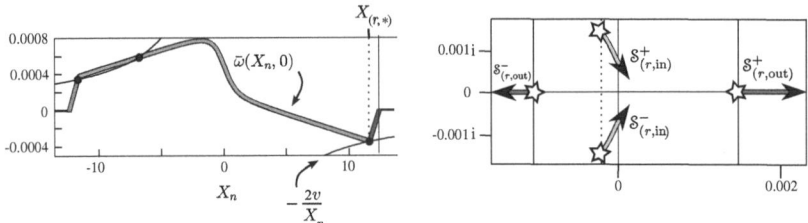

Figure 6.7
(left) Nontangential initial contact between $\bar{\omega}(X_n, 0)^T$ and $-2v/X_n$ for weights in the pursuer-pursuer quadrant. (right) Eigenvalue trace for the salient eigenvalues. The limit of $\mathcal{S}_{(r,*)}^{\pm}$ as the weights converge to $\partial\mathcal{R}_r$ from above are marked by open ☆'s. The salient eigenvalues $\mathcal{S}_{(r,\text{in})}^{\pm}$ form a complex conjugate pair with negative real part, as the weights rise up from $\partial\mathcal{R}_r$. Their traces are colored cyan. The sign of their real part increases and they cross the imaginary axis. The salient eigenvalues $\mathcal{S}_{(r,\text{out})}^{\pm}$ move apart along the positive and negative axis as the weights rise up from $\partial\mathcal{R}_r$. Their traces are colored magenta. Parameter values $P = 12$, $\psi = \pi/9$, $v = 1/500$, and $(w_l, w_r) \approx (0.006071, -0.006019)$.

node-like bifurcation. Numerical computation confirms this and shows furthermore that the salient eigenvalues, $\mathcal{S}_{(r,\text{out})}^{\pm}$, move outward along the purple paths from the green stars as illustrated in figure 6.6.

We now proceed to consider what happens to the salient eigenvalues after a nontangential initial contact. For the case shown in figure 6.7, the weights are in the pursuer-pursuer quadrant. The function $\bar{\omega}_n(X_n, 0)^T$ has a local minimum at $X_{(r,*)} = P - M$, while the function $-2v/X_n$ is concave down along the positive X_n-axis. The initial contact occurs at the local minimum and is nontangential. At the contact point the derivative of the function $-2v/X_n$ does not equal either the left or right derivatives of $\bar{\omega}_n(X_n, 0)^T$.

Shortly after the initial contact, there are two intersection points: $X_{(r,\text{in})}$ just below $P - M$, and $X_{(r,\text{out})}$ just above $P - M$ (comparable intersection points can be seen in the right panel of figure 6.4,). We consider the intersection at $X_{(r,\text{out})}$ first.

As in the tangential case, when X_n passes through $X_{(r,\text{out})}$ from left to right the hyperbola $-2v/X_n$ goes from being above the graph of $\bar{\omega}(X_n, 0)^T$ to being below it (like in figure 6.5). So the slope of the tangent to the hyperbola at the intersection point is less than the slope of the tangent to the graph of $\bar{\omega}_n(X_n, 0)^T$. Consequently, $\Delta m_{(r,\text{out})}$ is negative, the salient eigenvalues must have opposite signs, and the salient eigenvalue $\mathcal{S}_{(r,\text{out})}^+$ must be positive.

Numerical computation confirms this and shows that both of the salient eigenvalues, $\mathcal{S}_{(r,\text{out})}^{\pm}$, move outward from the white stars on the real axis as shown in figure 6.7 (the stars are shown as open ☆'s to indicate that the eigenvalues are not defined at the initial contact, but just after it). The traces of $\mathcal{S}^{\pm}_{(r,\text{out})}$ are basically the same in the tangential and nontangential cases. $\mathcal{S}^{\pm}_{(r,\text{out})}$ moves outwards in both cases, and so the relative equilibrium $\mathcal{E}_{(r,\text{out})}$ is nonattracting.

The situation is more complicated for $\mathcal{E}_{(r,\text{in})}$. As in the tangential case, when X_n passes through $X_{(r,\text{in})}$ from left to right the hyperbola goes from being below the graph of $\bar{\omega}(X_n, 0)^T$ to being above it. The slope of the tangent to the hyperbola at the intersection point is greater than the slope of the tangent to the graph of $\bar{\omega}_n(X_n, 0)^T$. So $\Delta m_{(r,\text{out})}$ is positive, the product of the salient eigenvalues is positive, and the real parts of $\mathcal{S}^{\pm}_{(r,\text{in})}$ have the same sign.

However, in the nontangential case, the salient eigenvalues $\mathcal{S}^{\pm}_{(r,\text{in})}$ do not begin at $\mathcal{T}_{(r,\text{in})}$ and 0. The slopes of the graphs of $\bar{\omega}_n(X_n, 0)^T$ and $-2v/X_n$ are not the same at the initial contact point, and numerical computation shows that the salient eigenvalues begin instead as a complex conjugate pair with negative real part (unless the weights are near the origin of W_{same}). This is shown in figure 6.7, where $\mathcal{S}^{\pm}_{(r,\text{in})}$ are marked with open ☆'s connected by a vertical dotted line. The relative equilibrium $\mathcal{E}_{(r,\text{in})}$ is observed in numerical integrations to be an attracting set shortly after the initial contact just as it is in the tangential case.

Despite the differences between the tangential and nontangential cases, the behaviors of the two bifurcations are similar. To see this, note that if we smoothed the function $\bar{\omega}_n(X_n, 0)^T$ just within a very small neighborhood of $P-M$ so that the initial contact would be tangential, the salient eigenvalues would trace out paths like those in figure 6.6. The salient eigenvalue $\mathcal{S}^{-}_{(r,*)}$ would begin at $\mathcal{T}_{(r,*)}$ and $\mathcal{S}^{+}_{(r,*)}$ would begin at 0. Shortly afterward, $\mathcal{S}^{\pm}_{(r,\text{in})}$ would move inwards, meet at $\mathcal{T}_{(r,\text{in})}/2$ and become a pair of complex numbers, eventually reaching the positions of the open ☆'s connected by a vertical dotted line in figure 6.7. As noted, $\mathcal{S}^{\pm}_{(r,\text{out})}$ behave the same in the tangential and nontangential cases, beginning at $\mathcal{T}_{(r,\text{out})}$ and 0 and then moving outwards.

The extra complication in the nontangential case is the result of using a piecewise linear scaling function in equation (2.7) to avoid complications in other aspects of the analysis. This bifurcation is degenerate but it often still resembles a saddle node bifurcation in that two relative equilibria appear, one of which is attracting, $\mathcal{E}_{(r,\text{in})}$, and one of which is nonattracting, $\mathcal{E}_{(r,\text{out})}$.

We now turn to the Hopf-like bifurcation that occurs after the saddle node-like bifurcation. This is where $\mathcal{E}_{(r,\text{in})}$ ceases to be attracting and a new attracting set, $\mathcal{P}_{(r,\text{in})}$, appears. This is also where the agents begin to meander around each other. The way this Hopf-like bifurcation occurs is similar in the pursuer-pursuer and avoider-avoider quadrants.

Further changing the weights after the initial contact between the hyperbola and the graph of $\bar{\omega}_n(X_n, 0)^T$ leads to an increase in the real part of $S_{(r,\text{in})}^{\pm}$ (see figures 6.6 and 6.7). The salient eigenvalues cross the imaginary axis and the relative equilibrium $\mathcal{E}_{(r,\text{in})}$ ceases to be attracting. Numerical integration confirms this and indicates that an attracting relative periodic orbit appears when the salient eigenvalues cross the imaginary axis.

So long as the relative equilibrium $\mathcal{E}_{(r,\text{in})}$ exists the value of $X_{(r,\text{in})}$ is positive. So the sign of the normal eigenvalue $\mathcal{T}_{(r,\text{in})}$ equals the sign of $\mu_{(r,\text{in})}$. As the weights are further varied, the value of $\mu_{(r,\text{in})}$ increases toward 0 and when $\mu_{(r,\text{in})}$ passes through 0, all three normal eigenvalues cross the imaginary axis. The relative equilibrium $\mathcal{E}_{(r,\text{in})}$ ceases to be an attracting set.

At the bifurcation, we have the degenerate condition that all six eigenvalues are purely imaginary and 0 has a multiplicity of 2. This is why we have called it a degenerate Hopf-like bifurcation. Despite the degeneracy, numerical integration indicates that this bifurcation has much in common with a Hopf-like bifurcation, which is familiar from the analysis of the BZ reaction.

We designate the region of W_{same} where $S_{(r,\text{in})}^{\pm}$ have positive real part $\mathcal{M}_r$ (see the rose-colored regions of figures 6.1 and 6.10). The bifurcation condition $\mu_{(r,\text{in})} = 0$ occurs for (w_ℓ, w_r) on the outer boundary of $\mathcal{M}_r$.

Since $X_{(r,\text{in})}$ is in the interval $[0, P-M]$ (see appendix A), the values of the distances $d_{(\ell,\text{in})}$ and $d_{(r,\text{in})}$ are less than P. So by equation (2.6), the slope of the scaling function is the same at $d_{(\ell,\text{in})}$ and $d_{(r,\text{in})}$ and this slope is less than P.

$$f'(d_{(\ell,\text{in})}) = f'(d_{(r,\text{in})}) = -\frac{1}{P}.$$

Substituting this into equation (B.1) from appendix B and rearranging shows that the bifurcation condition $\mu_{(r,\text{in})} = 0$ is equivalent to:

$$(w_\ell, \ w_r) \bullet (d_{(\ell,\text{in})}, -d_{(r,\text{in})}) = 0.$$

Since $X_{(r,\text{in})}$ varies little with changes in the weights, the distances $d_{(\ell,\text{in})}$ and $d_{(r,\text{in})}$ between the sensors and agents in the relative equilibrium $\mathcal{E}_{(r,\text{in})}$ vary little with the weights. So $(d_{(\ell,\text{in})}, -d_{(r,\text{in})})$ is approximately constant, and the equation above is nearly the equation for a line through the origin. So the condition $\mu_{(r,\text{in})} = 0$ occurs approximately when (w_r, w_ℓ) lies on such a line. This is also what we observe in our numerical integrations, as can be seen in figure 6.10. The outer boundary of $\mathcal{M}_r$ is approximately a line through the origin.

In section 6.7, we discuss the implications of this bifurcation analysis on the agents' representations of each other. In section 6.8, we analyze the meandering paths produced by the relative periodic orbits that emerge from the Hopf-like bifurcation.

6.7　Agents' Representations of Revolving and Meandering Behaviors

We have now seen the main attracting sets that occur for weights in W_{same}. They produce circular and meandering paths in the rest frame. In this section, we analyze these behaviors from the perspective of the agents themselves. In addition to shedding further light on the dynamics and bifurcations associated with $\mathcal{E}_{(r,\text{in})}$ and $\mathcal{P}_{(r,\text{in})}$, it illustrates the value of the open dynamical systems framework. This framework allows us to compare the internal behaviors of the agents considered on their own as separate systems with their collective behavior when interacting in a total dynamical system (recall chapter 4). It also allows us to interpret their representational dynamics in the context of this collective behavior.

The way the agents behave individually was discussed in chapter 2, and summarized in Figure 2.3. A single agent behaves as a pursuer when its weights are positive; it behaves as an avoider when its weights are negative; it behaves in a lateralized fashion (pursuing to one side; avoiding to the other) when the weights have opposite signs. This pattern largely persists in the weight space W_{same} but it is altered for some types of avoiders that are separated at the appropriate distance. A pair of pursuers often behave collectively as pursuers, but surprisingly, so can a pair of avoiders. A pair of avoiders can revolve or meander about each other, and go through a similar bifurcation sequence as occurs for pairs of actual pursuers (as discussed in section 6.6 and as shown in figure 6.10).

The value of an agential perspective has been recognized by some (e.g., Philippides et al. [2012]), although it runs counter to an anti-representational tendency in embodied approaches to cognitive science. Theorists in this area tend to treat dynamical systems models and "agent-environment systems" as a way to overcome more traditional approaches to cognition, according to which internal representations are states that carry information about the environment, and which control and mediate behavior (see section 4.2). We have argued that an open dynamical systems framework allows for a mixed perspective, which acknowledges the importance of representations in cognition, but also makes it possible to study how representational processes change when an agent is embedded in an environment (see Hotton and Yoshimi [2010, 2011]).

We combine mathematical and numerical analysis in this section. We will focus on the particular physical parameter values: $\psi = \pi/4$, $P = 12$, $v = 1/50$ and analyze the dynamics for (w_ℓ, w_r) in $[-15/4, 15/4]^2 \subset W_{\text{same}}$. This region of W_{same} is shown in figures 6.1 and 6.10.

Figure 6.8 shows the traces of the salient eigenvalue $\mathcal{S}^{\pm}_{(r,\text{in})}$ and open phase portraits for an agent before and after the main bifurcations discussed earlier.

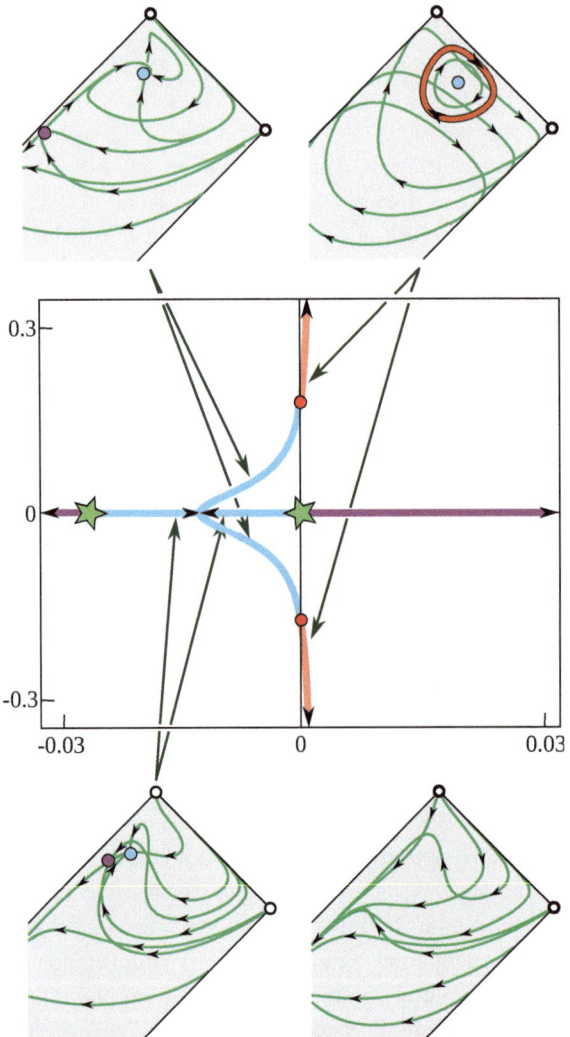

Figure 6.8
The eigenvalue traces of $\mathcal{S}^{\pm}_{(r,*)}$ and corresponding open phase portraits in the agent space, shown in the surrounding panels. The salient eigenvalues begin at the green stars. Prior to this the agent's state simply decays toward $(0,0)^T$ (bottom right panel). After this the traces of $\mathcal{S}^{\pm}_{(r,\text{in})}$ and $\mathcal{S}^{\pm}_{(r,\text{out})}$ are colored cyan and magenta respectively. The partially fixed points $\pi_n(\mathcal{E}_{(r,\text{in})})$ and $\pi_n(\mathcal{E}_{(r,\text{out})})$ are marked by cyan and magenta dots, respectively (left panels). They correspond to representations of one agent by the other when the two agents are revolving. The eigenvalues $\mathcal{S}^{\pm}_{(r,\text{in})}$ then meet, form a complex conjugate pair, and go on to cross the imaginary axis at the red dots. After this, the traces of $\mathcal{S}^{\pm}_{(r,\text{in})}$ are rose colored. The partially fixed point $\pi_n(\mathcal{E}_{(r,\text{in})})$ is no longer attracting, and an attracting simple closed path $\pi_n(\mathcal{P}_{(r,\text{in})})$ appears around it, colored orange (upper right panel). This corresponds to the fact that in a meandering path the agents represent each other as moving in closed loops.

The weights (w_ℓ, w_r) are varied along the dotted line in the avoider-avoider quadrant in figure 6.10. This is the same path as is shown in the magnified region in the lower left of figure 6.1. The weights begin in $\mathcal{O}_r \cap \mathcal{O}_\ell$ and move in the upper left direction of W_{same}. This corresponds to going clockwise through the panels in figure 6.8 from the bottom right to the top right. We see changes in the open phase portraits that resemble the familiar sequence of saddle node and Hopf bifurcations found in two-dimensional dynamical systems.

The colors in figure 6.10 correspond to figure 6.1: beige for no attracting relative equilibria, blue for attracting revolving type relative equilibria, and rose color for relative periodic orbits. The panels in figure 6.8 show the representations that occur in the open phase portraits of the agents before and after the main bifurcations. The green stars show where the saddle node-like bifurcation occurs. The red dots show where the Hopf-like bifurcations occur. The pair of salient eigenvalues help characterize the open phase portraits in a way that is similar to how pairs of eigenvalues for fixed points help characterize the phase portraits for classic two-dimensional dynamical systems (see sections 3.5 and 3.6).

At first, ϕ_τ has no attracting sets, but the open phase portrait does contain the attracting fixed point $(0, 0)^T$ (not visible in figure 6.8). This agent state corresponds to the agents traveling out of view of each other, which is what we observe in numerical integrations with weights in $\mathcal{O}_r \cap \mathcal{O}_\ell$.

The vertex $(0, 0)^T \in S_\alpha$ is partially fixed. The two agents can begin by facing each other while out of range, but then move within range of one another. In this case the agents' states begin at $(0, 0)^T$ and stay there for a time, until the agents come into range, at which point the activations become positive and so the agents' states leave $(0, 0)^T$. Therefore the preimage of $(0, 0)^T$ is not an invariant set of the total dynamical system and $(0, 0)^T$ is not a fully fixed point in S_{α_n}. However, for other initial conditions, the system does stay in $(0, 0)^T$ for all time. For example, the two agents can travel along parallel lines that are further apart than the sensor range. In this case, the entire orbit in S_τ projects to $(0, 0)^T$. Thus $(0, 0)^T$ is a partially fixed point. Depending on what is happening in the total system, the agent state goes in and out of $(0, 0)^T$, or stays in it for all time.

For weights in $\mathcal{O}_r \cap \mathcal{O}_\ell$ the agent dynamics simply decays to $(0, 0)^T$ representing "no agent." Although this vertex of S_{α_n} is not visible in the panels of figure 6.8, the bottom right panel does show how in this case agent states tend towards the partially fixed point $(0, 0)^T$.

As the weights progress along the dotted lines in figure 6.1 they enter the region $\mathcal{R}_r$ at the points marked by the green stars. This corresponds to moving from the bottom right to the bottom left panel in figure 6.8. We see a pair of partially fixed points appear from nothing in the open phase portraits, which resembles a saddle node bifurcation of fixed points for a two-dimensional dynamical system.

For the total dynamical system ϕ_τ a pair of relative equilibria come into existence along with their salient eigenvalues $\mathcal{S}^\pm_{(r,*)}$. The salient eigenvalues begin at the points marked by the green stars in W_{same} shown in figure 6.8 (also shown in figure 6.6). At this bifurcation, one of the salient eigenvalues is 0, as in a saddle node bifurcation of fixed points for a two-dimensional dynamical system, although this bifurcation is actually degenerate since the multiplicity of 0 is two. The zero eigenvalue has eigenvectors in the tangent space and in the normal space.

Just after this bifurcation, the salient eigenvalues $\mathcal{S}^\pm_{(r,\text{in})}$ move towards each other on the negative axis while $\mathcal{S}^\pm_{(r,\text{out})}$ move away from each other. Since $\mathcal{T}_{(r,\text{in})}$ is the sum of $\mathcal{S}^\pm_{(r,\text{in})}$, all three normal eigenvalues for the relative equilibrium $\mathcal{E}_{(r,\text{in})}$ have negative real part, and so it is attracting. The relative equilibrium $\mathcal{E}_{(r,\text{in})}$ projects to the attracting partially fixed point $\pi_n(\mathcal{E}_{(r,\text{in})})$ in S_{α_n}. And since $\mathcal{S}^+_{(r,\text{out})}$ emerges from 0 into the positive real axis the relative equilibrium $\mathcal{E}_{(r,\text{out})}$ is nonattracting. It projects to the nonattracting partially fixed point $\pi_n(\mathcal{E}_{(r,\text{out})})$ in S_{α_n}.

For the attracting relative equilibrium, the two avoiders will revolve around each other for many initial conditions. As they revolve, they stay on opposite sides of their circular path so they remain motionless relative to each other. In this situation, each agent represents the other with the attracting partially fixed point $\pi_n(\mathcal{E}_{(r,\text{in})})$ (marked by a cyan colored dot).

To see that $\pi_n(\mathcal{E}_{(r,\text{in})})$ is partially fixed, consider the state $\pi_2(\mathcal{E}_{(r,\text{in})})$ in S_{α_2}. In this case, the preimage of $\pi_2(\mathcal{E}_{(r,\text{in})})$ contains $\mathcal{E}_{(r,\text{in})}$ which contains many orbits of ϕ_τ. For any of these orbits, vehicle 2 represents vehicle 1 by the same agent state $\pi_2(\mathcal{E}_{(r,\text{in})})$ for all time. Now suppose we change the first vehicle's heading by 180 degrees. Since agent 2's state does not depend on the heading of agent 1, this configuration is also in the preimage of $\pi_2(\mathcal{E}_{(r,\text{in})})$. In this case, vehicle 2 will begin in the state $\pi_2(\mathcal{E}_{(r,\text{in})})$ but the vehicles will move away from each other and agent 2's state will move away from $\pi_2(\mathcal{E}_{(r,\text{in})})$. This shows that the preimage of $\pi_2(\mathcal{E}_{(r,\text{in})})$ is not an invariant set. Thus $\pi_2(\mathcal{E}_{(r,\text{in})})$ is a partially fixed point. In some cases, vehicle 2 represents vehicle 1 in the same fixed point for all times; in other cases, it can represent it as starting in that point and moving away from it, depending on what is happening in the total system. Similar reasoning applies to $\pi_1(\mathcal{E}_{(r,\text{in})})$ in S_{α_1}.

As (w_ℓ, w_r) moves further into $\mathcal{R}_r$, the salient eigenvalues $\mathcal{S}^\pm_{(r,\text{in})}$ continue to move towards each other. They go on to meet each other within the negative real axis, and become a complex conjugate pair with negative real part. The partially fixed point $\pi_n(\mathcal{E}_{(r,\text{in})})$ continues to be attracting. This corresponds to moving from the bottom left to the top left panel in figure 6.8.

In a two-dimensional dynamical system, the pair of negative eigenvalues of an attracting fixed point can also meet and form a complex conjugate pair. When this happens, states near the attracting fixed point cease to go more or less straight towards it and instead begin to spiral into it. This is not a bifurcation (no topological change in the orbits occurs), but it is a qualitative change in behavior comparable to the difference between an over-damped and under-damped oscillator. This is visible in the agent space: there is a shift from a tendency for the paths to go straight towards the attracting partially fixed point $\pi_n(\mathcal{E}_{(r,\text{in})})$ in the bottom left panel and to spiral into it in the top left panel.

The attracting partially fixed point $\pi_n(\mathcal{E}_{(r,\text{in})})$ has two negative salient eigenvalues while the nonattracting partially fixed point $\pi_n(\mathcal{E}_{(r,\text{out})})$ has one positive and one negative salient eigenvalue. This is like a classic saddle node bifurcation of two fixed points for a two-dimensional dynamical system that produces an attracting fixed point with two negative eigenvalues and a saddle point with one positive and one negative eigenvalue. These open phase portraits and the classic phase portrait of a dynamical system mainly differ in that paths in the open phase portrait can cross each other whereas the orbits do not cross each other in the phase portrait of a classical dynamical system.

As the weights (w_ℓ, w_r) continue to be varied, they enter the region $\mathcal{M}_r$, at the point marked by a crimson dot in figure 6.1. The region $\mathcal{M}_r$ is rose colored in figures 6.1 and 6.10. This corresponds to moving from the top left to the top right panel in figure 6.8. The salient eigenvalues $\mathcal{S}^{\pm}_{(r,\text{in})}$ cross the imaginary axis at the points marked by the red dot in figure 6.8.

As explained in section 6.6, the relative equilibrium $\mathcal{E}_{(r,\text{in})}$ ceases to be an attracting set at this point, and numerical integration indicates that an attracting relative periodic orbit, $\mathcal{P}_{(r,\text{in})}$, emerges from $\mathcal{E}_{(r,\text{in})}$ and the vehicles begin to meander, typically moving along quasiperiodic orbits in $\mathcal{P}_{(r,\text{in})}$.

When the real part of $\mathcal{S}^{\pm}_{(r,\text{in})}$ becomes positive, the partially fixed point $\pi_n(\mathcal{E}_{(r,\text{in})})$ ceases to be attracting. As shown in the upper right panel of figure 6.8, the point $\pi_n(\mathcal{E}_{(r,\text{in})})$ is now surrounded by the attracting path $\pi_n(\mathcal{P}_{(r,\text{in})})$ colored orange.

From the vehicle's standpoint, the meandering behavior of the total system corresponds to a periodic succession of representations. Each agent now represents the other as moving along a simple closed curve instead of being motionless. Thus the Hopf-like bifurcation of the total system resemble a Hopf bifurcation of a classic two-dimensional system that occurs when the eigenvalues of an attracting fixed point cross the imaginary axis. The salient eigenvalues play an analogous role in accounting for the open phase portraits in S_{α_n} as the two Hopf eigenvalues play in accounting for the two-dimensional closed phase portraits observed in a Hopf bifurcation. Again, the main difference is that paths can cross in open phase portraits whereas in a closed phase portrait they cannot.

Finally, numerical integration indicates that there are no attracting sets when the weights (w_ℓ, w_r) rise above the rose-colored region $\mathcal{M}_r$ in the avoider-avoider quadrant. It appears that the relative periodic orbit $\mathcal{P}_{(r,\mathrm{in})}$ that generates the meandering paths ceases to be attracting. However it also appears that the nonattracting relative periodic orbit $\mathcal{P}_{(r,\mathrm{in})}$ continues to play a role in the dynamics because there are long transients when the agents detect each other on their right-hand sides. During these long transients, the agents move along meandering-like paths in the rest frame. Eventually though they move out of range of their sensors and remain out of range of each other.

6.8   The Geometry of Meandering Paths for W_{same}

In the previous section we analyzed the types of partially fixed points and paths traced out in the agent state space when the vehicles revolve or meander around each other. In this section we shift the emphasis from paths in the agent space to the geometry of the physical paths traced out by the agents in the rest frame. We mainly focus on meandering paths generated when the total system is in a relative periodic orbit such as $\mathcal{P}_{(r,\mathrm{in})}$ or $\mathcal{P}_{(\ell,\mathrm{in})}$. We will continue to use the same values for the parameters P, ψ, and v as in section 6.7.

We will begin with a short review of epicyclic motion, which provides an approximate description for meandering paths in many types of systems. We then introduce the idea of the "total curvature of a periodic arc" in a meandering path which we denote by $\breve{\kappa}$. The number $\breve{\kappa}$ can be used to classify meandering paths and facilitates a comparison between Braitenberg vehicles and several types of physical systems.

Next we will describe the meandering paths of avoider-avoider systems, whose loops are inwardly directed, and which display dissipative behaviors similar to spiral tip meander in the BZ reaction. We then describe the meandering paths of pursuer-pursuer systems, whose loops are outwardly directed, and which display conservative behaviors similar to those of billiard balls and spherical pendulums. We will also see parallels between the weight space for the vehicles and the parameter spaces for the BZ reaction and for spherical pendulums. At the end of the section we discuss the behavior of the vehicles near W_{same} in the pursuer-pursuer quadrant, and the attracting set $\mathcal{Q}_{\mathrm{eq}}$, which highlights relationships between meandering paths in the revolving case and meandering paths in the translating case, discussed in chapter 7.

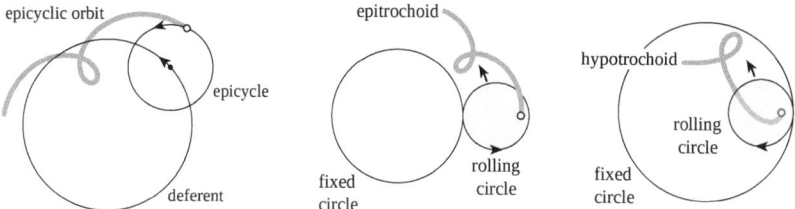

Figure 6.9
Two construction methods for the same types of curve. (left) Epicyclic orbits are constructed with epicycles and deferents, both of which are circles. The epicycle spins as
its center travels along the deferent. The point marked with ○ on the epicycle traces
out an epicyclic orbit, which is shown as a thick gray curve. (middle) An epitrochoid
(shown as a thick gray curve) is made by rolling a circle on the outside of a fixed circle.
(right) A hypotrochoid (shown as a thick gray curve) is made by rolling a circle on the
inside of a fixed circle. The interior of the rolling circles are shaded light gray. The
points marked with ○ are attached to the interior of the rolling circles.

Epicyclic motion involves one circle (called an *epicycle*) whose center
moves around another circle (called a *deferent*), as shown in the left panel
of figure 6.9. We call the path traced out by a point on the epicycle an *epicyclic
orbit* [3] since Ptolemy used such paths to represent the movement of planets in
the solar system. The rate at which the epicycle goes around the deferent and
the rate at which the epicycle spins are both fixed. Although Ptolemy's geo-
centric model for the motion of the planets has been replaced by heliocentric
models, epicyclic motion remains a good model for many of the moons that
orbit the planets. In modern terminology, we get a *prograde* epicyclic orbit if
the epicycle goes around the deferent in the same direction that it spins. If the
epicycle goes around the deferent in the opposite direction, we get a *retrograde*
epicyclic orbit.

"Epitrochoids" and "hypotrochoids" are curves that can be constructed by
rolling a circle (without slipping) around another circle that remains in place.
For epitrochoids, the rolling circle is outside of the fixed circle as shown in the
middle panel of figure 6.9. For hypotrochoids, the rolling circle is inside the
fixed circle as shown in the right panel of figure 6.9. In either case, the tracing
point need not be on the circumference of the rolling circle.

Although the method of construction for epitrochoids and hypotrochoids
is slightly different from that for epicyclic orbits, the two methods generate
the same curves. Epitrochoids and hypotrochoids are the same as prograde

3. Not to be confused with a dynamical system orbit.

and retrograde epicyclic orbits, respectively. For more on this topic, see Hotton (2016). Epicyclic motion has a long history in classical mechanics, while epitrochoids and hypotrochoids are more recent. We use the terms epitrochoids and hypotrochoids in this book.

Epitrochoids are known for having inwardly directed loops, like the curves in the left and middle panels of figure 6.9 and the left panels of figures 6.10 and 6.15, while hypotrochoids are known for having outwardly directed loops, like the curves in the right panel of figure 6.9 and the right panels of figures 6.10 and 6.15. These behaviors can also be seen in figures 1.2, 2.9, and 6.18.

Looking ahead to the translating case discussed in chapter 7, we see that the agents can also generate paths that resemble trochoids, which are curves constructed by rolling a circle (without slipping) along a line segment. Trochoids are known for having parallel loops, like the curves in the top left and bottom right panels of figure 7.4.

Epicyclic orbits can be partitioned into a sequence of arcs that are properly congruent to each other. This also happens with the physical paths generated by relative periodic orbits such as $\mathcal{P}_{(\ell,\mathrm{in})}$ and $\mathcal{P}_{(r,\mathrm{in})}$. Recall from section 5.7 that when the total dynamical system ϕ_τ begins in a relative periodic, orbit it travels through the proper congruence classes of $\mathbf{SE}(2)$ in S_τ in a periodic fashion. As the total system repeatedly passes through a succession of proper congruence classes, a succession of properly congruent configurations of the vehicles occurs over and over again.

Since the total dynamical system is equivariant with respect to the group of proper congruences, $\mathbf{SE}(2)$, the behavior of the agents is essentially the same each time it passes through a proper congruence class. Physical paths that begin from properly congruent configurations are themselves properly congruent. The result is that a meandering path generated by a relative periodic orbit can be partitioned into a sequence of properly congruent arcs called *periodic arcs* (Hotton 2016). Examples of periodic arcs are highlighted in black in figures 6.10, 6.11, 6.15, and 6.17.

In fact, there are an infinite number of partitions of a meandering path into properly congruent periodic arcs. Figure 6.11 gives an example of a closed curve partitioned into periodic arcs in two different ways. The different partitions are determined by where in the meandering path we choose to begin constructing the partition.

Meandering paths often have periodic arcs in the form of a *loop*, that is, an arc that crosses itself at a single point. For example the periodic arcs of the curve on the left of figure 6.11 are loops. The loops are positioned radially about a central point. Loops like this have often been called "petals" in the study of meandering spiral waves in excitable media such as the BZ reaction,

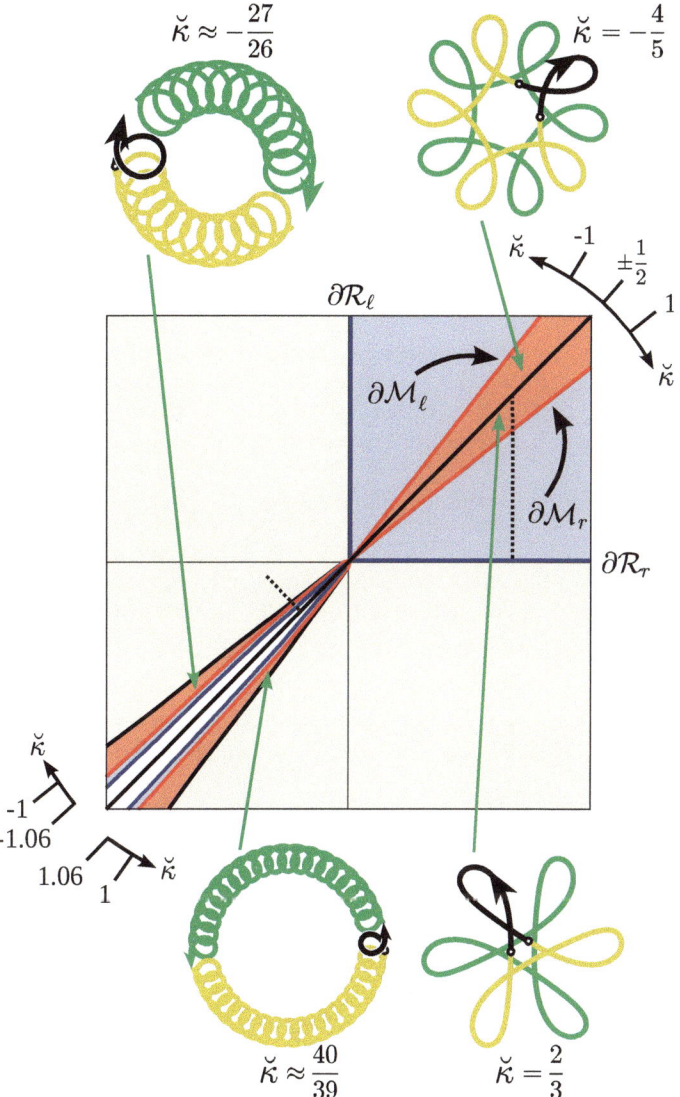

Figure 6.10
A bifurcation diagram in $[-15/4, 15/4]^2$ and four corresponding physical paths. The white region surrounding the half line $w_\ell = w_r \leq 0$ is $\mathcal{O}_r \cap \mathcal{O}_\ell$. The curves $\partial\mathcal{R}_r, \partial\mathcal{R}_\ell$ are colored blue, and the adjacent regions are colored lavender where ϕ_τ has an attracting relative equilibrium. The curves $\partial\mathcal{M}_r, \partial\mathcal{M}_\ell$ are colored red, and the adjacent rose-colored regions are where ϕ_τ exhibits an attracting set which bifurcated from a relative equilibrium. The beige-colored regions are where this attracting set has lost its stability. In each panel, one agent's path is colored gold, while the other agent's path is colored sea green except for a single periodic arc, which is colored black. The black periodic arcs each begin pointing upwards. The green arrows show the location of the corresponding weights (w_ℓ, w_r). The dotted line on the lower left corresponds to a variation of parameters described in several places in this chapter, including figure 6.8.

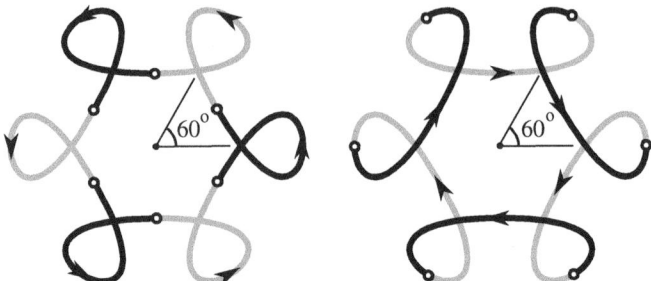

Figure 6.11
Two partitions of a symmetrical curve into six periodic arcs. The endpoints of the periodic arcs are marked by ○ and successive periodic arcs are colored alternately black and gray. The figure has an inter-petal angle of 60°. (left) The periodic arcs are loops. (right) The periodic arcs do not self-intersect. In this case, the self-intersection points of the whole curve arise from the intersection of distinct periodic arcs.

because their union can resemble the petals of a flower. The angle between adjacent loops or the *inter-petal angle* of a meandering path is determined by the parameters of the excitable medium (Winfree 1991). As we will see, the inter-petal angle can be computed from the total curvature of a periodic arc, $\check{\kappa}$. This makes it possible to compare the structure of the weight space W_{same} for the Braitenberg vehicles and the structure of the parameter spaces for the BZ reaction and for the spherical pendulum, which we do at the end of this section.

The *local curvature* for a sufficiently smooth curve is the instantaneous rate of change in the direction of the curve's tangent vector (in radians) with respect to the arc length of the curve. It characterizes the shape of a curve at each point. A standard formula for the local curvature of a planar curve $(x_n(t), y_n(t))^T$ parameterized by time t is:

$$\kappa(t) = \frac{\dot{x}_n(t)\ddot{y}_n(t) - \ddot{x}_n(t)\dot{y}_n(t)}{(\dot{x}_n(t)^2 + \dot{y}_n(t)^2)^{3/2}}. \tag{6.16}$$

Two properly congruent curves must have the same local curvature at corresponding points.

The *total curvature* for a sufficiently smooth curve of finite length is the total amount the tangent vector turns from the beginning to the end of the curve (Milnor 1950). Two properly congruent curves of finite length, such as the periodic arcs of a meandering path, have the same total curvature.

The quantity $\check{\kappa}$ is defined to be the total curvature of a periodic arc. When the speed v at which a curve is traversed is constant and the local curvature $\kappa(t)$ has a minimal period (i.e., the smallest positive amount of time before the

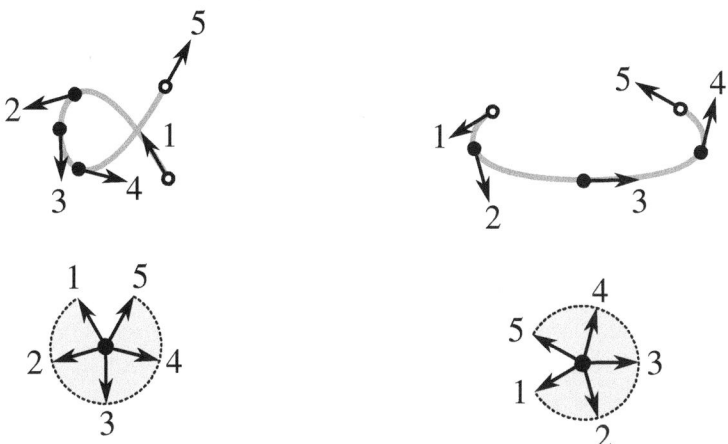

Figure 6.12

(top) Two types of periodic arcs for the same curve are shaded gray. Five tangent vectors are shown on each periodic arc, numbered **1, 2, 3, 4, 5**, from start to finish. (top left) The periodic arc from the left side of the left panel in figure 6.11. (top right) The periodic arc from the top of the right panel in figure 6.11. (bottom) The tangent vectors are repositioned without changing their direction so they have a common base point. The tangent vectors sweep out 5/6 in both cases so that $\breve{\kappa} = 5/6$ for these periodic arcs.

function's values begin to repeat), then we can obtain the total curvature of a periodic arc from the formula:

$$\breve{\kappa} = \frac{v}{2\pi} \int_{t_1}^{t_1+T} \kappa(\tau)\, d\tau. \tag{6.17}$$

where T is the minimal period of $\kappa(t)$ (Hotton 2016). The purpose of dividing by 2π is to measure the total amount by which the tangent vector turns along the length of a periodic arc relative to a single complete turn.

The value of $\breve{\kappa}$ is independent of the choice for $t_1 \in \mathbf{R}$ (i.e., the point where the periodic arc begins). The total curvature of a periodic arc is the same regardless of how we partition a curve into periodic arcs. Figure 6.12 shows two examples of periodic arcs for the curve in figure 6.11. We can see that the tangent vector turns by the same amount for both periodic arcs.

There are two kinds of curves with constant curvature in a Euclidean plane: lines and circles. Constant functions are technically periodic functions, but they have no minimal period. Any amount of time could be chosen as the period of the curvature for a line or circle and this would lead to different values for $\breve{\kappa}$. Since there is no minimal period for $\kappa(t)$ the value of $\breve{\kappa}$ is not well defined in these cases. Lines and circles are the type of physical paths traced out by the Braitenberg vehicles when the total system is in a relative equilibrium.

We only define the value of $\check{\kappa}$ when the local curvature of the physical paths has a minimal period. For the Braitenberg vehicles this happens when the state of the total system is one of the relative periodic orbits $\mathcal{P}_{(\ell,\text{in})}$ or $\mathcal{P}_{(r,\text{in})}$. These relative periodic orbits exist when the weights are in $\mathcal{M}_\ell$ or $\mathcal{M}_r$, that is, the rose-colored regions of the pursuer-pursuer and avoider-avoider quadrants of W_{same}, shown in figure 6.10. Positive values of $\check{\kappa}$ correspond to each vehicle turning overall to the left (anticlockwise) in a periodic arc, while negative values correspond to the vehicles turning overall to the right (clockwise) in a periodic arc. The larger $|\check{\kappa}|$ is, the more the vehicles turn in each periodic arc.

If $\check{\kappa}$ is not an integer, then the curve has a center of symmetry, that is, a point the curve can be rotated around so that it is mapped back to itself. Furthermore, if $\check{\kappa}$ is a non-integral rational number then the curve must be closed, that is, the coordinate functions, $x_n(t)$ and $y_n(t)$, of the curve are periodic functions with a common minimal period, and the point $(x_n(t), y_n(t))^T$ repeatedly traces over the curve's image. If $\check{\kappa}$ is irrational, then the coordinate functions $x_n(t)$ and $y_n(t)$ are quasiperiodic functions of t. For instance, a hypotrochoid that densely fills an annulus is quasiperiodic.

When the meandering paths are closed curves, the quantity $\check{\kappa}$ is closely related to the Whitney turning number for the curve (Whitney 1937). For a sufficiently smooth closed curve, the tangent vector must sweep out an integer number of complete turns between its starting point and final point, because these are the same point, with the same tangent vector. This integer is called *Whitney's turning number* for a closed planar curve and its value is a consequence of the topological properties of the curve. For instance, if a closed curve does not cross itself, then its Whitney turning number must be ± 1 as in the top left and right panels of figure 6.13.

Whitney's turning number can be used to interpret $\check{\kappa}$ for closed curves. As long as a curve has periodically varying curvature, we can associate the value $\check{\kappa}$ for each periodic arc in the curve to the curve as a whole. The image of a symmetrical closed curve can be partitioned into a finite number of periodic arcs, so the quantity $\check{\kappa}$ must be a rational number for a closed curve, whose numerator and denominator have straightforward interpretations. The numerator (in lowest terms) for $\check{\kappa}$ is the Whitney turning number for the closed curve, that is, the number of complete turns the tangent vector makes on the curve's image. The denominator (in lowest terms) for $\check{\kappa}$ is the number of periodic arcs in the image of the closed curve.

The closed curve in figure 6.11 and the bottom right panel of figure 6.13 provides an example. It is a closed curve with the symmetry of a hexagon. In this

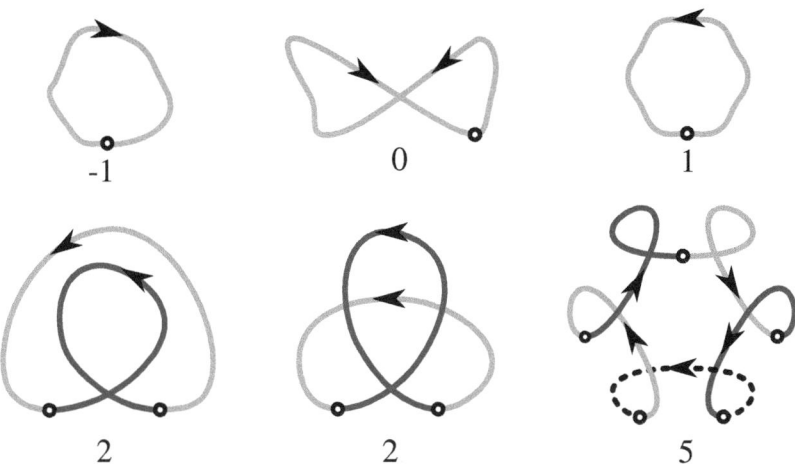

Figure 6.13
The Whitney turning numbers for six closed curves are displayed below each curve. (top left) The curve makes one complete right turn. (top middle) The curve makes right and left turns, which cancel each other out. (top right) The curve makes one complete left turn. (Bottom left, middle) The curves make two complete left turns. (bottom right) The curve makes five complete left turns. The closed curves in the bottom row are partitioned into arcs that each make one complete left turn rather than into periodic arcs.

case $\breve{\kappa} = 5/6$, which is also the value of $\breve{\kappa}$ for any of its periodic arcs (see figure 6.12). The numerator (the Whitney turning number) is 5, that is, the tangent vector makes five complete turns on the curve's image. The denominator is 6, that is, there are six periodic arcs that can be joined together to make the whole closed curve, and they are mapped into each other by successive rotations by 1/6 of a turn or 60°. This is the inter-petal angle shown in figure 6.11.

Conversely, we can reconstruct a closed planar curve for any non-integral rational $\breve{\kappa}$ from a periodic arc. In this case, given that $\breve{\kappa} = 5/6$, we know the total curvature for the union of six periodic arcs is 5. In fact the union is a closed curve whose Whitney turning number is 5. The inter-petal angle in degrees is obtained by dividing 360° by the denominator of $\breve{\kappa}$.

However, meandering paths are typically quasiperiodic with irrational $\breve{\kappa}$. In the quasiperiodic case, a meandering path cannot actually be drawn in its entirety because it is infinite in length, but we can draw a finite number of its periodic arcs. The value of $\breve{\kappa}$ in these cases can be interpreted in terms of rational number approximations to $\breve{\kappa}$. For every irrational number there are infinite sequences of rational numbers that converge to it. So although a quasiperiodic

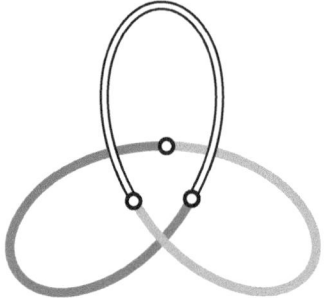 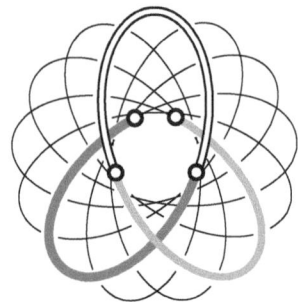

Figure 6.14
An illustration of how rational $\check{\kappa}$ can be used to approximate irrational $\check{\kappa}$. (left) A closed hypotrochoid with $\check{\kappa} = 2/3$. It is partitioned into three periodic arcs whose endpoints are marked by ○. One periodic arc is shaded dark gray, another is shaded light gray, and the third is white with a black outline. (right) A closed hypotrochoid with $\check{\kappa} = 8/13$. It is depicted with a thin back curve except for three periodic arcs which are shaded the same as in the left panel. The hypotrochoid in the left panel approximate the corresponding periodic arcs in the right panel. The hypotrochoid in the right panel approximates a hypotrochoid with $\check{\kappa} = 2/(1 + \sqrt{5})$.

curve is not closed, it can be approximated by a sequence of closed curves of increasing length. In the limit the quasiperiodic curve densely fills an annulus.

For example, Kepler showed that the infinite sequence formed by the ratios of consecutive Fibonacci numbers: 2/3, 3/5, 5/8, 8/13, ..., converges to the the reciprocal of the golden number, that is, the irrational number $2/(1 + \sqrt{5})$ (Herz-Fischler 1987). Figure 6.14 shows two closed hypotrochoids with periodic arcs whose total curvature comes from this sequence. The hypotrochoid on the left approximates the hypotrochoid on the right over three of its periodic arcs. While we can not draw the entire hypotrochoid with $\check{\kappa} = 2/(1 + \sqrt{5}) \approx 0.618$, we can approximate it over thirteen of its periodic arcs by the trochoid with $\check{\kappa} = 8/13 \approx 0.615$.

For irrational $\check{\kappa}$, the subset of angles that preserve the quasiperiodic curve is dense in the set of all angles. As we go through the sequence of closed curve approximations, the number of periodic arcs in the closed curve increases without bound. Simultaneously, the inter-petal angle of the closed curves becomes smaller. The finite subsets of angles that preserve the successive closed curve approximations converges to a dense subset of angles that preserve the quasiperiodic curve.

Note that in some cases there is a difference between the sign of $\check{\kappa}$ and how the curve winds around its center of symmetry. For the curve in figure 6.11, the local curvature is positive everywhere since the curve is always turning to the left (anticlockwise), and therefore $\breve{\kappa}$ is positive. Yet, as time progresses, the curve actually winds around its center in the clockwise direction.

Typically the value of $\breve{\kappa}$ changes as the weights vary inside $\mathcal{M}_\ell \cup \mathcal{M}_r$. However, $\breve{\kappa}$ remains constant along certain curves in $\mathcal{M}_\ell \cup \mathcal{M}_r$. Winfree (1991) coined the term *isogon* or *isogonal contour* to denote curves like this in parameter spaces for the BZ reaction.

For the Braitenberg vehicles, the isogonal contours nearly radiate outward from the origin of W_{same}, conforming with the overall shape of $\partial\mathcal{M}_\ell$ and $\partial\mathcal{M}_r$. Braitenberg vehicles whose weights lie on the same isogonal curve produce meandering paths with the same value for $\breve{\kappa}$. The protractor-like rulings at the corners of the avoider-avoider and pursuer-pursuer quadrants in figure 6.10 mark isogonal contours associated with specific values of $\breve{\kappa}$.

We now look at how the meandering paths change as we vary the weights across isogonal contours. In particular, we vary the weights along the dotted lines shown in figures 6.1 and 6.10. The Hopf-like bifurcation occurs at the boundary, $\partial\mathcal{M}_r$, between the blue to rose-colored regions in the figures. The Hopf-like bifurcations are marked by the red dots in the avoider-avoider and pursuer-pursuer quadrants shown in figure 6.1.

Recall that before the Hopf-like bifurcation, the agents revolve around a common center on opposite sides of a circular path. After the Hopf-like bifurcation, the revolving type relative equilibrium $\mathcal{E}_{(r,\text{in})}$ becomes nonattracting, and the attracting relative periodic orbit $\mathcal{P}_{(r,\text{in})}$ emerges. The midpoint between the agents remains the same as they transition from revolving to meandering.

Shortly after the Hopf-like bifurcation, the minimum and maximum radii of the meandering path are close and the meandering paths only veer slightly from the circular paths the agents once followed. As the weights enter further into $\mathcal{M}_r$, the minimum and maximum radii separate further apart. How the meandering paths expand into the enlarging annular region depends on whether the agents' weights are in the avoider-avoider quadrant or the pursuer-pursuer quadrant.

Figure 6.15 provides a summary of the main types of meandering paths that occur after the Hopf-like bifurcation. The left column shows meandering paths for the avoider-avoider quadrant, which are inwardly directed, and the right column shows meandering paths for the pursuer-pursuer quadrant, which are outwardly directed.

As the weights move vertically along the dotted lines in figures 6.1 and 6.10, we go from the bottom panels in figure 6.15 to the top panels. The meandering paths that occur right after the Hopf-like bifurcation are shown in the bottom

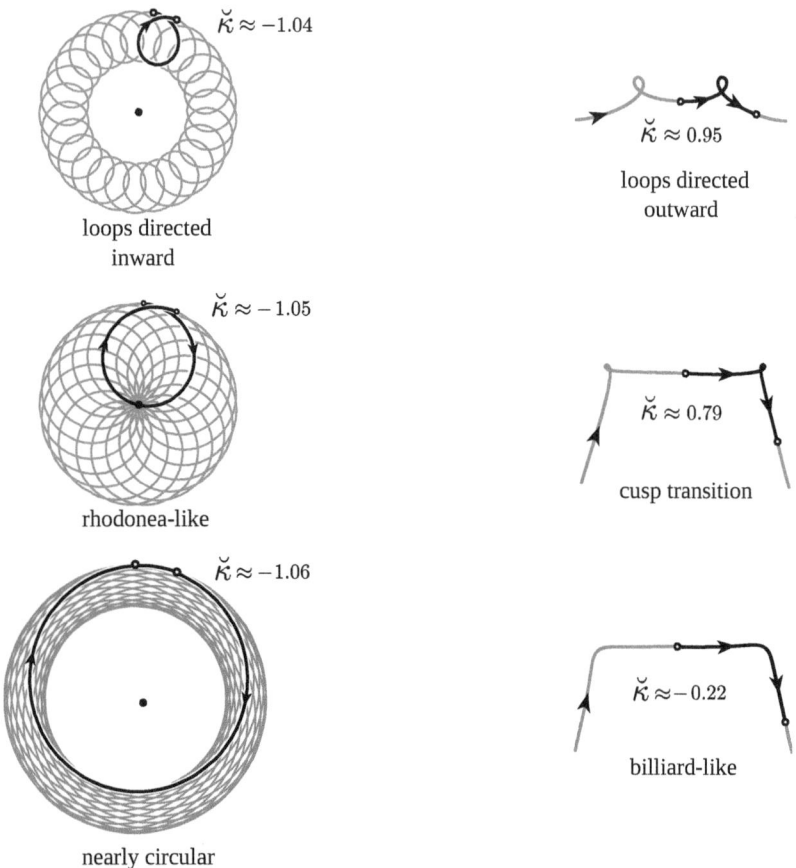

Figure 6.15
Meandering paths after the bifurcation at $\partial\mathcal{M}_r$. A single periodic arc is highlighted in black within each path; their endpoints are marked with ○. (left) Paths for weights in the avoider-avoider quadrant go through a rhodonea transition. (right) Paths for weights in the pursuer-pursuer quadrant go through a cusp transition.

row. Important transitions in the shape of the paths are indicated in the middle row, and the meandering paths after these transitions are shown in the top row.

The periodic arcs in a given meandering path are confined between a minimum and maximum radius. The black loops on the left column of figure 6.15 begin at the maximum radius, pass through the minimum radius, and end back at the maximum radius. In the three cases shown, the loops are directed inward and so resemble epitrochoids. The periodic arcs in the right column of the figure begin at the minimum radius, pass through the maximum radius, and

rhodonea
curve

$\breve{\kappa} = -20/19$

rhodonea-like
meandering path

$\breve{\kappa} \approx -1.05$

Figure 6.16
(left) A closed rhodonea curve. (right) A meandering path traced out by a Braitenberg vehicle with nearly the same value for $\breve{\kappa}$. The weights are $(w_\ell, w_r) \approx (-2.8310, -2.4021)$.

end back at the minimum radius. In the upper two cases shown, the loops are directed outward and so resemble hypotrochoids.

We first consider the avoider-avoider case (left side of figure 6.15). In this case, as the weights enter into $\mathcal{M}_r$, we have a transition from the loops being nearly circular and winding around the center (bottom left), to loops passing right through the center (middle left, where the minimum radius is 0), to loops not winding around the center but rather traveling outside the center (top left). The space filled in by the meandering paths in the physical space goes from being an annulus, to a disk, and back to an annulus.

In the case where the minimum radius is 0, the abstract Braitenberg vehicles pass through the center simultaneously (and thus through each other). Numerical integration indicates that the meandering paths look as though they have the same topology as that of a family of curves known as "rhodoneas" or "roselike" epitrochoids.[4] An example is shown in figure 6.16 that has a fairly simple formula in terms of $\breve{\kappa} = -20/19$:

$$\begin{pmatrix} x(t) \\ y(t) \end{pmatrix} = \cos(t) \begin{pmatrix} \cos((2\breve{\kappa}+1)t) \\ \sin((2\breve{\kappa}+1)t) \end{pmatrix}.$$

This transition through a rhodonea-like meandering path has been observed in dissipative systems such as the meandering tips of spiral waves in the BZ

4. Studied by Guido Grandi around 1723 and many others since. A rhodonea can be a hypotrochoid or an epitrochoid. In this case, they are epitrochoids.

reaction, as chemical concentrations are varied (cf. figure 3 in Li et al. [1996] and figure 6 in Skinner and Swinney [1991]). The rhodonea transition has been observed in models of the BZ reaction (cf. figure 4 in Jahnke, Skaggs, and Winfree [1989]; figure 15 in Winfree [1991]; and in Hotton [2016]). The transition has also been observed in a model for the meandering tips of spiral waves in heart tissue (cf. figure 11 in Efimov, Krinsky, and Jalife [1995]).

We now consider the pursuer-pursuer case (right side of figure 6.15). As the weights move vertically in $\mathcal{M}_r$, there is a transition from billiard-like dynamics (bottom right), through a cusp transition (middle right), to outwardly directed loops (top right), that are increasingly elongated as the weights move towards W_{eq}, and that come to resemble the paths made by a spherical pendulum.

The paths below the cusp transition correspond to a very thin region just above $\partial \mathcal{M}_r$, where the values for $\check{\kappa}$ are slightly greater than -1, and a very thin region just below $\partial \mathcal{M}_\ell$ where the values for $\check{\kappa}$ are slightly less than 1. (These regions are too thin to be marked in figure 6.10). For weights in these regions, the physical paths resemble billiard-like hypotrochoids. Rañó has also observed Braitenberg vehicles following meandering paths that resemble billiard-like hypotrochoids (Rañó 2010).[5]

Billiard-like hypotrochoids have relatively long arcs whose curvature is close to 0. These relatively long arcs alternate with relatively short arcs whose curvature is far from 0. The relatively long arcs correspond to the straight paths followed by a billiard ball, while the relatively short arcs correspond to a billiard ball bouncing off the sides of a circular billiard table.

As the weights move further into $\mathcal{M}_r$ in the pursuer-pursuer quadrant, the curvature of the short arcs of the meandering paths increases without bound and in numerical simulations they appear to form cusps. Afterward outwardly directed loops emerge from the cusps so that the meandering paths resemble hypotrochoids.

After the Hopf-like bifurcation at $\partial \mathcal{M}_r$, the value of $\check{\kappa}$ varies continuously with the weights until the cusps form. Numerical simulations indicate that the value of $\check{\kappa}$ jumps up by 1 as the weights pass through the cusp stage (in the figure it changes from -0.22 to 0.79; the values are taken from slightly before and after the cusp transition). The cusp transition also occurs in $\mathcal{M}_\ell$ but the value of $\check{\kappa}$ jumps down by -1.[6]

---

5. Also see Ahmed and Teahan (2021) for billiard-like behavior in a "hybrid Braitenberg vehicle."
6. These jumps in the value of $\check{\kappa}$ resembles the jump by ± 1 in the Whitney turning number of a closed curve when it develops a cusp as it is deformed. In fact, there is a relationship between the jump in $\check{\kappa}$ and in the jump of the Whitney turning number. Recall that if $\check{\kappa}$ is a non-integral rational number p/q, for coprime integers p, q and $q > 0$, then the whole curve is closed with q periodic arcs and its Whitney turning number is p. If a rotationally symmetric closed curve forms cusps as it

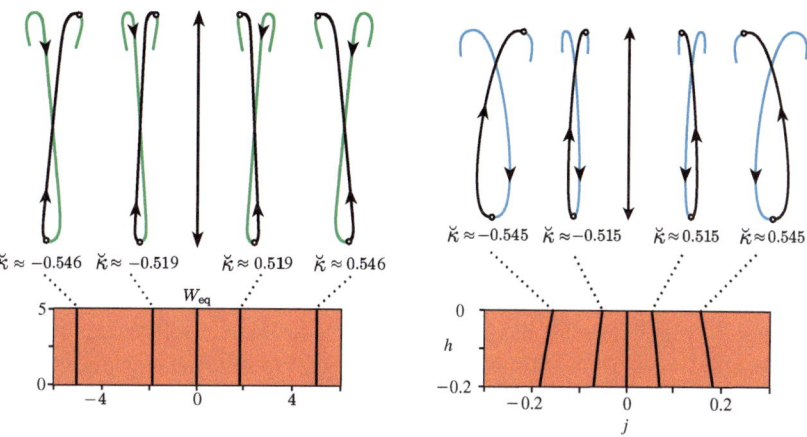

Figure 6.17
(top left) Physical paths for nearly equal weights are fairly straight, punctuated by sharp turns. A periodic arc in each path is highlighted in black with its endpoints marked by ○. Each path's value for $\check{\kappa}$ is shown below it. The straight line segment in the middle indicates that the agents engage in uniform linear motion when their weights are equal. (bottom left) The rose-colored rectangle is anchored at $(2.295, 2.295) \in W_{\text{eq}}$, which is near the middle of the pursuer-pursuer quadrant shown in figures 6.1 and 6.10. Isogonal contours are shown as black curves. Dotted lines match the value of $\check{\kappa}$ with an isogonal curve. The value of $\check{\kappa}$ becomes ambiguous ($\pm 1/2$) as the weights approach W_{eq}. (top right) Physical paths seen from above a spherical pendulum. A periodic arc in each path is highlighted in black with its endpoints marked by ○. Each path's value for $\check{\kappa}$ is shown below it. The periodic arcs can be approximated with arcs of ellipses when the magnitude of the angular momentum, j, and the energy, h, are around their minimal values $(0, -1)$. The straight line segment in the middle panel is the path generated when the angular momentum of the pendulum is 0 and it swings within a plane. (bottom right) Rose-colored region in the energy-momentum space for a spherical pendulum with isogonal contours shown as black curves. The dotted lines match the value of $\check{\kappa}$ with an isogonal curve. The value of $\check{\kappa}$ becomes ambiguous ($\pm 1/2$) as the angular momentum approaches 0.

After the cusp transition, the meandering paths produce outwardly direct loops that are elongated as the weights in $\mathcal{M}_r$ approach W_{eq}. In the limit, the value of $\check{\kappa}$ approaches $1/2$ from above (see figure 6.17). As the weights in $\mathcal{M}_\ell$ approach W_{eq}, the value of $\check{\kappa}$ approaches $-1/2$ from below. We can think of

---

is deformed while preserving its symmetry (and the curve does not pass through its center), then there will be q cusps in the whole curve. Each cusp will contribute to a change in the Whitney turning number either by 1 or by –1. Since there are q cusps, the Whitney turning number of the whole curve will change by $\pm q$. So the value of $\check{\kappa}$ will have to change by ± 1 on each periodic arc.

the agents as approaching a limiting case where they move in straight lines (in $\mathcal{Q}_{eq}$ they do in fact move along single-file paths). As we will see, meandering paths in the translating case approach this limit in a similar way (see the end of section 7.6).

The elongated, outwardly directed loops of the physical paths generated by the agents when their weights are in this region resemble those made by a spherical pendulum with low energy and angular momentum (Cushman 1983; Hotton 2016). The parameter spaces of the two systems also resemble each other with the isogonal contours near W_{eq} organized in a similar way to the isogonal contours in the energy-momentum space for the spherical pendulum shown in figure 6.17. To illustrate the analogy, the small diagonally oriented rectangle in W_{same} has been selected for figure 6.17. It is the convex hull of the four points:

$$\{ (2.29200, 2.29800), (2.29800, 2.29200),$$

$$(2.29223, 2.29827), (2.29827, 2.29223) \}. \tag{6.18}$$

This rectangle is bisected by W_{eq} and for clarity in the figure it has been stretched away from W_{eq} and rotated so that W_{eq} is vertical. The precise transformation from the weights to the horizontal and vertical coordinates of the rectangle in figure 6.17 is given in table 6.1.

For the Braitenberg vehicles, the value of $\breve{\kappa}$ is undefined for weights precisely in W_{eq}. For the spherical pendulum, the value of $\breve{\kappa}$ is undefined on the line of zero angular momentum. In both cases, the value of $\breve{\kappa}$ approaches 1/2 from one side of the line and $-1/2$ from the other side. However, there is a small difference in the behavior between the two cases. The Braitenberg vehicles with equal weights engage in uniform linear motion, whereas a spherical pendulum with zero angular momentum swings back and forth within a vertical plane.

Horizontal	$10 \, \text{sign}(w_\ell - w_r) \, \left| w_\ell - w_r \right|^{(1/10)}$
Vertical	$10^4 \, (w_\ell + w_r - 4.59)$

Table 6.1
The transformation of (w_ℓ, w_r) to the coordinates for the rose-colored rectangle in figure 6.17.

6.9   A Saddle Node-Like Bifurcation of Relative Periodic Orbits

Numerical simulation suggests that an additional pair of relative periodic orbits can arise from a saddle node-like bifurcation in W_{same}, where an attracting and repelling relative periodic orbit come into existence and co-exist with the main relative periodic orbits described earlier, and produce similar meandering paths. We denote the attracting set by $\mathcal{P}_{(\ell,-)}$ and the repelling set by $\mathcal{P}_{(\ell,+)}$. Examples of these paths and their positions relative to $\mathcal{P}_{(\ell,\text{in})}$ and $\mathcal{E}_{(\ell,\text{in})}$ are shown in figure 6.18.

To study this behavior the sensor range, P, was lowered to 10 and the speed, v, was slowed down to $1/100$. The weights were varied in the pursuer-pursuer quadrant from near the positive w_r-axis to near W_{eq}. This additional bifurcation can be found around $(w_\ell, w_r) = (0.90, 1.14)$.

By symmetry, there are relative periodic orbits $\mathcal{P}_{(r,\pm)}$ that behave in the same way but with opposite chirality. This occurs for weights in the pursuer-pursuer quadrant between the w_ℓ-axis and W_{eq}.

After the Hopf-like bifurcation produces the attracting relative periodic orbit $\mathcal{P}_{(\ell,\text{in})}$, the revolving relative equilibrium $\mathcal{E}_{(\ell,\text{in})}$ persists, but it is no longer attracting. Initially, the maximum radius of the meandering paths generated by the relative periodic orbit $\mathcal{P}_{(\ell,\text{in})}$ and the radius of the revolving paths generated by the relative equilibrium $\mathcal{E}_{(\ell,\text{in})}$ are about same. As (w_ℓ, w_r) moves closer to W_{eq}, the maximum radius of the meandering paths generated by $\mathcal{P}_{(\ell,\text{in})}$ shrink to about a quarter of their initial size while the radius of the revolving path generated by $\mathcal{E}_{(\ell,\text{in})}$ remains about the same (see figure 6.18).

When $\mathcal{P}_{(\ell,\pm)}$ first appear, the maximum radius of the meandering paths they generate is intermediate in size between the maximum radius of the meandering paths generated by $\mathcal{P}_{(\ell,\text{in})}$ and the radius of the revolving paths generated by $\mathcal{E}_{(\ell,\text{in})}$ (figure 6.18). As with the relative periodic orbit $\mathcal{P}_{(\ell,\text{in})}$, the midpoint between the agents remains fixed when the state of the total dynamical system is in either of the relative periodic orbits $\mathcal{P}_{(\ell,\pm)}$, and it is the center of symmetry of the meandering paths they generate.

Despite the difference in size between the meandering paths generated by the invariant sets $\mathcal{P}_{(\ell,\text{in})}$ and $\mathcal{P}_{(\ell,\pm)}$, there is a geometric resemblance between them. Both paths have outwardly directed loops, and both wind anticlockwise around the center even though the agents themselves are always turning to the right. In both cases, the outwardly directed loops and the retrograde motion causes the meandering paths to resemble hypotrochoids. The values of $\check{\kappa}$ for the meandering paths generated by $\mathcal{P}_{(\ell,\text{in})}$ and $\mathcal{P}_{(\ell,\pm)}$ in figure 6.18 are about the same: -0.96. This generates an inter-petal angle of about $14°$ as can be seen

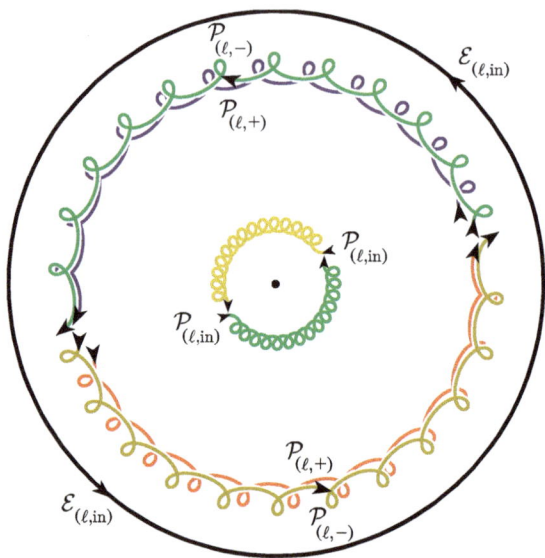

Figure 6.18

The meandering paths traced by the two agents for the attracting relative periodic orbits $\mathcal{P}_{(\ell,\text{in})}$ and $\mathcal{P}_{(\ell,-)}$ are colored gold and sea green as in previous figures. The meandering paths traced by the two agents for the repelling relative periodic orbit $\mathcal{P}_{(\ell,+)}$ are colored orange and violet. The large black circle on the outside is a revolving path for the repelling relative equilibrium $\mathcal{E}_{(\ell,\text{in})}$. The meandering paths for the attracting set $\mathcal{P}_{(\ell,-)}$ and repelling set $\mathcal{P}_{(\ell,+)}$ are just inside of the revolving path. Their maximum radii are about the same after an apparent saddle node-like bifurcation.

in the figure. This suggests that the relative periodic orbits for the invariant sets $\mathcal{P}_{(\ell,\text{in})}$ and $\mathcal{P}_{(\ell,\pm)}$ in S_τ have the same topology and dynamics.

6.10 W_{eq} as Approached from within W_{same}

The subspace W_{eq} is the intersection of W_{same} and W_{rev} (see section 2.5), where all four sensor weights for the two vehicles are equal. As we have seen, when the weights approach W_{eq} in the pursuer-pursuer quadrant of W_{same}, the meandering paths generated by $\mathcal{P}_{(\ell,\text{in})}$ and $\mathcal{P}_{(r,\text{in})}$ have increasingly flattened loops (figure 6.17). When the weights are directly on W_{eq}, there is a bifurcation in which the agents engage in uniform linear motion in a single-file configuration.

For positive weights in W_{eq}, the invariant set $\mathcal{Q}_{\text{eq}}$ is observed numerically to be attracting. However, it is the union of individually nonattracting relative equilibria. Each of these relative equilibria corresponds to a specific distance between the agents as they move single file. Set the agents at some single-file distance from each other, and they will maintain this distance indefinitely. If the agents are perturbed slightly in any direction, they will return to the invariant set Q_{eq} but usually with a different displacement and thus on a different relative equilibrium contained by Q_{eq}.

If the agents are separated further apart while remaining single-file, so that the agent in back becomes out of view of the agent in front, there is a transition from Q_{eq} to a trivial translating type relative equilibria (discussed in section 7.1). The agents still engage in uniform linear motion, but the trivial translating type relative equilibria is not part of Q_{eq}. In this case, perturbing the heading of the agent in front causes the agents to head in different directions indefinitely.

Figure 6.17 shows that the meandering behaviors generated by $P_{(\ell,in)}$ and $P_{(r,in)}$ turn into the single-file behaviors generated by Q_{eq}, as the meandering paths become a sequence of increasingly flattened loops. This occurs when we approach W_{eq} within the pursuer-pursuer quadrant of W_{same}. The numerical evidence indicates that the positive part of W_{eq} is a bifurcation curve in the pursuer-pursuer quadrant of W_{same}. Within W_{same} the slightest perturbation of the weights from equality results in the vehicles reversing their direction indefinitely instead of undergoing uniform linear motion. So there is no weight pair (w_l, w_r) in the parameter space W_{eq} with an open neighborhood of W_{same} in which all of the dynamical systems are equivalent.

As we will see in the next chapter, something similar occurs as we approach the positive part of W_{eq} within W_{rev} (compare figures 6.17 and 7.8). A slight perturbation of the weights from equality within the pursuer-pursuer quadrant also results in the vehicles reversing their direction indefinitely instead of undergoing uniform linear motion. In this case, the vehicles reverse direction by following oval paths that become increasingly flattened as the weights approach W_{eq}. The oval paths are generated by attracting sets that we call "counterrotating relative periodic orbits." These attracting sets also appear to turn into Q_{eq} when the weights reach the positive part of W_{eq} from within the pursuer-pursuer quadrant of W_{rev}.

Thus W_{eq}, (the intersection of W_{same} and W_{rev}), can be thought of as linking the analysis of the meandering paths in the revolving case with the analysis of the meandering paths in the translating case. This linkage can be seen by comparing figures 6.1 and 7.1: in both cases the main bifurcation sequences can be seen to terminate with Q_{eq}.

The positive part of W_{eq} is a bifurcation curve within the pursuer-pursuer quadrants of W_{same} and W_{rev}. Within W_{same} and W_{rev}, this bifurcation curve appears as the boundary between open regions of topologically equivalent dynamical systems. However, within the three-dimensional subspace W_{same} + W_{rev}, the positive part of W_{eq} can only be a small part of the boundary between any open regions of equivalent dynamical systems. In this larger context, the attracting set $\mathcal{Q}_{eq}$ is involved in an unusual type of bifurcation. The exceptional character of this bifurcation may play a role in the prevalence of lateralization in various types of animals (Bisazza, Rogers, and Vallortigara 1998; Cooper et al. 2011; Frasnelli 2013; Workman and Andrew 1986; Templeton et al. 2012; Hunt et al. 2014).

7 Translating Type Relative Equilibria and Their Bifurcations

In this chapter, we study translating type relative equilibria in which both vehicles undergo uniform linear motion together, as well as relative periodic orbits in which the vehicles meander along parallel paths. These types of invariant sets occur when the sensor weights of the two agents are exactly reversed from each other. For two pursuers to maintain the same heading, the pursuer on the left needs to have a left-turning bias and the pursuer on the right needs to have a right-turning bias. Otherwise they would move towards each other. Similarly, to keep two avoiders moving in the same direction, it is necessary for the avoider on the left to have a right-turning bias and the avoider on the right to have a left turning bias. Either way, the agents need to have their turning biases exactly reversed from each other to obtain a translating type relative equilibrium. The set of all reversed weights forms the subspace W_{rev} of the total weight space, W_{total}, as defined in equation (2.11).

The bifurcations that occur in W_{rev} have some parallels to the bifurcations in W_{same} discussed in chapter 6. To emphasize the resemblance between the two subspaces W_{rev} and W_{same}, we continue to let the ordered pair (w_ℓ, w_r) denote the sensor weights for both agents, but in this chapter we set

$$(w_\ell, w_r) \cong w_\ell \, (1, 0, 0, 1)^T + w_r \, (0, 1, 1, 0)^T .$$

So, if $(w_\ell, w_r) = (1/2, -1/2)$, then the weights for agent 1 are $(w_{(\ell,1)}, w_{(r,1)}) = (1/2, -1/2)$ and the weights for agent 2 are $(w_{(\ell,2)}, w_{(r,2)}) = (-1/2, 1/2)$.

As with W_{same}, there are Hopf-like bifurcations in W_{rev} but instead of a saddle node-like bifurcation of relative equilibria, an attracting relative equilibrium emerges from a non-standard bifurcation. These bifurcations divide W_{same} up into dynamical regimes that correspond to distinct behaviors. These resemble corresponding regions in W_{same}, and they too are symmetrically positioned in W_{rev} about the subspace W_{eq} (the intersection of W_{same} and W_{rev}), as shown in figure 7.1.

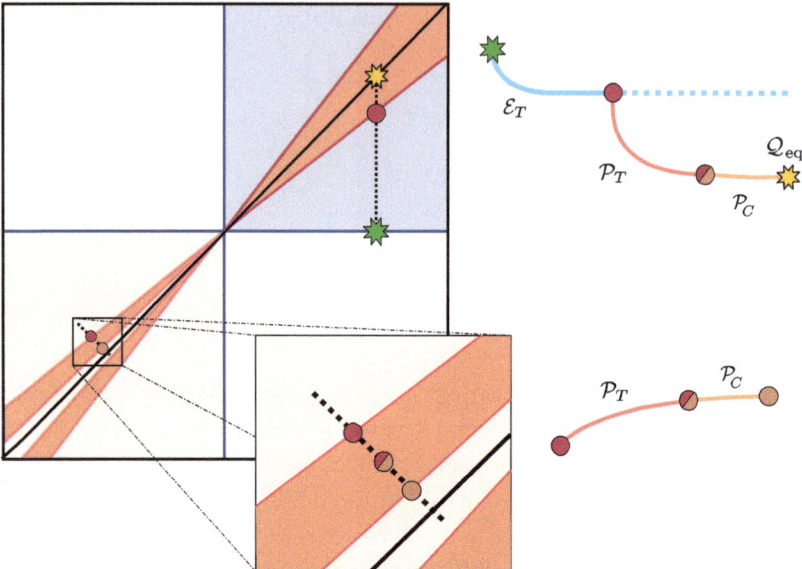

Figure 7.1
A summary of the invariant sets and bifurcations discussed in this chapter. The region $[-15/4, 15/4]^2$ in W_{rev} is shown on the left with a blowup of a smaller region in the avoider-avoider quadrant. The black dotted lines show two paths along which we vary the weights. The eight-pointed green star and colored disks on these dotted lines show where important bifurcations occur. Diagrams for the bifurcations are shown on the right. The green star corresponds to a nonstandard bifurcation from which a single relative equilibrium, $\mathcal{E}_T$, emerges. The crimson dot in the pursuer-pursuer quadrant corresponds to a Hopf-like bifurcation in which $\mathcal{E}_T$ loses stability and an attracting relative periodic orbit $\mathcal{P}_T$ appears. The crimson/orange disk corresponds to a bifurcation from $\mathcal{P}_T$ to $\mathcal{P}_C$. This subsequently bifurcates to the attracting set $\mathcal{Q}_{eq}$ indicated by the yellow star. The solid curves on the right indicate that the labeled invariant sets are attracting, while the dotted line on the right indicates that $\mathcal{E}_T$ is not attracting.

Symmetry plays a slightly different role for agents whose weights are in W_{same} versus agents whose weights are in W_{rev}. In W_{same}, there are regions where both agents can turn left together and regions where they can both turn right together. Where these regions overlap in W_{same}, the agents are capable of turning left together and of turning right together. By contrast, in W_{rev}, the agents either do not turn or when they do turn they do so in opposite directions.

Figure 7.1 summarizes the main invariant sets of the total system ϕ_τ for weights in W_{rev}, along with the bifurcations that occur when traveling along the two paths shown with dotted lines. For weights in the lateralized quadrants (the white regions), there are no relative equilibria and no attracting sets have been observed. The agents simply move out of view of one another. We denote the union of the closed pursuer-pursuer and avoider-avoider quadrants by $\mathcal{R}_T$ (see also figure 7.4).

In the blue regions, there is a single attracting relative equilibrium, $\mathcal{E}_T$, where the two agents move side by side in straight lines. This side-by-side relative equilibrium emerges from a nonstandard bifurcation marked by a green eight-pointed star. As the weights cross the w_ℓ-axis, an infinite number of relative equilibria momentarily come into existence. Afterward there is just a single attracting relative equilibrium $\mathcal{E}_T$.

When the weights cross from a blue region to the rose-colored region in the pursuer-pursuer quadrant, a Hopf-like bifurcation occurs in which the attracting relative periodic orbit $\mathcal{P}_T$ appears, while $\mathcal{E}_T$ persists as a non-attracting invariant set. In this region, the vehicles move in parallel meandering paths. Less is known about how $\mathcal{P}_T$ becomes an attracting set in the avoider-avoider quadrant.

As the weights are varied towards W_{eq} in either the pursuer-pursuer or avoider-avoider quadrants, numerical evidence indicates that an additional bifurcation occurs in which the the meandering paths generated by the relative periodic orbit $\mathcal{P}_T$ become counterrotating paths generated by another type of relative periodic orbit we call $\mathcal{P}_C$. When this happens, the vehicles stop parallel meandering and begin following a pair of symmetrically positioned simple closed curves that are nearly on top of one another (shown in the bottom left and top right panels of figure 7.4). This bifurcation is marked by red/orange disks in figure 7.1, consistent with their intermediate status between the meandering paths generated by $\mathcal{P}_T$ and the single-file paths generated by $\mathcal{Q}_{eq}$.

As the weights approach W_{eq}, in the pursuer-pursuer quadrant of W_{rev} something similar happens to the counterrotating paths as happens to the meandering paths as the weights approach W_{eq} in the pursuer-pursuer quadrant of W_{same}. In both cases, periodic arcs become increasingly flattened, and eventually turn into the single-file behaviors of $\mathcal{Q}_{eq}$.

We do not provide a discussion of agent space representations in this chapter analogous to the discussion of agential representations in section 6.7, because the agent representations that occur in the translating case are very similar to those that occur in the revolving case. When the agents move out of view of each other, as they do for weights in the lateralized quadrants (white regions in figure 7.1), the agents' representations decay to the "alone state" $(0, 0)^T$ representing no agent. For the pursuer-pursuer and avoider-avoider quadrants (the nonwhite regions, $\mathcal{R}_T$, in figure 7.1), there are partially fixed points in S_{α_n} that correspond to one agent representing the other. Since the relative location of the agents remains fixed when the agents are in the side-by-side relative equilibrium $\mathcal{E}_T$ of ϕ_τ, each agent's state is at a partially fixed point in S_{α_n}, where one agent represents the other at a specific location relative to themselves. When the weights enter the rose-colored region $\mathcal{M}_T$, the agents move along meandering paths, and a simple closed curve appears in each agent's state space that represents the other agent's periodically varying relative position.

We discuss several other trivial and minor relative equilibria that occur in W_{rev}. In this chapter, we again treat the physical parameters P, ψ, v_1, and v_2 as given features of the Braitenberg vehicles and the pair of sensor weights (w_ℓ, w_r) as adjustable parameters. We also drop the subscript from v_1 and v_2, but we do *not* drop the subscripts from $\bar{\omega}_n$ as we did in chapter 6 because in this chapter the turning functions are not the same for both agents.

7.1   Trivial and Nontrivial Translating Type Relative Equilibria

Trivial translating type relative equilibria are abundant when $v_1 = v_2$. They ordinarily arise from the vehicles being too far apart. If the vehicles are out of range of each other's sensors and they happen to be heading in the same direction, then technically the orbit that is being followed by the total dynamical system is part of a translating type relative equilibrium. We will not emphasize these trivial relative equilibria. It is clear that the vehicles move in straight paths when they are out of range, and under such circumstances it is not especially significant that they happen to be heading in the same direction. Such translating type relative equilibria are not attracting sets since a slight perturbation can cause the vehicles to move further apart.

Trivial translating type relative equilibria can occur in other ways. If both sensor weights are zero, then the Braitenberg vehicles always behave as though they were physically out of range of each other's sensors. If exactly one of the sensor weights is zero and each vehicle is only within range of the sensor with the zero weight, then again they behave as though they were physically out of range of both sensors. Under these circumstances, if both vehicles also happen to be heading in the same direction, then the state of the system is in a translating type relative equilibrium. These relative equilibria are like those that occur when the vehicles are physically beyond each other's sensor range. They are not attracting sets since a small perturbation can send the vehicles moving apart.

If one vehicle is only within range of a single sensor on the other vehicle and that sensor has a non-zero weight, then the other vehicle's angular velocity cannot be zero and the state of the total dynamical system cannot be in a translating type relative equilibrium.

The only remaining way the agents can be in a translating type relative equilibrium is for at least one sensor weight to be non-zero and for both vehicles to be in range of both sensors on the other vehicle. The definition for a *nontrivial* translating type relative equilibrium can be concisely stated as:

$$(w_\ell, w_r) \neq (0, 0) \quad \text{and} \tag{7.1}$$

$$(X_n, Y_n)^T \in (D_{(\ell,n)} \cap D_{(r,n)}) \quad \text{for} \quad n = 1, 2.$$

Recall that the region $D_{(\ell,n)} \cap D_{(r,n)}$ is the central lune[1] where the sensors' fields of view overlap. This is shown in green in figure 2.4 and in light gray in figure 7.3. Non-trivial translating type relative equilibria can be attracting sets, and they will be our focus for most of this chapter.

7.2 Necessary Conditions for Nontrivial Translating Type Relative Equilibria

Here we begin our analysis of nontrivial translating type relative equilibria. We consider necessary conditions on the weights and necessary conditions on the agents' configurations, which are summarized in equation (7.5). We then classify the nontrivial translating type relative equilibria into three cases: side-by-side, oblique, and single-file.

If (w_ℓ, w_r) is in the interior of the lateralized quadrants of W_{rev}, then for all $(X_n, Y_n)^T$ in the central lune, the turning function $\bar{\omega}_n$ is either strictly positive or strictly negative. This means the agent must turn in some direction when the other agent is in view of both sensors. So the system cannot be in a translating type relative equilibrium.

If (w_ℓ, w_r) is on the coordinate axes of W_{rev} [except at $(0, 0)$] then for all (X_n, Y_n) in the central lune, one of the terms in equation (2.8) for the function $\bar{\omega}_n$ is either positive or negative while the other term is zero. So again the agent must turn when the other agent is in view of both sensors and the system cannot be in a translating type relative equilibrium. If the other agent was in view of just one sensor, the one with the zero weight, then the system could be in a trivial translating type relative equilibrium, but not in nontrivial translating type relative equilibrium.

Even though we allow one of the sensor weights to be zero in the definition of nontrivial translating type relative equilibria, in fact it is necessary for the weights to be in the interior of the pursuer-pursuer or avoider-avoider quadrants in order for them to exist. We denote the union of the pursuer-pursuer and avoider-avoider quadrants by $\mathcal{R}_T$.

---

1. Recall that a lune is a planar region bounded by the arcs of two circles.

The necessary conditions on the agents' configurations for translating type relative equilibria to exist are that they move in the same direction and do not turn. We can express these conditions concisely in terms of the following equations:

$$\theta_1 = \theta_2$$
$$\bar{\omega}_1(X_1, Y_1)^T = 0 \tag{7.2}$$
$$\bar{\omega}_2(X_2, Y_2)^T = 0.$$

We can simplify equation (7.2) by using a simple relationship between $\bar{\omega}_1$ and $\bar{\omega}_2$. First, note that if $\theta_1 = \theta_2$ then in general:

$$R_{(\pi/2-\theta_2)} \begin{pmatrix} x_1 - x_2 \\ y_1 - y_2 \end{pmatrix} = -R_{(\pi/2-\theta_1)} \begin{pmatrix} x_2 - x_1 \\ y_2 - y_1 \end{pmatrix}.$$

So it follows from equation (2.4) for $(X_n, Y_n)^T$ that:

$$(X_2, Y_2)^T = -(X_1, Y_1)^T \tag{7.3}$$

when $\theta_1 = \theta_2$. It follows from equation (2.5) that:

$$d_{(\ell,2)}(X_2, Y_2)^T = d_{(r,1)}(-X_2, Y_2)^T$$
$$d_{(r,2)}(X_2, Y_2)^T = d_{(\ell,1)}(-X_2, Y_2)^T.$$

This is a consequence of the sensors being symmetrically positioned about the agent's heading. It then follows from these two equations, equation (2.8) for $\bar{\omega}_n$, and the fact that the sensor weights are exactly reversed that:

$$\bar{\omega}_2(X_2, Y_2)^T = -\bar{\omega}_1(-X_2, Y_2)^T. \tag{7.4}$$

From equation (7.3), we get $-X_2 = X_1$ and $Y_2 = -Y_1$. Substituting this into the right-hand side of equation (7.4) gives us:

$$\bar{\omega}_2(X_2, Y_2)^T = -\bar{\omega}_1(X_1, -Y_1)^T.$$

This is the simple relationship between $\bar{\omega}_1$ and $\bar{\omega}_2$ we sought. From this it follows that conditions (7.2) are equivalent to the conditions:

$$\theta_1 = \theta_2$$
$$\bar{\omega}_1(X_1, Y_1) = 0 \tag{7.5}$$
$$\bar{\omega}_1(X_1, -Y_1) = 0.$$

We have succeeded in eliminating two of the variables in equation (7.2). Conditions similar to (7.5) hold with the subscripts 1 and 2 reversed, but we do not need them since once we have $(X_1, Y_1)^T$ we can use (7.3) to get $(X_2, Y_2)^T$.

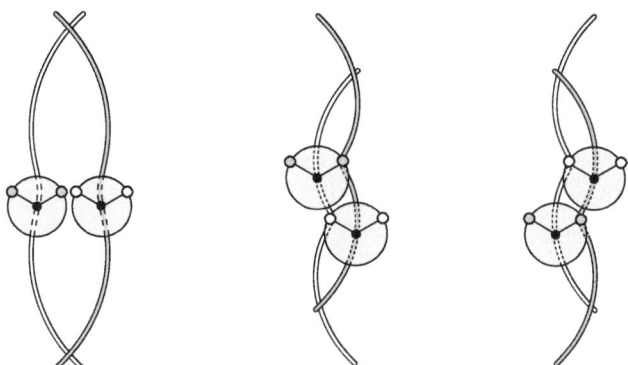

Figure 7.2
Side-by-side and oblique configurations for translating type relative equilibria along with the zero-sets of their turning functions $\bar{\omega}_n(X_n, Y_n)$, that is, locations for agent $\neg n$ where agent n won't turn. The agents' heading in the rest frame are directly upward. The sensors for agent 1 are shaded gray, and the sensors for agent 2 are shaded white. The zero set of $\bar{\omega}_n(X_n, Y_n)$ is shaded to match the color of agent n's sensors. The outline of the zero-sets are dashed where they overlap an agent's body. The location of each agent is on the zero set of the other agent, so neither of them changes its heading. In this figure $\psi = \pi/3$, and $(w_l, w_r) = (1, 3/2)$. (left) The agents' configuration for the side-by-side relative equilibrium $\mathcal{E}_T$ ($P = 6.45$). (middle, right) There are two mirror symmetric configurations for oblique relative equilibria ($P = 4.72$).

If $(X_1, Y_1)^T$ satisfies conditions (7.5), then so does $(X_1, -Y_1)^T$. This is illustrated in figure 7.2. In each of the three panels of the figure, agent 1 is on the left with its sensors shaded gray and agent 2 is on the right with its sensors shaded white. In the right panel, agent 2 is located at $(X_1, Y_1)^T$ in the body frame of agent 1 with $Y_1 > 0$. In the middle panel, agent 2 is located at $(X_1, -Y_1)^T$ in the body frame of agent 1 with $-Y_1 < 0$. So configurations that satisfy condition (7.5) with $Y_1 \neq 0$ come in pairs.

Nontrivial translating type relative equilibria always involve some balancing of the sensor activations whereby the value of the turning function for each agent ends up being 0. Configurations in the rest frame for which this occurs are shown in figure 7.2. Each of the outlined gray curves is the zero-set[2] of the turning function, $\bar{\omega}_1(X_1, Y_1)$, for agent 1 and each of the outlined white curves is the zero-set of the turning function, $\bar{\omega}_2(X_2, Y_2)$, for agent 2. As agent 1 moves in the rest frame, it carries its zero-set along with it. Agent 1 will not turn whenever agent 2 is located on this zero-set. As agent 2 moves in the rest

---

2. The set of points where the function's value is 0.

frame, it carries its zero-set along with it. Agent 2 will not turn whenever agent 1 is located on this zero-set.

For *both* agents to maintain a fixed heading, they must continually be positioned on the zero set of each other's turning function. This occurs in a few specific ways: side-by-side (shown in the left panel of figure 7.2), in an oblique configuration, involving specific vertical separations (shown in the middle and right panels of figure 7.2), or in a single-file configuration (not shown in figure 7.2).

7.3   Side-by-Side Translating Type Relative Equilibria

The *side-by-side* relative equilibria correspond to solutions (X_1, Y_1) for conditions (7.5) such that $X_1 \neq 0$ and $Y_1 = 0$. These two equations mean the agents are located beside each other. This is shown in the left panel of figure 7.2.

The configurations for side-by-side relative equilibria are symmetric under the reflection about the perpendicular bisector of the agents. This reflection swaps the locations of those sensors that have the same weights. So the locations in the plane associated to the weights are preserved by this reflection.

In order for the relative equilibria to be nontrivial, $(X_1, 0)$ has to be in the central lune. The intersection of the X_1-axis with this lune is the open line segment:

$$I = \left(- \left(\sqrt{P^2 - N^2} - M \right), \ \sqrt{P^2 - N^2} - M \right) \times \{0\}.$$

The line segment I is shown in figure 7.3. A necessary condition for translating type relative equilibria to be side-by-side is that $(X_1, 0) \in I$.

To show that such solutions exist, we first show that for weights in $\mathcal{R}_T$ the turning function $\bar{\omega}_1$ has opposite values at the two endpoints of I. By the intermediate value theorem (and the fact that the turning function is continuous), this implies that there is at least one point in I where the turning function is 0, that is, one solution to (7.5) with $(X_1, 0) \in I$. To show this, we first compute the values of $\bar{\omega}_1$ at the endpoints of I. By symmetry, the distance between the left endpoint of I and agent 1's right sensor is the same as the distance between the right endpoint of I and its left sensor. They are both equal to P. So one sensor's activation in these two cases is 0. Since $\bar{\omega}_1 = w_l \, a_l - w_r \, a_r$, this implies that the value of $\bar{\omega}_1$ is proportional to the activation of the other sensor.

We let d_l denote the distance between the left endpoint of I and the left sensor. By symmetry, this is the same as the distance between the right endpoint of I and the right sensor. We obtain the values for $\bar{\omega}_1$ at the endpoints of I from

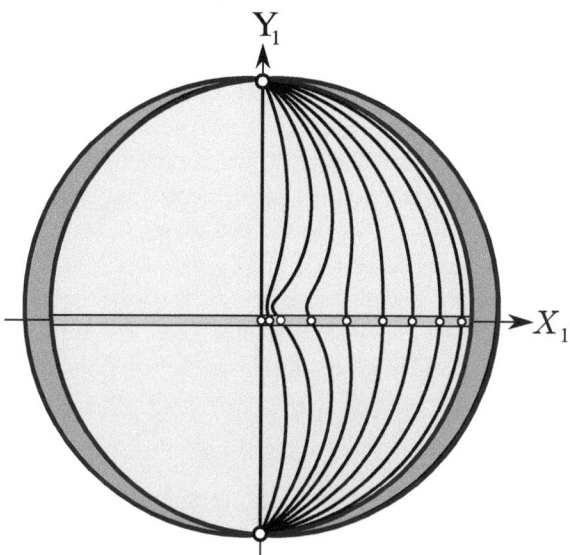

Figure 7.3
The central lune is shaded light gray. The horizontal line segment I is shaded medium gray and outlined in black. The left and right lunes are shaded dark gray. The zero sets of $\bar{\omega}_1(X_1, Y_1)$ inside the lune are shown as nine black curves for nine values of (w_ℓ, w_r) from the vertical dotted line in figure 7.1. As (w_ℓ, w_r) moves upwards in figure 7.1, a succession of zero sets emerges from right to left, beginning with the curve next to the central lune's right circular boundary. Eventually these curves cease to be arcs of a strictly convex curve. When (w_ℓ, w_r) reaches W_{eq} the zero set becomes the vertical line segment connecting the central lune's vertices. In this figure, $P = 12$ and $\psi = \pi/4$.

equations (2.6) and (2.8):

$$\bar{\omega}_1 \begin{pmatrix} -\left(\sqrt{P^2 - N^2} - M\right) \\ 0 \end{pmatrix} = w_\ell \left(1 - d_I/P\right)$$

$$\bar{\omega}_1 \begin{pmatrix} \sqrt{P^2 - N^2} - M \\ 0 \end{pmatrix} = -w_r \left(1 - d_I/P\right) \quad .$$

Their product is:

$$-w_\ell \, w_r \left(1 - d_I/P\right)^2 .$$

For (w_ℓ, w_r) in the interior of $\mathcal{R}_T$, the weights have the same sign, and so the product as a whole must be negative. The only way this can happen is for the turning function to have opposite signs at the two end points. So, by the intermediate value theorem, the turning function's value must pass through 0 at least once somewhere in I. Therefore, there is at least one solution to equation (7.5) with $(X_1, 0)$ somewhere in I.

To address the issue of uniqueness for solutions to equation (7.5) with $(X_1, 0) \in I$, we numerically computed the zero sets of $\bar{\omega}_1$ within the central lune, for example the curves shown in the central lune of figure 7.3. Each of these curves connects the vertices of the central lune and passes through the X_1-axis just once. For each pair of weights $(w_\ell, w_r) \in \mathcal{R}_T$, we denote the unique solution in I for equation (7.5) by $(X_T, 0)$.

We will study side-by-side relative equilibria in greater detail in sections 7.5 and 7.6. It will be useful to have a particular solution to differential equation (4.14) of a side-by-side relative equilibrium in the rest frame. This solution is:

$$\phi_\tau \left((-X_T, 0, \pi, X_T, 0, \pi)^T / 2 \right) =$$

$$\begin{pmatrix} x_1(t) \\ y_1(t) \\ \theta_1(t) \\ x_2(t) \\ y_2(t) \\ \theta_2(t) \end{pmatrix} = \frac{1}{2} \begin{pmatrix} -X_T \\ 2vt \\ \pi \\ X_T \\ 2vt \\ \pi \end{pmatrix}. \tag{7.6}$$

For this particular solution, the perpendicular bisector of the agents is the y-axis of the rest frame. We can perform a quick check that equation (7.6) does provide a solution to differential equation (4.14). Differentiating both sides of (7.6) gives:

$$\begin{pmatrix} \dot{x}_1(t) \\ \dot{y}_1(t) \\ \dot{\theta}_1(t) \\ \dot{x}_2(t) \\ \dot{y}_2(t) \\ \dot{\theta}_2(t) \end{pmatrix} = \begin{pmatrix} 0 \\ v \\ 0 \\ 0 \\ v \\ 0 \end{pmatrix}.$$

Substituting equation (7.6) into the right-hand side of differential equation (4.14) gives:

$$\begin{pmatrix} 0 \\ v \\ \omega_1(-X_T/2, vt, \pi/2, X_T/2, vt, \pi/2)^T \\ 0 \\ v \\ \omega_2(-X_T/2, vt, \pi/2, X_T/2, vt, \pi/2)^T \end{pmatrix}.$$

where we have used $\theta_1 = \theta_2 = \pi/2$ and $v_1 = v_2 = v$ to simplify the expression. We now just need to show that the third and sixth components are zero. By equation (4.13):

$$\omega_1 \begin{pmatrix} -X_T/2 \\ vt \\ \pi/2 \\ X_T/2 \\ vt \\ \pi/2 \end{pmatrix} = \bar{\omega}_1 \left(\mathcal{C}_1 \begin{pmatrix} -X_T/2 \\ vt \\ \pi/2 \\ X_T/2 \\ vt \\ \pi/2 \end{pmatrix} \right) = \bar{\omega}_1(X_T, 0).$$

This is zero since $(X_T, 0)$ is defined to be a solution to equation (7.5). Similar reasoning along with equation (7.4) shows that the sixth component also equals zero. This confirms that the function in equation (7.6) is a solution to differential equation (4.14).

This also shows that for $(X_T, 0) \in I$ the equation $\bar{\omega}_1(X_T, 0) = 0$ is not only a necessary condition for the existence of a side-by-side relative equilibrium but also a sufficient condition. We denote this side-by-side relative equilibrium by $\mathcal{E}_T$.

The relative equilibrium $\mathcal{E}_T$ comes into existence in a nonstandard bifurcation as (w_ℓ, w_r) enters $\mathcal{R}_T$ from the lateralized quadrants. As (w_ℓ, w_r) crosses the coordinate axes, an infinite number of trivial translating type relative equilibria momentarily appear. The relative equilibrium $\mathcal{E}_T$ emerges out of the union of these relative equilibria. It is attracting when (w_ℓ, w_r) enters the interior of the pursuer-pursuer quadrant but not when it enters the avoider-avoider quadrant. This is illustrated in figures 7.1 and 7.4 and discussed in section 7.5.

Section 7.5 discusses how $\mathcal{E}_T$ ceases to be attracting in a Hopf-like bifurcation that produces an attracting relative periodic orbit that we call $\mathcal{P}_T$. This bifurcation resembles the Hopf-like bifurcations of the relative equilibria $\mathcal{E}_{(r,\mathrm{in})}$ and $\mathcal{E}_{(\ell,\mathrm{in})}$ that produce the relative periodic orbits $\mathcal{P}_{(r,\mathrm{in})}$ and $\mathcal{P}_{(\ell,\mathrm{in})}$ respectively.

7.4 Oblique and Single-File Relative Equilibria

Oblique relative equilibria correspond to solutions (X_1, Y_1) in equation (7.5) with $X_1 \neq 0$ and $Y_1 \neq 0$. The agents are obliquely positioned as they travel along parallel lines. Examples are shown in the middle and right panels of figure 7.2. Since $Y_1 \neq 0$, oblique relative equilibria occur in pairs corresponding to the two solutions $(X_1, \pm Y_1)$ to equation (7.5). However, we have not observed them to be attracting in numerical integrations of equation (4.14), so our analysis will be brief.

Oblique relative equilibria only occur when the portion of the zero set in the central lune is not an arc of a strictly convex curve (the "wavy" curves in figure 7.3). Numerically it appears that this curve converges to a boundary circle of the lune as (w_ℓ, w_r) approaches either of the coordinate axes of W_{rev} from within $\mathcal{R}_T$. This is illustrated in figure 7.3. For weights close to a coordinate axis, the portion of the zero set in the lune is an arc of a strictly convex curve and there are no oblique relative equilibria. As the weights are varied away from the coordinate axes, this curve ceases to be strictly convex, and a pair of oblique relative equilibria appear.

The configuration for an oblique relative equilibrium has no nontrivial symmetries. However, there is a glide reflection (defined in section 5.1) that maps one oblique configuration to the other. These oblique configurations occur in chiral pairs neither of which appear to be attracting.

Single-file relative equilibria correspond to solutions to equation (7.5) with $X_1 = 0$. One agent is in front of the other, and they move single file in the plane. In this configuration the sensor weights must be equal to prevent the agents from turning.

As $(w_\ell, w_r) \in W_{\text{rev}}$ approaches W_{eq}, the portion of the zero set of $\bar{\omega}_1$ in the central lune becomes less curved, and when the weights reach W_{eq} the zero set in the lune becomes the open line segment connecting the lune's vertices (see figure 7.3). Every point in this line segment gives us a single-file relative equilibrium.

Individually the single-file relative equilibria are not attracting. For a single-file configuration, either agent can be displaced slightly forwards or backwards without altering its heading to obtain another single-file relative equilibrium.

The union of all of the single-file relative equilibria forms a connected four-dimensional invariant set $\mathcal{Q}_{\text{eq}}$ (introduced in section 6.10). The invariant set $\mathcal{Q}_{\text{eq}}$ appears numerically to be attracting in the pursuer-pursuer quadrant of W_{rev} but not in the avoider-avoider quadrant.

We can summarize the results thus far by considering the zero set as it moves from the middle of the central lune out to its right boundary (see figure 7.3). At each step along the way, the shape of the zero set determines which relative equilibria exist. The middle of the central lune ($X_1 = 0$) is associated with a collection of single-file relative equilibria, which determine the four-dimensional invariant set $\mathcal{Q}_{\text{eq}}$. This line segment is a convex but not strictly convex curve. As the zero set moves to the right, it ceases to be convex (it becomes "wavy" inside the lune), and we get chiral pairs of oblique relative equilibria. Whenever the zero-set in the lune is outside of line segment $X_1 = 0$, there is a side-by-side relative equilibrium. We now turn to the analysis of these side-by-side relative equilibria.

7.5 Linear Stability Analysis for Side-by-Side Relative Equilibria

In this section, we linearize the side-by-side relative equilibria in a translating frame, within which the agents are stationary with respect to each other. To counteract the translating motion we use the one-parameter subgroup $t \mapsto T_{(-vt)} \subset \mathbf{SE}(2)$, where:

$$T_{(-vt)}(x, y)^T = (x, y - vt)^T$$

for $(x, y) \in \mathbf{R}^2$. The action of the proper congruences $T_{(-vt)}$ on the total state space S_T is:

$$s''(t) = \begin{pmatrix} x_1'' \\ y_1'' \\ \theta_1'' \\ x_2'' \\ y_2'' \\ \theta_2'' \end{pmatrix} = G_{T_{(-vt)}} \begin{pmatrix} x_1 \\ y_1 \\ \theta_1 \\ x_2 \\ y_2 \\ \theta_2 \end{pmatrix} = \begin{pmatrix} x_1 \\ y_1 - vt \\ \theta_1 \\ x_2 \\ y_2 - vt \\ \theta_2 \end{pmatrix}.$$

The coordinates of the translating frame are written with double primes, whereas the coordinates of the rotating frame were written with single primes.

To obtain the differential equation in the translating frame, we differentiate both sides of the equation above and express the result in the translating frame:

$$\begin{pmatrix} \dot{x}_1'' \\ \dot{y}_1'' \\ \dot{\theta}_1'' \\ \dot{x}_2'' \\ \dot{y}_2'' \\ \dot{\theta}_2'' \end{pmatrix} = \begin{pmatrix} \dot{x}_1 \\ \dot{y}_1 - v \\ \dot{\theta}_1 \\ \dot{x}_2 \\ \dot{y}_2 - v \\ \dot{\theta}_2 \end{pmatrix} = \begin{pmatrix} v\cos(\theta_1) \\ v\sin(\theta_1) - v \\ \omega_1(x_1, y_1, \theta_1, x_2, y_2, \theta_2)^T \\ v\cos(\theta_2) \\ v\sin(\theta_2) - v \\ \omega_2(x_1, y_1, \theta_1, x_2, y_2, \theta_2)^T \end{pmatrix}$$

$$= \begin{pmatrix} v\cos(\theta_1'') \\ v\sin(\theta_1'') - v \\ \omega_1(x_1'', y_1'', \theta_1'', x_2'', y_2'', \theta_2'')^T \\ v\cos(\theta_2'') \\ v\sin(\theta_2'') - v \\ \omega_2(x_1'', y_1'', \theta_1'', x_2'', y_2'', \theta_2'')^T \end{pmatrix} = F \begin{pmatrix} x_1'' \\ y_1'' \\ \theta_1'' \\ x_2'' \\ y_2'' \\ \theta_2'' \end{pmatrix} + \begin{pmatrix} 0 \\ -v \\ 0 \\ 0 \\ -v \\ 0 \end{pmatrix} = F'' \begin{pmatrix} x_1'' \\ y_1'' \\ \theta_1'' \\ x_2'' \\ y_2'' \\ \theta_2'' \end{pmatrix}. \quad (7.7)$$

Here we have used the fact that $\omega_n(s''(t)) = \omega_n(s(t))$. This follows from equations (4.2) and (4.13), as in section 6.4.

It can be checked that substituting any $t \in \mathbf{R}$ into solution (7.6) for differential equation (4.14) gives a fixed point for differential equation (7.7). We focus on the fixed point for $t = 0$, which is:

$$s_T = (-X_T, 0, \pi, X_T, 0, \pi)^T / 2. \quad (7.8)$$

We also introduce notation for the matrix of the derivative of F'' evaluated at the fixed point s_T. This is:

$$\mathbf{J}_T = DF''(s_T).$$

The entries of $\mathbf{J}_T$ and its eigenvalues are functions of the weights for the total dynamical system ϕ_τ. This allows us to study bifurcations of ϕ_τ as the weights are varied.

To simplify subsequent formulas we introduce symbols for three quantities:

$$\nu_T = -\frac{\partial}{\partial X_2}\,\bar{\omega}_2\,(X_T,\,0)$$

$$\mu_T = -\frac{\partial}{\partial Y_2}\,\bar{\omega}_2\,(X_T,\,0) \tag{7.9}$$

$$\mathcal{T}_T = \mu_T\,X_T\,.$$

We can use the results from the revolving case (computed in appendix B) to facilitate the computation of the total derivative in the translating case. Since $F'' - F$ is constant, $DF'' = DF$, and since $F' = F$ when $\Omega_{(r,*)} = 0$ it follows that $DF'' = DF'$ for $\Omega_{(r,*)} = 0$. The only differences symbolically between s_T [equation (7.8)] and $s_{(r,*)}$ [equation (6.8)] are that $X_{(r,*)}$ has been replaced by X_T and $-\pi$ has been replaced by π. So here we can quickly compute $\mathbf{J}_T$ from $\mathbf{J}_{(r,*)}$ in equation (6.10):

$$\mathbf{J}_T = \begin{pmatrix} 0 & 0 & -\nu & 0 & 0 & 0 \\ 0 & 0 & 0 & 0 & 0 & 0 \\ \nu_T & \mu_T & \mathcal{T}_T & -\nu_T & -\mu_T & 0 \\ 0 & 0 & 0 & 0 & 0 & -\nu \\ 0 & 0 & 0 & 0 & 0 & 0 \\ \nu_T & \mu_T & 0 & -\nu_T & -\mu_T & \mathcal{T}_T \end{pmatrix}.$$

We can study the bifurcations of $\mathcal{E}_T$ by considering the eigenvalues of the matrix $\mathbf{J}_T$ that represents its linearization at s_T. We show in appendix C that the set of eigenvalues for $\mathbf{J}_T$ is $\{0,\,\mathcal{T}_T,\,\mathcal{S}_T^\pm\}$ where 0 has a multiplicity of at least three and

$$\mathcal{S}_T^\pm = \frac{\mathcal{T}_T \pm \sqrt{\mathcal{T}_T^2 - 8\nu\nu_T}}{2}. \tag{7.10}$$

The eigenspace for 0 contains the tangent space of $\mathcal{E}_T$ at s_T. As with revolving type relative equilibria the tangential eigenvalues describe how the system behaves inside the relative equilibrium, with parallel orbits that neither attract nor repel one another. As in the revolving case, the tangential eigenvalues will not play an important role in the bifurcation analysis.

We will focus on the three normal eigenvalues $\{\mathcal{T}_T, \mathcal{S}_T^\pm\}$, which describe how the system behaves in a small neighborhood of the relative equilibrium. If the real part of all of the normal eigenvalues are negative, then all states sufficiently close to the relative equilibrium tend towards it. If the real part of any of the normal eigenvalues is positive then, there are states arbitrarily close the relative equilibrium that move away from it. Thus we can use the normal eigenvalues to determine whether $\mathcal{E}_T$ is an attracting set or not.

As in the revolving case, there are redundancies in the eigenvalues. The tangential eigenvalues are all 0. For the normal eigenvalues, it follows from (7.10) that:

$$\mathcal{T}_T = \mathcal{S}_T^+ + \mathcal{S}_T^-.$$

Thus, if the real parts of $\mathcal{S}_T^\pm$ are both negative, then $\mathcal{T}_T$ is negative and so all three normal eigenvalues each have negative real part. So we can deduce the stability of $\mathcal{E}_T$ just from the two eigenvalues $\mathcal{S}_T^\pm$ which is why we call them the *salient eigenvalues* of $\mathcal{E}_T$.

The side-by-side type relative equilibrium comes into existence through a nonstandard bifurcation that somewhat resembles the saddle node-like bifurcation of relative equilibria in the revolving case. The side-by-side relative equilibria begin when the weights cross the boundary $\partial\mathcal{R}_T$, which is the union of the coordinate axes in W_{rev}. The revolving type relative equilibria appear as the weights cross at the boundaries $\partial\mathcal{R}_r$ and $\partial\mathcal{R}_\ell$, which in the pursuer-pursuer quadrant are close to the coordinate axes. On the other hand, in the translating case, infinitely many trivial translating type equilibria appear at the boundary $\partial\mathcal{R}_T$, where the graph of the turning function $\bar{\omega}_n$ overlaps a line segment in the X_n-axis. As the weights enter the interior of $\mathcal{R}_T$, a single side-by-side equilibrium $\mathcal{E}_T$ appears. It is attracting in the pursuer-pursuer quadrant but not in the avoider-avoider quadrant.

The behavior of $\mathcal{S}_T^\pm$ for weights in the pursuer-pursuer quadrant is analogous to the case for $\mathcal{S}_{(r,\mathrm{in})}^\pm$ shown in figure 6.7. After the nonstandard bifurcation, $\mathcal{S}_T^\pm$ appear at two points with negative real part in the complex plane. As the weights are varied further, $\mathcal{S}_T^\pm$ move toward the imaginary axis. When the salient eigenvalues cross the imaginary axis a Hopf-like bifurcation produces an apparent relative periodic orbit $\mathcal{P}_T$ which generates meandering paths.

The real part of the salient eigenvalues is $\mu_T X_T/2$, and X_T is not zero so long as $w_\ell \neq w_r$. Thus $\mu_T = 0$ is the condition for a Hopf-like bifurcation that determines the boundary $\partial\mathcal{M}_T$. By the same reasoning as at the end of section 6.6, this boundary is approximately a line through the origin, as can be seen in figures 7.1 and 7.4

7.6   The Geometry of Meandering Paths for W_{rev}

In this section, as in section 6.8, we analyze the geometry of the paths traced out by the agents when the state of the total dynamical system appears to be near a relative periodic orbit. We combine the linear stability analysis with numerical analysis of representative examples of agents with reversed weights. The physical parameters will have the same values as in section 6.8 but the weights will be in the space W_{rev} instead of W_{same}. To emphasize the resemblance between the parameter spaces W_{same} and W_{rev}, we examine the dynamics of ϕ_τ for (w_ℓ, w_r) inside a square region of W_{rev} with the same coordinates, $[-15/4, 15/4]^2$, as in section 6.8.

We call the region where we observe meandering behavior $\mathcal{M}_T$. It consists of three sector shaped regions positioned symmetrically about W_{eq} with two regions in the avoider-avoider quadrant and one in the pursuer-pursuer quadrant. They are the rose-colored regions in figure 7.4. $\mathcal{M}_T$ resembles the union of the meandering regions, $\mathcal{M}_\ell \cup \mathcal{M}_r$, of W_{same} as shown in figure 6.10 of section 6.8.

We showed in section 7.3 that the side-by-side relative equilibrium $\mathcal{E}_T$ exists if and only if the weights are in the pursuer-pursuer or avoider-avoider quadrants. We saw numerically in section 7.5 that for weights in the pursuer-pursuer quadrant, $\mathcal{E}_T$ is attracting near the coordinate axis of W_{rev}. It ceases to be attracting in a Hopf-like bifurcation near W_{eq}, which appears to produce an attracting relative periodic orbit. This bifurcation occurs when $\mu_T = 0$, which demarcates the boundary of $\mathcal{M}_T$ in the pursuer-pursuer quadrant.

In the avoider-avoider quadrant $\mathcal{E}_T$ is never observed to be attracting, though there appear to relative periodic orbits that are attracting. It is not clear to what extent $\mathcal{E}_T$ is involved in their change in stability at the borders of $\mathcal{M}_T$ in the avoider-avoider quadrant. The boundary of $\mathcal{M}_T$ in the avoider-avoider quadrant of W_{rev} does *not* correspond to a Hopf-like bifurcation, even though it resembles the boundary of the union of $\mathcal{M}_\ell$ and $\mathcal{M}_r$ in the avoider-avoider quadrant of W_{same}, which does correspond to a Hopf-like bifurcation.

For weights in W_{rev}, the value of the total curvature for the periodic arcs of a meandering path appears numerically to always be very close to ± 1 or to $\pm 1/2$, depending on whether the agents go through a complete turn or a half turn within a single periodic arc. The first case corresponds to the relative periodic orbit $\mathcal{P}_T$; the second case corresponds to the counterrotating relative periodic orbit $\mathcal{P}_C$.

In general, whenever $\check{\kappa}$ is ± 1 then either the congruence between adjacent periodic arcs is a translation, in which case the curve has translational symmetry, or else the path is a simple closed curve (as in an oval without rotational

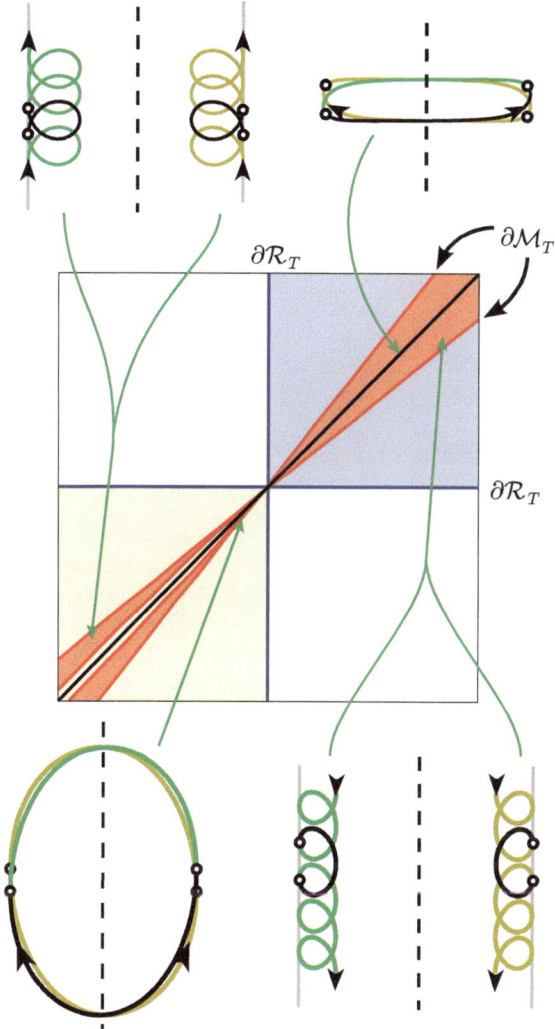

Figure 7.4

The bifurcation diagram in $[-15/4, 15/4]^2$. $\mathcal{R}_T$ is the union of the pursuer-purser and avoider-avoider quadrants; its boundary is the coordinate axes colored blue. The region $\mathcal{M}_T$ is rose-colored and its boundary is colored red. Within the pursuer-pursuer quadrant the complement of $\mathcal{M}_T$ in $\mathcal{R}_T$ is lavender colored. Within the avoider-avoider quadrant the complement of $\mathcal{M}_T$ in $\mathcal{R}_T$ is colored beige. The white region is the complement of $\mathcal{R}_T$. The panels show four pairs of physical paths corresponding to the weights indicated by the green arrows. The dashed lines are the axis of reflectional symmetry. Selected periodic arcs are colored black and their end points are marked by ∘. These endpoints have been placed where the tangent line is parallel to the axis of reflection. The bounding lines for the trochoid shaped paths are colored gray.

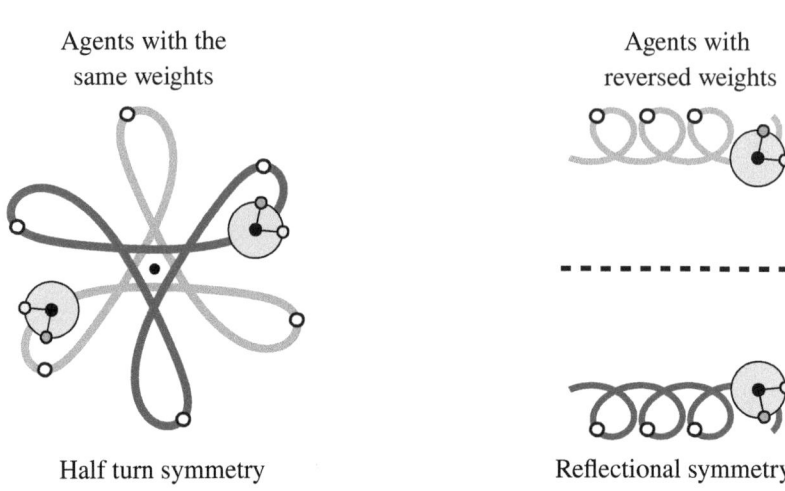

Agents with the same weights

Agents with reversed weights

Half turn symmetry

Reflectional symmetry

Figure 7.5
(left) A configuration of two agents with the same pair of weights. Their configuration is symmetric under a half turn about their fixed midpoint and they trace out paths that are symmetric under the same half turn. (right) A configuration of two agents with reversed weights that is symmetric under a reflection about their perpendicular bisector. They trace out paths that are symmetric under the same reflection. Points of maximum curvature are marked with ○ (the size of the agents are not drawn to scale).

symmetry). In our observations, whenever $\check{\kappa} \approx \pm 1$ the paths have translational symmetry and are not simple closed curves. These meandering paths correspond to the relative periodic orbit $\mathcal{P}_T$. The agents trace out a pair of parallel curves that resemble trochoids—that is, curves generated by a circle rolling on a line—as shown in the top left and bottom right panels of figure 7.4.

Similar behavior occurs in the BZ reaction in which a pair of counter rotating spirals undergo "linear drift" (e.g., Schmidt and Müller [1997]; Manz, Ginn, and Steinbock [2003]). This is illustrated in the right panel of figure 8.1. The wave is propagating to the upper right. The upper left spiral tip turns anticlockwise, while the lower right spiral tip turns clockwise. The wave also drifts, in this case in the upper right direction. The meandering spiral tips trace trochoid shaped paths (shown in black), while the wave undergoes linear drift.

When we observe the agents tracing out simple closed curves, then $\check{\kappa} \approx \pm 1/2$. The two simple closed curves are symmetrical under a half turn, and they can be partitioned into two periodic arcs. These meandering paths correspond to the counterrotating relative periodic orbits $\mathcal{P}_C$. The agents counterrotate along a pair of simple closed curves such as the pairs of ovals shown in the bottom left and top right panels of figure 7.4. As can be seen in the figure, the union of the two simple closed curves is symmetrical under the same half turn.

There is a basic difference between the symmetry of the meandering paths for weights in W_{rev} and W_{same}. This difference corresponds to the symmetry of the two agents' configuration in the two cases. In section 6.2, we saw that for a revolving type relative equilibrium, the configuration of a pair of agents is symmetrical under a half turn about the midpoint of their locations. This half turn also preserves the locations of the agents' sensors. On the other hand, it follows from section 7.2 that for a side-by-side relative equilibrium the configuration for the agents is symmetrical under the reflection about the perpendicular bisector of their locations. This reflection also preserves the locations of the agents' sensors.

Numerical integration shows that these symmetries for the physical paths appear to persist even when the system is following a relative periodic orbit. Examples are shown in figure 7.5. For weights in W_{same}, the two meandering paths are symmetric under a half turn about the agent's midpoint which remains fixed over time. For weights in W_{rev}, the two meandering paths are symmetric under a reflection about the agent's perpendicular bisector, which remains fixed over time. The revolving and translating cases are both shown in figure 7.5, where the sensors are shaded to make it easier to confirm their symmetries (lines of reflectional symmetry are also shown in the panels of figure 7.4). Notice that in the translating case there is reflectional symmetry but no rotational symmetry. The trochoid-like paths form a chiral pair of curves.

In section 6.8 we saw that meandering paths in the revolving case produced inward or outwardly directed loops, and that the paths are confined between a minimum and maximum radius. The situation is similar with meandering paths in the translating case. The trochoid-like paths in the translating case also have a "center" of symmetry, although this center of symmetry is a line, not a point. We say a trochoid-like path is *inwardly directed* if the loops point toward their line of symmetry; they are *outwardly directed* if the loops point away from it.

The trochoid-like paths are confined between a pair of parallel lines that intersect the meandering paths tangentially infinitely many times at a discrete set of points (see figure 7.6). These parallel lines play an analogous role to the circle of maximum radius in the $\mathcal{M}_\ell \cup \mathcal{M}_r$ case. These intersection points can be taken as the end points of periodic arcs. Within a periodic arc, one endpoint is its starting point and the other one is its final point. The starting points are marked by light gray dots and the final points are marked by dark gray dots in figure 7.6. The starting and final points of periodic arcs can be used to provide an alternative characterization of inwardly versus outwardly directed trochoid-like paths. If the direction of the velocity vector at its final point is pointed away from the starting point of that periodic arc, then the trochoid-like path is inwardly directed. If the direction of the velocity vector at the final point is pointed towards the starting point, then the trochoid-like path is outwardly directed.

For weights in $\mathcal{M}_T$ of the avoider-avoider quadrant and sufficiently far away from W_{eq}, the meandering paths have inwardly directed loops, while for weights in the pursuer-pursuer quadrant the meandering paths have outwardly directed loops. This matches the revolving case in which the meandering paths have inwardly directed loops for weights in the avoider-avoider quadrant of W_{same} and outwardly directed loops for weights in the pursuer-pursuer quadrant of W_{same}. The revolving case is shown in figure 6.10, and the translating case is shown in figure 7.4.

The counterrotating case between the inwardly and outwardly directed trochoid-like paths generates simple closed curves. As shown in figure 7.4, this case occurs in both the pursuer-pursuer and avoider-avoider quadrants for weights in $\mathcal{M}_T$ near W_{eq}. As shown in figure 7.6, we can think of this as a transition in which the starting point of a periodic arc passes through its final point. Note that in these cases the value of $\breve{\kappa}$ jumps from ± 1 to $\pm 1/2$ so that the meandering paths have rotational symmetry and there are actually two periodic arcs in each of the simple closed curves, instead of just one.

The counterrotating case persists within a subregion of $\mathcal{M}_T$. The final point of the second periodic arc coincides with the starting point of the first periodic arc as the weights are varied. This is consistent with $\mathcal{P}_C$ being a distinct type of relative periodic orbit from $\mathcal{P}_T$. We focus on $\mathcal{P}_C$ in the next section.

7.7   The Geometry of Counterrotating Paths near W_{eq}

In this section, we consider the geometry of the physical paths as the weights are varied towards W_{eq} within the pursuer-pursuer quadrant of W_{rev}. So far, we have seen that in this quadrant a nonstandard bifurcation produces the attracting side-by-side relative equilibrium $\mathcal{E}_T$, and that a Hopf-like bifurcation produces the attracting relative periodic orbit, $\mathcal{P}_T$, which generates trochoid like meandering paths. As the weights get closer to W_{eq}, another bifurcation appears to occur in which the relative periodic orbit $\mathcal{P}_T$ is transformed into the another type of relative periodic orbit that we call $\mathcal{P}_C$. It generates counterrotating oval paths in the plane.

Numerical evidence indicates that the orbits comprising $\mathcal{P}_C$ are periodic. Numerically, the configuration of the vehicles after one period is not only congruent to the starting configuration but it is virtually identical to it. Thus $\mathcal{P}_C$ appears to be a union of periodic orbits. By contrast, all of the orbits comprising $\mathcal{P}_T$ are unbounded. Because of this change in the topology of the orbits in

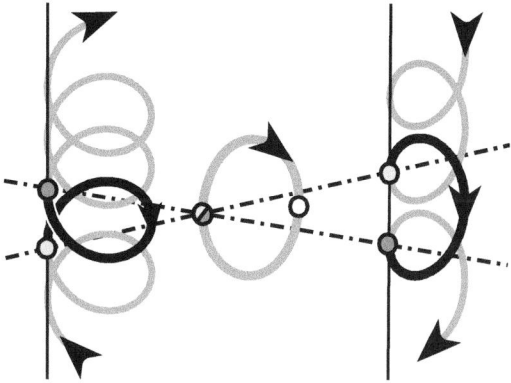

Figure 7.6
An illustration of the relationship between inwardly directed loops, counterrotating paths, and outwardly directed loops. The physical paths of a single right-turning agent in three panels of figure 7.4 are shown here in gray. A selected periodic arc for the trochoid-like paths is colored black. The starting point of the periodic arc is marked by a light gray dot, and the final point is marked by a dark gray dot. The path on the left is from the top left panel of figure 7.4. The velocity of its final point is directed away from the starting point. These endpoints bound an inwardly directed loop. The path on the right is from the bottom right panel of figure 7.4. The velocity of its final point is directed towards the starting point. These end points mark the tips of an outwardly directed loop. The path in the middle is from the bottom left panel in figure 7.4. The dot that is half light, half dark marks the starting point, which is the same as the final point. The path is a simple closed curve composed of two periodic arcs whose endpoints are the light/dark gray dot and the white dot. This type of path occurs in the counterrotating case.

the attracting sets, the transition between $\mathcal{P}_T$ and $\mathcal{P}_C$ constitutes a bifurcation of ϕ_τ.

We will consider the counterrotating paths of $\mathcal{P}_C$ in further detail and show that geometrically they resemble the meandering paths generated by the relative periodic orbits $\mathcal{P}_{(\ell,\text{in})}$ and $\mathcal{P}_{(r,\text{in})}$ for weights near W_{eq} within the pursuer-pursuer quadrant of W_{same}. We will conclude this chapter on the exceptional character of another type of bifurcation that occurs at W_{eq} and that involves the attracting set $\mathcal{Q}_{\text{eq}}$.

The trochoid-like paths generated by $\mathcal{P}_T$ do not possess any rotational symmetries, either individually or jointly. On the other hand, the counterrotating paths generated by $\mathcal{P}_C$ are symmetrical under a half turn, both individually and jointly (as can be seen in the bottom left and top right panels in figure 7.4). The bifurcation between $\mathcal{P}_T$ and $\mathcal{P}_C$ changes the symmetry of each of the two physical paths. This allows the bifurcation to be detected by the jump in the value of $\check{\kappa}$ for the two paths (between ± 1 and $\pm 1/2$).

The trochoid-like paths and oval paths may appear to be very different types of curves so that the bifurcation between $\mathcal{P}_T$ and $\mathcal{P}_C$ can seem rather abrupt but, as explained in the previous section, the counterrotating paths can be regarded as an intermediate case between the trochoid-like paths with inwardly and outwardly directed loops. This is summarized in figure 7.6.

We can use the bounding lines of the meandering paths to further elaborate on how the trochoid-like paths generated by $\mathcal{P}_T$ relate to the oval paths generated by $\mathcal{P}_C$. As the weights are varied within $\mathcal{M}_T$, the distance between the starting and final points of the periodic arc changes. As this distance goes to 0, the endpoints of a periodic arc pass through each other on the bounding line. The velocity at the final point changes continuously from pointing away from the starting point to pointing towards it. This is indicated by the dash-dot lines in figure 7.6. The loops of the trochoid-like paths go from being inwardly directed to being outwardly directed. In this sense the counterrotating paths are an intermediate case between inwardly and outwardly directed trochoid-like paths.

Because this intermediate case involves the starting and final points of a periodic arc passing through each other it might seem that counterrotating paths should be exceptional. However, numerical evidence indicates they are robust to small changes in the weights within W_{rev} and that the relative periodic orbit $\mathcal{P}_C$ occurs for a relatively small open subset of $\mathcal{M}_T$.

In the pursuer-pursuer quadrant, the bifurcation between $\mathcal{P}_T$ and $\mathcal{P}_C$ is observed to take place very close to W_{eq}. In the avoider-avoider quadrant, the bifurcation appears to occur along a pair of symmetrically positioned curves (*not* shown in figure 7.4) in the two sectors of $\mathcal{M}_T$ inside the avoider-avoider quadrant. Each bifurcation curve appears to separate the sector it is in into two regions: one away from the origin and one that is close to the origin. In the the avoider-avoider quadrant, $\mathcal{P}_T$ occurs in the region away from the origin and $\mathcal{P}_C$ occurs in the region close to the origin.

So the attracting set $\mathcal{P}_C$ is an intermediate case in two different ways. First, the physical paths produced by $\mathcal{P}_C$ are intermediate in form between the inwardly and outwardly directed trochoid-like paths of $\mathcal{P}_T$ as shown in figure 7.6. Second, the regions in W_{rev} where $\mathcal{P}_C$ occurs is geometrically in between the regions where $\mathcal{P}_T$ produces the inwardly and outwardly directed loops in its trochoid-like paths.

There are also some common aspects between the relative periodic orbit $\mathcal{P}_C$ for weights close to W_{eq} in the pursuer-pursuer quadrant of W_{rev} and the relative periodic orbits $\mathcal{P}_{(\ell,\text{in})}$ and $\mathcal{P}_{(r,\text{in})}$ for weights close to W_{eq} in the pursuer-pursuer

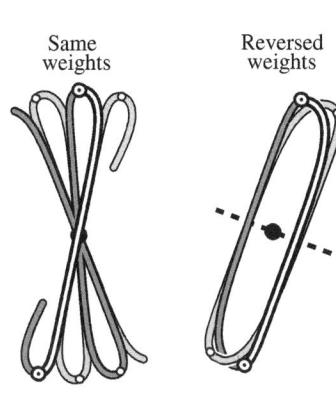

Same weights

Reversed weights

Figure 7.7
An illustration of how counterrotating paths in the translating case are related to meandering paths in the revolving case. (left) Weights in W_{same}. (right) Weights in W_{rev}. Both weight pairs are very close to $(2.295, 2.295)$ in W_{eq}. The paths for the two agents are shaded light and dark gray and outlined in black except for a single periodic arc in the left and right panels, which are white and outlined in black. Points of maximum absolute curvature are marked with $\bigcirc$ except for the endpoints of the white periodic arcs, which are marked with $\odot$. The white periodic arcs are nearly congruent. The center of rotational symmetry is marked by $\bullet$. On the right, the agent's perpendicular bisector is marked by the dashed line.

quadrant of W_{same}. This is illustrated in figure 7.7. The paths shown on both sides of the figure have rotational symmetry. [3] The meandering paths on the left side of the figure are generated by $\mathcal{P}_{(r,\text{in})}$, and they only have rotational symmetry without any reflectional symmetry. On the other hand, the counterrotating paths on the right side are generated by $\mathcal{P}_C$, and their union has reflectional symmetry as well as rotational symmetry. These counterrotating paths form a pair of simple closed curves with a common center of symmetry and opposite chirality.

As the weights in $\mathcal{M}_T$ move towards W_{eq}, the periodic arcs of the closed paths take on shapes that resemble the periodic arcs of the meandering paths for weights in $\mathcal{M}_\ell \cup \mathcal{M}_r$ near W_{eq}. A single left-turning periodic arc is indicated with white curves outlined in black in the left and right panels of figure 7.7. They are bounded by points of maximum absolute curvature. The local curvature along these two periodic arcs are close to being the same so that the periodic arcs are nearly congruent.

Furthermore the total absolute curvature of the periodic arcs, that is, $|\check{\kappa}|$, is close to 1/2 near W_{eq} regardless of whether we approach W_{eq} from within W_{rev} or from within W_{same}. This is illustrated in figures 6.17 and 7.8. The panels in both figures show how the shapes of the meandering paths change as the weights approach W_{eq}. In figure 7.8, a small diagonally oriented rectangle has

---

3. Recall from section 6.8 that the rotational symmetry is for the whole curve in which time goes infinitely far backwards and forwards and that not all of the curve can be drawn.

been selected from W_{rev} to illustrate how the shape of the counterrotating paths generated by $\mathcal{P}_C$ changes as the weights approach W_{eq}. To facilitate the comparison of the rose-colored rectangles in figures 6.17 and 7.8, the rectangle in figure 7.8 is the convex hull of a set of points with the same coordinates as in equation (6.18) except here the coordinates (w_ℓ, w_r) refer to points in W_{rev}. For clarity, the rectangle in figure 7.8 has also been stretched away from W_{eq} and rotated so that W_{eq} is vertical.

The coordinates for the midpoint of the rectangle's bottom edge is $(2.295, 2.295)$, which is in W_{eq}. This is also the midpoint of the bottom edge of the rectangle in figure 6.17. These two rectangles are congruent and intersect orthogonally inside of the three-dimensional subspace $W_{\text{same}} + W_{\text{rev}}$. There are no isogonal curves in figure 7.8, since it appears $\breve{\kappa} = \pm 1/2$ for all weights in this rectangle.

In the counterrotating case when the weights are above W_{eq} in $\mathcal{M}_T$, the value of $\breve{\kappa}$ is $-1/2$ for agent 1 and $1/2$ for agent 2. This is shown in in figure 7.8. When the weights are below W_{eq} in $\mathcal{M}_T$ the value of $\breve{\kappa}$ is $1/2$ for agent 1 and $-1/2$ for agent 2. As the weights in $\mathcal{M}_T$ pass through W_{eq} the sign of $\breve{\kappa}$ reverses for both agents. In the revolving case, the sign of $\breve{\kappa}$ also reverses for both agents as the weights pass through W_{eq} within $\mathcal{M}_\ell \cup \mathcal{M}_r$. The value of $\breve{\kappa}$ converges to $-1/2$ for both agents as the weights approach W_{eq} from within $\mathcal{M}_\ell$ and it converges to $1/2$ for both agents as the weights approach W_{eq} from within $\mathcal{M}_r$. This is shown in figure 6.17.

For weights in W_{rev} that are near W_{eq}, the union of the two physical paths are symmetrical under a rotation by exactly one half turn. For weights in W_{same} that are near W_{eq} (but not in W_{eq}), the union of the two physical paths are symmetrical under a rotation by almost a half turn. As the weights in $\mathcal{M}_T$ approach W_{eq} from either side, the paths become narrower, just as they do for weights in $\mathcal{M}_\ell \cup \mathcal{M}_r$. In both cases, the paths become an alternating succession of nearly straight arcs connected by hairpin turns, and in both cases the closer the weights are to W_{eq} the closer the nearly straight arcs are to each other. When the weights reach W_{eq}, the agents stop turning; instead the agents undergo uniform linear motion. It appears that $\mathcal{Q}_{\text{eq}}$ is the limiting case of $\mathcal{P}_C$ as well as of $\mathcal{P}_{(\ell,\text{in})}$ and $\mathcal{P}_{(r,\text{in})}$.

The orbits of $\mathcal{P}_C$ are periodic while the orbits of $\mathcal{Q}_{\text{eq}}$ are unbounded. So the positive part of W_{eq} is a bifurcation curve in the pursuer-pursuer quadrant of W_{rev}. It seems the bifurcation which turns $\mathcal{P}_T$ into $\mathcal{P}_C$ helps prepare for the bifurcation which turns $\mathcal{P}_C$ into $\mathcal{Q}_{\text{eq}}$. The value of $\breve{\kappa}$ is not defined for weights in W_{eq}, because when the agents are undergoing uniform linear motion the curvature of the paths has no minimal period. When the value of $\breve{\kappa}$ is defined for a curve and its value is not an integer, then the curve must have rotational

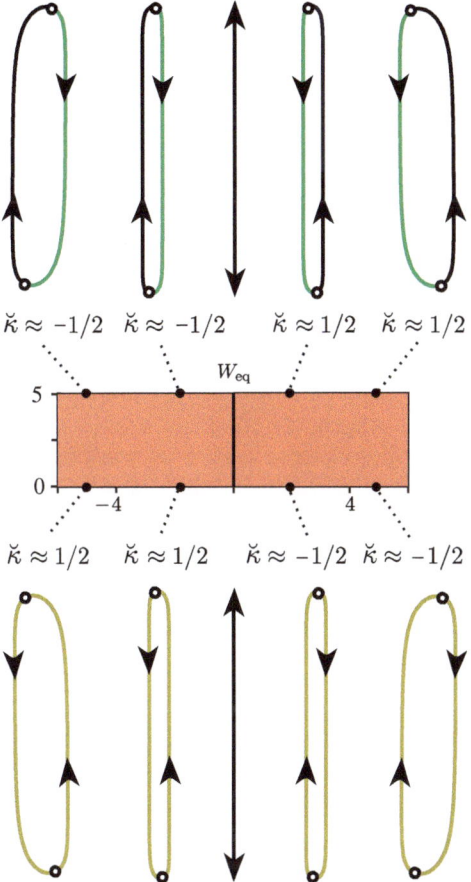

Figure 7.8
An illustration of how $\breve{\kappa}$ changes for counterrotating paths when the weights are in W_{rev} near W_{eq}. (middle) The rose-colored rectangle is anchored at $(2.295, 2.295)$ in W_{rev}. The coordinate axes are given by the formulas in table 6.1 to facilitate comparison with figure 6.17. (top) The paths for agent 1. A periodic arc in each path is highlighted in black. (bottom) The paths for agent 2. Points of maximum absolute curvature are marked by ○ in the top and bottom. The paths are labeled with their value for $\breve{\kappa}$. Dotted lines match the weights to the value they produce for $\breve{\kappa}$. The straight paths in the top and bottom indicate that the agents engage in uniform linear motion when their weights are in W_{eq}.

symmetry (Hotton 2010, 2016). So when the value of $\breve{\kappa}$ is defined for a curve without rotational symmetry, it must be an integer. For the trochoid-like paths of $\mathcal{P}_T$, the value of $\breve{\kappa}$ is ± 1. As the weights within W_{rev} get close to W_{eq}, they also get close to points in W_{same} where the value of $\breve{\kappa}$ is nearly $\pm 1/2$. So the trochoid-like paths have to undergo some type of transformation if the value of $\check{\kappa}$ is to approach the value of $\check{\kappa}$ for nearby points in W_{same}.

The bifurcation that occurs in the positive part of W_{eq} is exceptional within the subspace $W_{\text{same}} + W_{\text{rev}}$. The positive part of W_{eq} is also a bifurcation curve in the pursuer-pursuer quadrant of W_{same} (recall section 6.8). Numerical evidence indicates that the attracting sets $\mathcal{P}_{(r,\text{in})}$ and $\mathcal{P}_{(\ell,\text{in})}$ exist for open regions of the two-dimensional subspace W_{same}, and the attracting set $\mathcal{P}_C$ exists for open regions of the two-dimensional subspace W_{rev}. The positive part of W_{eq} is the boundary of these open regions within the pursuer-pursuer quadrants of W_{same} and W_{rev}. However the positive part of W_{eq} lacks enough dimensions to be the boundary of an open region in the three-dimensional subspace $W_{\text{same}} + W_{\text{rev}}$.

In a single-file configuration, each agent must be directly in front of or directly behind the other agent. So to obtain single-file relative equilibria, it is necessary for the sensor weights of each individual agent to be equal. It is not necessary for the two agents to have all of their sensor weights equal, that is, the quadruple of weights do not have to be in W_{eq}. It is sufficient that the left and right sensor weights of each individual agent be equal. The sensor weights of the agent in front can be different from the agent in back. In other words, it is necessary and sufficient for the sensor weights of the two agents to not break the bilateral symmetry of their bodies. As explained in section 2.5, the subspace of W_{total} for which the weights do not break the bilateral symmetry of agents' bodies is $W_{\text{opp}} + W_{\text{eq}}$. This subspace is shown in figure 2.10.

The intersection of the subspaces $W_{\text{same}} + W_{\text{rev}}$ and $W_{\text{opp}} + W_{\text{eq}}$ is W_{eq}. So the only place within $W_{\text{same}} + W_{\text{rev}}$ where $\mathcal{Q}_{\text{eq}}$ can exist is W_{eq}. If we perturb a quadruple of weights in W_{eq} to any point in $W_{\text{same}} + W_{\text{rev}}$ that is outside of W_{eq}, we no longer get a single-file relative equilibrium and the invariant set $\mathcal{Q}_{\text{eq}}$ ceases to exist.

We have seen that the value of $\check{\kappa}$ is close to $\pm 1/2$ for weights in the "paddle wheel," $W_{\text{same}} \cup W_{\text{rev}}$, that are close to W_{eq}. For these weights, the physical paths consist of an alternating succession of nearly straight arcs and hairpin turns. It is possible that within $W_{\text{same}} + W_{\text{rev}}$ other types of behavior occur near W_{eq}, but outside of $W_{\text{same}} \cup W_{\text{rev}}$, for which the value of $\check{\kappa}$ is close to $\pm 1/2$ and the physical paths consist of nearly straight arcs and hairpin turns. In the bifurcation at the positive part of W_{eq}, the uniform linear motion of $\mathcal{Q}_{\text{eq}}$ may consistently arise from nearly straight arcs becoming perfectly straight and the agents ceasing to turn.

The left and right sensor weights being equal might be regarded as the normal situation, but their strengths can never be perfectly equal in a physical implementation of Braitenberg vehicles. Slight perturbations from perfect equality produce diverse behaviors such as counterrotating paths and hypotrochoid-like meandering paths. Such diverse behavior may be relevant to how agents interact with each other in the natural world, illustrating once again Braitenberg's idea that simple agents can produce complicated behaviors.

8 Relation to Naturally Occurring Systems

Braitenberg conceived of his vehicles as a way to demonstrate that simple systems can engage in complicated behavior. He was not attempting to model the movement of any specific organism. Yet the paths traced out by Braitenberg vehicles resemble the paths traced out by many physical systems and simple organisms. The final section of Braitenberg's book ("Biological Notes on the Vehicles") discusses applications to natural systems including compound eyes, olfactory orientation, object fixation, and acoustic form perception. We conclude our book in a similar way, relating our work to a range of natural phenomena, to give a sense of the possible applications of this analysis. The panels in figure 8.1 show some physical systems whose behaviors resemble the behaviors of Braitenberg vehicles, and the panels in figure 8.2 show some biological systems whose behaviors resemble the behaviors of Braitenberg vehicles.[1] In each case, physical or biological, the system traces out planar curves that resemble those produced by Braitenberg vehicles over some range of parameter values.

One reason Braitenberg vehicles can recapitulate the movement of so many types of phenomena is that they have a broad behavioral repertoire, which spans both conservative and dissipative systems. A conservative system looks basically the same whether it is run forwards or backwards in time. For instance our solar system can be accurately modeled as a conservative system in Newtonian mechanics. If the velocities of the planets could be reversed at some moment in time, then they would all go around the Sun along the same paths as they currently do but in the opposite direction. On the other hand, a dissipative system can have an attracting set in its state space. The states neighboring the attracting set move toward it. Such a system looks different when run forwards or backward in time. For instance, an elastic ball bouncing on a table is a

---

1. See Shaikh and Rañó (2020) for further examples of Braitenberg vehicles as models of animal behavior, beyond those discussed in this chapter.

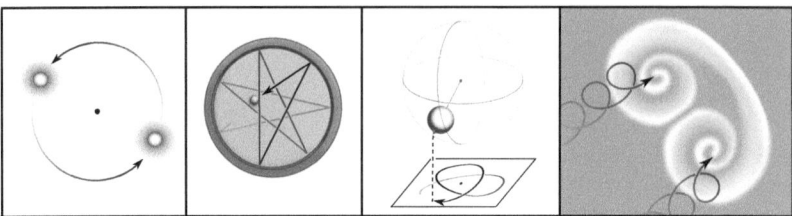

Figure 8.1
Illustrations of physical systems that follow paths resembling those followed by Braitenberg vehicles (numbered from left to right). (1) A binary star system in which the stars revolve around their common center of mass. This resembles the behavior of a pair of Braitenberg vehicles when the system is in a revolving type relative equilibrium. (2) A billiard ball rolling on a frictionless circular table ($\check{\kappa} \approx 3/7$). Braitenberg vehicles can produce physical paths which are like "rounded out" versions of these paths (Rañó 2010). (3) A spherical pendulum, when viewed from above, traces out a planar path that looks like a hypotrochoid (Hotton 2016). This resembles the path traced by a vehicle when the system is in a relative periodic orbit ($\check{\kappa} \approx -2/3$, figure 6.10). (4) Counter-rotating spirals of the BZ reaction, whose tips undergo linear drift (Schmidt and Müller 1997). This also resembles the behavior of a pair of Braitenberg vehicles when the system appears to be in a relative periodic orbit ($\check{\kappa} = \pm 1$, figure 7.4).

simple dissipative system. In forward time, the system loses energy with each bounce and the ball eventually comes to rest on the table. The attracting set in this case is just a single point in the state space that corresponds to the ball being at rest. However, in backward time the system gains energy with each bounce, so the ball bounces forever, reaching ever greater heights.

Systems with complicated dynamics can sometimes exhibit both conservative and dissipative behavior. One way this can happen is when the dynamical system has an attracting invariant manifold with several dimensions, such as with Braitenberg vehicles. When the initial state of such a system is far from the attracting set, the state can converge toward the attracting set in an irreversible fashion. However, when the initial state is in the attracting set, its behavior can look the same forward and backward in time.

8.1 Conservative Systems and Classical Dynamics

To a certain degree, Braitenberg vehicles behave like point particles subject to central forces (attractive for pursuers and repulsive for avoiders). Classical mechanics has long modeled the motion of celestial bodies as point masses moving under a central force such as gravity. It is not unusual for two gravitationally bound stars in a binary star system to have about the same mass

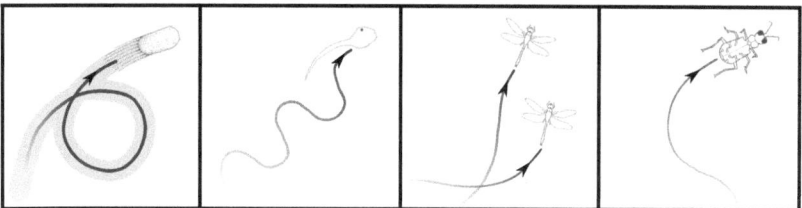

Figure 8.2
Illustrations of biological systems discussed in this chapter whose behaviors are similar to those of Braitenberg vehicles (numbered left to right). (1) A looping *Listeria* bacterium, (2) an ascidian larva moving sinusoidally, (3) a pair of dragonflies in a territorial pursuit, and (4) a tiger beetle chasing prey.

and revolve in nearly circular orbits about their common center of mass. This is like Braitenberg vehicles in a revolving type relative equilibrium. Although we no longer think of the planets in our solar system as performing epicyclic motion, the orbits for some of their moons can be fairly well approximated with epicyclic motion. The physical paths traversed by the Braitenberg vehicles can also be approximated with epicyclic motion. For instance, we confirmed the observation of Rañó (2010) of billiard-like epicyclic behavior for particular values of the sensor weights (section 6.8).

Unlike point particles though, Braitenberg vehicles have bodies that extend into the physical plane. We have treated Braitenberg vehicles as non-deformable bodies so that they have rigid body kinetics. Euler showed that a rigid body in Newtonian mechanics can precess in the absence of external forces (Euler 1776). Poinsot later showed that this precession is given by a cone rolling without slipping around another cone, which is a slight variation on epicyclic motion (Poinsot 1851). Klein subsequently studied spinning tops moving across a horizontal plane (Klein 1897). Klein verified the common observation that a top can perform epicyclic-like motion in the plane.[2] This research produced some of the earliest mathematical models for the movement of bodies in an environment. As in our analysis of the Braitenberg vehicles, these models make use of two frames of reference, one for the moving body and one for the environment.

As an example of conservative behavior by Braitenberg vehicles, Rañó (2010) shows, in his implementation of Braitenberg vehicles, that the agent

2. Of course, in the long run, it dissipates energy because of friction and comes to rest. A few years later, Klein addressed some of the complications in a top's behavior introduced by frictional forces (Klein and Sommerfeld 2012).

traces out physical paths that resemble billiards on a circular table, a classic example of a conservative dynamical system (Birkhoff 1927). In this dynamical system, the billiard ball bounces inelastically at the table's circular boundary. Its behavior in backwards time is the same as if the ball's initial velocity had the same magnitude but with the opposite direction. The paths in Rañó's version of Braitenberg vehicles have relatively long straight arcs alternating with relatively short sharply curved arcs. They can be seen as a slight "rounding out" of the paths taken by billiards on a circular table. In our implementation of Braitenberg vehicles, the agents can also trace out paths that resemble a rounding out of the paths traced by a billiard on a circular table, for example, in the bottom right of figure 6.15 (where $\breve{\kappa} \approx -1/5$). As explained in section 6.8, this motion occurs for weights near the outer boundary of the meander region $\mathcal{M}_\ell \cup \mathcal{M}_r$ within the pursuer-pursuer quadrant of W_{same}.

Section 6.10 presents another example of a conservative dynamical system from classical mechanics that behaves like the Braitenberg vehicles: the frictionless spherical pendulum. Whereas the billiard-like paths occur near one boundary of $\mathcal{M}_r$, the paths that are like those traced by a spherical pendulum occur near the other boundary of $\mathcal{M}_r$, near W_{eq}. It is similar for $\mathcal{M}_\ell$. As the weights approach W_{eq} within the pursuer-pursuer quadrant, the meandering paths again have relatively long straight arcs interspersed with relatively short sharply curved arcs. The value of $\breve{\kappa}$ tends to $1/2$ on one side of W_{eq} and $-1/2$ on the other side of W_{eq}. This is similar to what happens with the value of $\breve{\kappa}$ for the frictionless spherical pendulum as its angular momentum approaches the half-line of zero angular momentum within its energy-momentum space.

8.2 Dissipative Systems

Although Braitenberg vehicles can behave like a conservative system for weights in the pursuer-purser quadrant of W_{same}, they also have aspects in common with dissipative systems for weights in the avoider-avoider quadrant of W_{same}. As Braitenberg pointed out (Braitenberg 1984), his vehicles would be subjected to frictional forces that would tend to bring them to rest unless they had an energy supply.

The examples from classical mechanics describe conservative systems, whose energy remains constant over time. However, Braitenberg vehicles require energy to execute their prescribed behaviors and are therefore better thought of as dissipative systems. An important property of dissipative systems, as opposed to conservative systems, is the presence of attracting sets. Attracting sets correspond to the long-term behavior of a dynamical system and the analysis of dissipative systems often comes down to identifying the attracting sets of a system and classifying their types.

A well-known class of dissipative systems with attracting sets are excitable media which exhibit spiral waves, in particular the BZ chemical reaction (Winfree 1973). When confined to a thin layer, the spiral waves can move like a rigid body even though the medium is in a liquid state. This corresponds to an attracting relative equilibrium for the BZ reaction. The relative equilibrium appears from a saddle node-like bifurcation that produces an attracting and non-attracting pair of relative equilibria (Kiss et al. 2003). After a subsequent Hopf-like bifurcation, both relative equilibria are non-attracting and an attracting relative periodic orbit appears (Golubitsky, LeBlanc, and Melbourne 1997; Sandstede, Scheel, and Wulff 1999; Golubitsky and Stewart 2003). This can be seen in the tips of the spiral waves traversing epicyclic-like curves in the plane. As we have seen, a similar sequence of bifurcations occurs in the Braitenberg system. We discuss this example in more detail in the next section.

Some common features in the examples from classical mechanics, the BZ reaction, and the Braitenberg vehicles are that they are equivariant with respect to a group of proper congruences and that they can engage in quasiperiodic behavior. These two features combine to produce planar paths in which the local curvature varies periodically even when the path is not closed. So the paths can be subdivided into an infinite series of congruent arcs (see section 6.8). This allows us to apply the techniques in (Barkley and Kevrekidis 1994; Barkley 1994; Hotton 2010, 2016) to the study of Braitenberg vehicles.

In recent years, many other examples of spatially extended dissipative systems have been described that also produce spirals or meandering paths with periodic curvature, for example systems involving active media, self-propelled domains, and cell motility (Ogawa et al. 2006; Shenoy et al. 2007; Hiraiwa et al. 2010; Wen, Leung, and Chen 2012a, 2012b; Ishihara 2013; Bhalla, Griffith, and Patankar 2013; Dreher, Aranson, and Kruse 2014; Reufer et al. 2014; Brun et al. 2015; Wen, Leung, and Chen 2016; Tarama 2017; Ohta 2017; Ren et al. 2020; Young, Shelley, and Stein 2021; Ziepke et al. 2022). Many of these studies have relied on numerical techniques with little dynamical systems theory. Our work here can serve as an example of the kind of detailed mathematical analysis that might be achieved with these systems.

8.3   The BZ Reaction

At an abstract level, the behavior of the Braitenberg vehicles resemble the dissipative dynamics of the Belousov-Zhabotinsky chemical reaction (Belousov 1985; Zhabotinsky 1964). The BZ reaction is a well-studied autocatalytic oxidation-reduction reaction (Field, Körös, and Noyes 1972; Zhabotinsky et al. 1993). Unlike more familiar chemical reactions, the concentrations of the reagents in the BZ reaction oscillate as the system converges to its thermo-dynamic equilibrium state. As certain dissolved metal ions are oxidized, they catalyze their own production. So the rate of oxidation increases as the concentration of these oxidized metal ions increases. Negative ions accumulate as the oxidation reaction accelerates, and eventually the concentration of negative ions becomes large enough to inhibit the oxidation reaction. The oxidized metal ions are subsequently reduced back down to begin the cycle again. The oscillations occur with a period minutes long (Belousov 1985; Epstein and Showalter 1996) and the amplitude of the oscillations slowly decays over thousands of periods before the system nears its equilibrium state (Zhabotinsky 2007).

When the reagents of the BZ reaction are present in a region of sufficiently large extent, without being stirred, the chemical concentrations can vary spatially as well as temporally (Zaikin and Zhabotinsky 1970). The reaction only depends on how the reagents are distributed in relation to each other. When a BZ reaction is confined to a thin layer in physical space, the dynamics is, in principle, equivariant with respect to the group of proper congruences of the plane, $\mathbf{SE}(2)$ (Barkley 1994). If we could initialize the reaction throughout a thin layer of infinite extent to any chosen concentration distribution for the reagents, we should find that the system ends up in the same state regardless of whether we run the reaction for a specific amount of time and then apply a proper congruence to the thin layer or apply the proper congruence and then run the system for the same amount of time.

In a spatially distributed BZ reaction, the reaction mainly occurs within a narrow region that travels through the medium. The reaction is quiescent outside of this reaction zone. When the BZ reaction is confined to a thin layer, there are, in addition to the homogeneous equilibrium state, two commonly observed patterns: "target waves" and "spiral waves." Target waves have equally spaced concentric reaction zones that propagate outward. We do not present images of target waves here. We focus on spiral waves instead. With spiral waves, the reaction zone is shaped like a spiral whose coils are equally spaced. In the figures in this section, the reactions zones are indicated by lighter shades of gray. The coils propagate in such a way that the reaction zone appears to rotate (Winfree 1972; Keener and Tyson 1986).

It is not practical to initialize a BZ reaction in the form of a spiral wave. Instead, spiral waves appear after a transient period. One way a BZ reaction can be induced to form spiral waves is by blocking a section of a propagating reaction zone so that it breaks up into separate pieces. The broken ends of the reaction zone curl up, and the shape of the reaction zones converges to spirals. Figure 8.3 shows the formation of spiral waves from broken reaction zones.

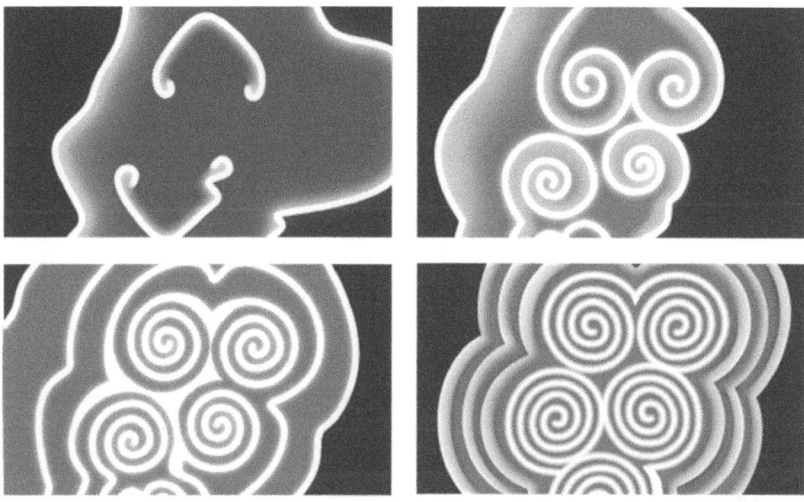

Figure 8.3
Convergence toward spiral waves of the BZ reaction. Chemical concentration is indicated with shades of gray. Lighter shades of gray correspond to reaction zones, whose tips trace out the curves of interest here. (top left) The broken ends of the wave fronts have begun to curl up. (top right) Irregularly shaped spirals have formed. The outer coils of two spiral waves on the right have just made contact. (bottom left) Four spiral waves have made contact with each other. (bottom right) The coils of the spiral waves have become nearly equally spaced within their domains. Adapted from panels (a), (e), (d), and (h) of plate II in Zhabotinsky and Zaikin (1973).

In this case, the behavior of the spiral waves is complicated by the interaction of their outer coils. The coils of neighboring spiral waves propagate outward until they meet each other. The reaction zones become quiescent after meeting. Eventually the coils of the individual spiral waves appear equally spaced, and each spiral wave appears to rotate within a limited domain. The broken ends of the reaction zones become the tips of the spiral waves.

Idealized mathematical models of the BZ reaction describe spiral waves that extend infinitely in space. To keep things mathematically simple, the focus is usually on single spiral waves. The motion of a single spiral tip is often similar to what we observe with each member of a pair of Braitenberg vehicles. A common version of this type of model is the Oregonator model of the BZ reaction (Field and Noyes 1974). In such a model, as a spiral wave rotates its tip generally traces out a circular path around the quiescent core, as shown in figure 8.4. Theoretically the reaction zone of a spiral wave can coil outward without end. Any point in the plane can be the location for the tip of a single spiral wave and every direction from that point can be the direction in which the reaction zone emanates from the tip. For any two distinct unbounded spiral waves whose reaction zones coil outward in the same direction there is a unique proper congruence that maps one to the other. The set of unbounded spiral waves that coil outward in the clockwise direction is a copy of **SE**(2) inside the space of all possible concentration distributions for the BZ reaction. The set of unbounded spiral waves that coil outward in the anticlockwise direction forms another copy of **SE**(2). Each of these copies of **SE**(2) is a relative equilibrium for the BZ reaction, much like the relative equilibria $\mathcal{E}_{(\ell,\text{in})}$ and $\mathcal{E}_{(r,\text{in})}$ for the Braitenberg vehicles. As with the Braitenberg vehicles, the relative equilibria of the BZ reaction can be attracting for certain parameter values.

The BZ reaction can also undergo a Hopf-like bifurcation as the parameters are varied. The relative equilibria cease to be attracting and relative periodic orbits appear. The spiral wave then performs a secondary oscillation in which the shape of the spiral wave varies in a periodic manner as it propagates and the spiral wave tip traces out a path that resembles an epicyclic orbit (see figure 8.4). The shape of the meandering path varies with the parameters of the BZ reaction.

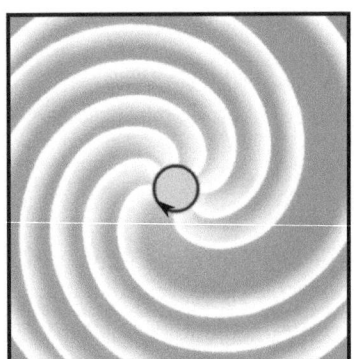 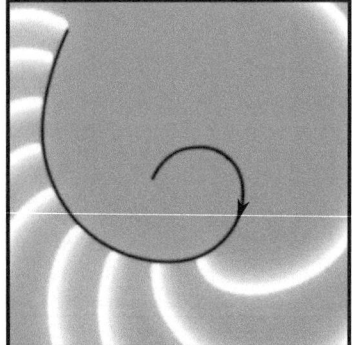

Figure 8.4
(left) A single spiral wave at seven moments in time superimposed in a single image. The spiral wave rotates clockwise and its tip traces out the black circle. (right) A single meandering spiral wave at nine moments in time superimposed in a single image. The spiral wave propagates in an overall clockwise direction, and its tip traces out the black path resembling an epicyclic orbit. Illustrations adapted from figure 1 in Kheowan et al. (2001), where the sides of the panels are roughly 6 mm in length.

Not only do the meandering paths of the spiral wave tips resemble epicyclic orbits, they also have the same types of symmetry as epicyclic orbits. As explained in section 6.8, the loops traced out by a meandering spiral wave tip have been called "petals" because of the way they resemble flowers (Zykov 1986; Winfree 1991). The loops for epicyclic orbits and wave tip paths are arranged radially around a central point, and the angle between neighboring loops, as measured from the central point, has been called the inter-petal angle for the meandering path. We explained in section 6.8 that when the inter-petal angle in degrees is rational it can be obtained from the total curvature for any periodic arc of the meandering path, that is, from $\check{\kappa}$, by dividing $360°$ by the denominator of $\check{\kappa}$ in its reduced form.

The value of $\check{\kappa}$ varies with the parameters of the BZ reaction. A subset of the parameter space that yields the same value for $\check{\kappa}$ is called an isogonal contour or an isogonal curve when the parameter space is two-dimensional. The portion of a two-dimensional parameter space where the BZ reaction undergoes spiral wave meander is partitioned into isogonal curves. The partition of a parameter plane into isogonal curves has been called a "flower garden" (Zykov 1986; Winfree 1991), each isogonal curve being like a row in a garden. Each flower in a row has the same number of petals, but the flowers are different sizes.

In section 5.6, we explained how we have adapted mathematical techniques developed in (Barkley 1994) for the analysis of spiral wave meander in the BZ reaction to the analysis of Braitenberg vehicles. So it's not surprising that the numerically computed isogonal curves in the avoider-avoider quadrant of W_{same} appear to be a warped image of the isogonal curves computed in a region of the parameter space of the Barkley model for the BZ reaction (Barkley 1994; Hotton 2016). The left panels of figure 6.10 show how the inwardly directed loops of the epitrochoid-like paths contract from winding around the center of symmetry, to passing through the center of symmetry, to winding outside the center of symmetry. We called this a rhodonea transition in section 6.8 (Hotton 2016). Section 6.8 lists several studies in which this transition is observed for the paths traced out by meandering spiral tips of the BZ reaction.

8.4 Actin-Based Motility

It may seem initially that the BZ reaction has little to do with the movement of individual agents. However, since the surface tension of the medium depends on the chemical concentrations in it, the chemical waves of the BZ reaction can transport small objects (Ichino et al. 2007; Tanaka et al. 2013). In addition to transporting objects, the BZ reaction can also produce vibrations in a gel. The BZ reaction has been modified to work with gels that oscillate synchronously with the chemical reaction (Maeda et al. 2008; Chen et al. 2011;

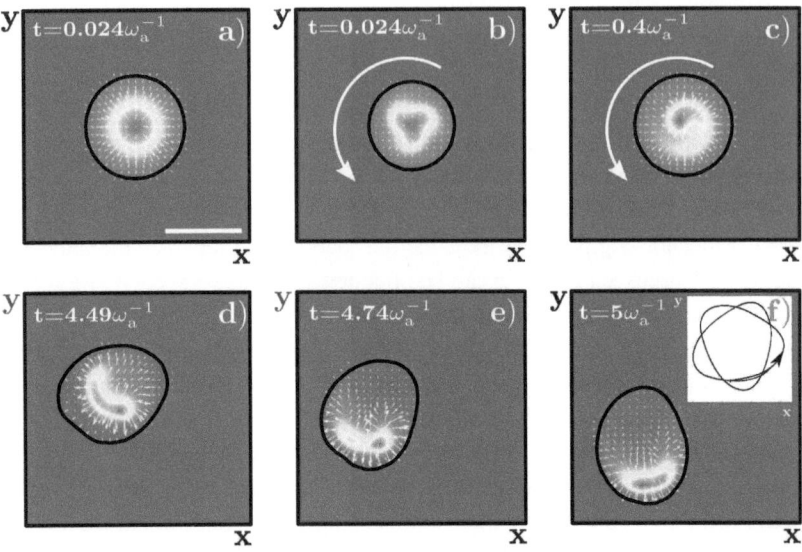

Figure 8.5
Actin density is indicated by a gray scale. As the density goes from low to high, the shading goes from dark to light and back to dark again. The thick black curves stand for the cell's boundary. The density is low outside of the cell and increases to local maxima inside the cell. The small white arrows indicate the direction of growth for the actin filaments, and the large white arrows indicate the rotation of an entire cell. Shown are: (a) steady state, (b) and (c) rotating circular cells, (d)–(f) a migrating asymmetric cell at different moments in time. The thin black curve inside the white inset in panel (f) shows the physical path of the cell's center of mass. It resembles a hypotrochoid with $\check{\kappa} \approx 2/5$ (adapted from figure 7 in Dreher, Aranson, and Kruse [2014]. Also see Ziebert and Aranson [2016]).

Levin, Deegan, and Sharon 2020; Ren et al. 2020). These are examples of chemo-mechanical transduction, where chemical energy is used to generate a mechanical force.

Chemo-mechanical transduction mechanisms in the cytoskeletons of cells are important for many forms of cell transport. For example, they are important in amoeboidal movement, where cells move by the extension and retraction of locomotory structures such as pseudopodia. The formation of pseudopodia involves the polymerization of cytoskeletal proteins such as actin into fila-mentous aggregates. Individual actin filaments grow by the addition of actin monomers at one end and shrink by the removal of the monomers at the other end, a process known as "treadmilling." Each treadmilling filament creates a small mechanical force, and the combined effect of many treadmilling actin filaments generates propulsive forces on the cell. These are responsible for amoeboidal motion and other forms of motion (Weiner et al. 2007; Doubrovinski and Kruse 2008; Aranson 2016). To produce these propulsive forces, the actin filaments self-organize in various ways, and there is evidence that actin filaments can form into spiral waves inside of cells (Vicker 2000; Whitelam, Bretschneider, and Burroughs 2009).

In Dreher, Aranson, and Kruse [2014] a set of partial differential equations is used to model actin polymerization and amoeboidal movement. The model produces spiral actin waves inside an idealized cell, which generate propulsive forces that transport the cell. For the initial parameter values, the system tends towards a steady state in which the cell assumes a circularly symmetric form whose location remains fixed (see figure 8.5). As the parameters are varied, a spiral wave can form within the cell that rotates while the cell's shape and location are largely unchanged. As the parameter values are further varied, the motion of the spiral wave becomes more complicated and the cell's shape deforms, sometimes in a periodic fashion. The periodic behavior of the actin wave inside the cell leads to the cell's center of mass moving along a path with periodically varying curvature.

The dynamics of this spatially extended system bears a resemblance to the BZ reaction. In this actin polymerization model, and in the BZ reaction, chemical concentrations take the form of spiral waves. The regions of high concentration have spiral shapes. For some parameter values, the regions of high concentration rotate like a rigid body even though they are not solid. For other parameter values, the shape of the spiral regions can vary periodically. The spiral waves can cause the cell to move along paths that resemble epicyclic orbits. It is likely that the transition from rotation to meandering in this model involves a Hopf-like bifurcation.

In this example of amoeboidal movement, force was generated by the collective treadmilling of actin *inside* of an agent. The actin rotates and moves the agent from within. However, actin polymerization can also operate *outside* the agent, pushing it around. Various microorganisms, such as the bacteria *Listeria monocytogenes* , can co-opt a host cell's actin polymerization mechanism to move in this way. When *L. monocytogenes* enters a cell it propels itself around the host's cytoplasm using the host's actin molecules (Tilney and Portnoy 1989). This allows the bacteria to migrate through their host while evading the host's immune system. Individual *L. monocytogenes* bacteria can travel in long meandering paths when confined within a thin layer of cytoplasmic extract (Shenoy et al. 2007).

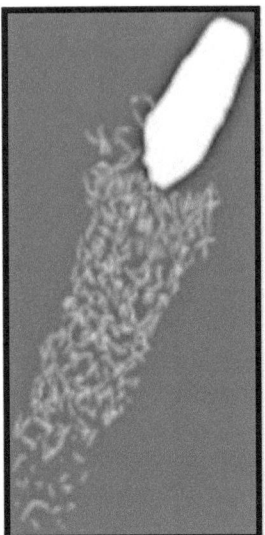 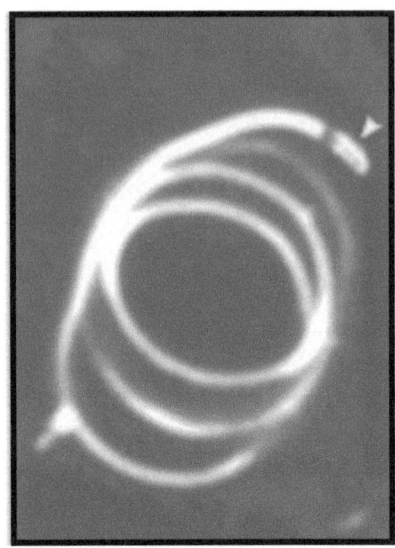

Figure 8.6
(left) Actin filaments trailing behind an *L. monocytogenes* bacterium in the upper right (adapted from figure 2 in Gouin, Welch, and Cossart 2005). (right) An actin comet tail left behind the *L. monocytogenes* bacterium shown next to the arrow in the upper right. The comet tail shows the epicyclic-like path traced by the bacterium (adapted from the youtube video "Exploring 'New Microbiology' with *Listeria monocytogenes*, part 2," Pascale Cossart).

This behavior can be reproduced by coating microscopic beads with compounds that catalyze the polymerization of actin and placing them in dishes containing actin-rich cytoplasmic extract (see figure 8.6). The filamentous aggregate left behind an individual bacterium, called a "comet tail," provides a record of the physical path it traversed (Rutenberg and Grant 2001; Soo and Theriot 2005). These paths sometimes resemble epicyclic orbits, as can be seen in the figure. A dynamical systems analysis of *L. monocytogenes* motility can be found in (Hotton 2010).

8.5 Feedback Mechanisms and Locomotion: The Cybernetic Thread

Another long-running research thread relevant to the study of Braitenberg vehicles is the use of feedback in the control of mechanical devices. An early example of feedback control was the Watt governor, which regulated the speed of steam engines to provide a consistent power supply (Farey 1827; Van Gelder 1995). When the engine ran too slow a valve would let more steam into the engine, and when it ran too fast the valve lowered the amount of steam. In the mid-nineteenth century, the physicist James Clark Maxwell developed a dynamical theory of governors, such as the Watt governor, for engineers and mathematicians (Maxwell 1868).

In the mid-twentieth century, the mathematical biologist Lotka described an example of feedback control in a mechanical device that he thought was relevant for living systems (Lotka 1956). This device was a small wind-up toy designed to roll across a table. The toy was moderately more complicated, as a mechanical device, than the spinning tops studied by Klein. The toy had a small bump that propped up its front end. When it reached the edge of the table, the bump went down allowing a third wheel to make contact with the table's surface and turn the device away from the edge. This feedback mechanism allowed the toy to run around the table without falling off the edge, unlike a spinning top.

Also in the mid-twentieth century, the cyberneticists Walter and Wiener each built phototropic robots inspired by animal behavior, which they named the "Tortoise" and "Moth," respectively. These robots had photo cells as part of an electromechanical feedback mechanism that could drive them toward or away from light sources (Walter 1950; Wiener 1954). These devices show hints of the meandering behavior we observe in pairs of Braitenberg vehicles. Walter described this behavior as "cycloidal" (see figure 8.7).

More recently, Long et al. presented physical and mathematical models of ascidian larval motility and proposed that their models be used in cognitive science as an alternative to Braitenberg vehicles (Long et al. 2004). Ascidian larva provide us with an example of a multicellular organism with a simple method of locomotion that follows epicyclic-like paths. Ascidians are primitive animals whose larva resemble small tadpoles. Each individual has a single eyespot and a simple nervous system. They are propelled by a tail whose orientation relative to its head varies in response to the amount of light the eyespot receives (McHenry 2001). When confined to a thin layer, ascidian larva can move in a complicated fashion in the presence of a small light source (Long et al. 2004).

The advantage to cognitive science that Long *et al.* see in their models is that the agents only have one sensor instead of two. Conversely, ascidian larva could potentially be modeled with Braitenberg vehicles by clamping the activation for one of the vehicle's sensors at an appropriate value. Their model differs from ours in a few important respects. First, their model has a single agent that detects the intensity of a spatially varying light field at the location of the sensor, instead of two agents that detect each other. Second, the form of the field is not always radial, so their agents cannot always be interpreted as pursuing or avoiding an object. Long et al. use the same variable names for the

Figure 8.7
Two of Walter's tortoises in the same box. They each have a light bulb in the front and a light sensor near the top. They are attracted to each other's light and engage in "mutual cycloidal" motion. The zigzag portion in the middle of the paths results from the two vehicles bumping into each other (drawing adapted from Walter [1950]).

states in their mathematical model (or "simulated robot") as we have here for Braitenberg vehicles. The equation of motion for the simulated robots is:

$$(\dot{x}, \dot{y}) = U(\cos(\theta), \sin(\theta)), \qquad \dot{\theta} = \omega,$$

which resembles the two halves of equation (4.14) that give the rate of change for the position and heading of the Braitenberg vehicles. Their parameter U is the fixed speed of the model larva and is essentially the same as our parameter v_n. Their function ω depends on the agent's state (x, y, θ) and the light field. Their ω contains four parameters, one of which, δ, is the angular displacement between the sensor and the agent's heading, much like our parameter ψ.

A detailed mathematical analysis, like the one performed for Braitenberg vehicles, has not yet been done for the ascidian larva model. However, the numerical analysis summarized in (Long et al. 2004) shows the simulated robots in a radial field engaging in behaviors similar to those we have presented here. Their agent pursues the field's center when $\delta = 0°$, but because U

is a fixed positive number the agent cannot simply stop at the center and it ends up revolving along a small circle about the center. This behavior is very much like Braitenberg vehicles when their state is in a revolving type relative equilibrium. When $\delta = 90°$, their agent's paths converge to epicyclic-like meandering paths and it's likely that the simulated robot undergoes a Hopf-like bifurcation as δ is varied between $0°$ and $90°$.

Pursuits between pairs of animals have been extensively studied, for example territorial pursuit between pairs of dragonflies. Typically one dragonfly chases another when it flies into the first dragonfly's territory. Members of these pairs have been referred to as "pursuers" and "evaders" (Lohmann, Corcoran, and Hedrick 2019), analogously to pursuer and avoider Braitenberg vehicles. Lohmann *et al.* observed that the pursuer frequently overshoots the evader and they proposed that this overshooting is an adaptive feature of these pursuits.

As shown in figure 8.8, the dragonflies turned left and right during the pursuit. In other words, the local curvature of their paths oscillated about zero. Because of the overshooting, the amplitude of the oscillation was greater for the pursuer than for the evader. On the other hand the total curvature, that is, the total amount the dragonflies turned, was about the same for both paths. This resulted in the path for the pursuer being longer than the path for the evader, so the pursuer needed to fly faster to keep up with the evader.

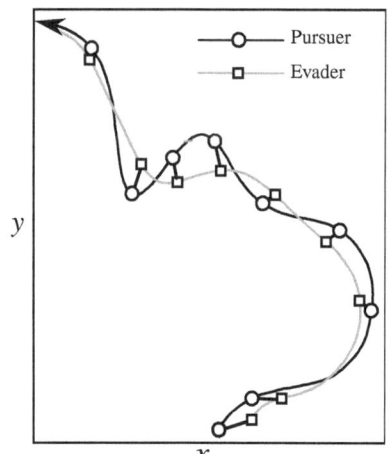

Figure 8.8
The paths traced by a pair of dragonflies in a territorial pursuit. The dragonflies traveled from bottom to top. The pursuer repeatedly overshoots the evader. It can be seen that the pursuer made tighter turns than the evader but that on average they turned by about the same amount. Also the path for the pursuer is longer than that for the evader. The simultaneous positions of the dragonflies are adjacent or connected by a small line (adapted from figure 2 in Lohmann, Corcoran, and Hedrick 2019).

Pursuer-avoider pairs of Braitenberg vehicles can behave like this when the pursuer moves slightly faster than the avoider. In the example shown in figure 2.10, the pursuer and avoider both turn left and right with the purser repeatedly overshooting the avoider. The local curvature of both paths varied in a periodic manner. The amplitude in the oscillations for the local curvature was much larger for the pursuer, but the total curvature for the periodic arcs, that is, the value for $\breve{\kappa}$, was nearly 0. Both vehicles moved overall in the same direction even though the pursuer had to repeatedly "correct" for overshooting the avoider.

Another pertinent example is provided by a model of tiger beetle pursuit (Haselsteiner, Gilbert, and Wang 2014). These small beetles prey on smaller insects. Tiger beetles have a compound eye on each side of their head, and the visual field can differ considerably between them. Although the beetles have six legs, the front and back legs on one side act in unison with the middle leg on the other side. The beetles move by alternatively standing on one triple of legs while bringing the other triple of legs forward. This method of walking is called a "tripod gait" and is used by many insects and robots (Cruse et al. 2009).

The pair of compound eyes detect prey and exercise control over the beetle's tripod gait through its nervous system, so that the beetle can catch its prey. This is comparable to a Braitenberg vehicle: the beetle's two compound eyes correspond to the Braitenberg vehicle's two sensors, and the beetle's two triples of legs correspond to the two wheels in Braitenberg's original conception.[3] The heading of the beetle can be varied by differentially changing the durations of the strides on either side (approximately the amount of time a forward-moving triple of legs is in the air) (Cruse et al. 2009) and by differentially altering the force exerted by the legs (Noest and Wang 2017). This corresponds to changing the speed of the motors driving the wheels in Braitenberg's original conception.

A rigid body resting on three or more points of a planar surface is in a stable static equilibrium, that is, it will not fall over when perturbed slightly. By contrast, a body balancing on one or two points is an unstable static equilibrium. Thus, the body will fall over when perturbed slightly. Consequently, tripod gait places less of a burden on an agent to distribute its weight across its legs than for agents engaging in tetrapod or bipedal gaits. The reduced need for proprioceptive feedback allows for a simpler neural network to control the pursuit (Cruse et al. 2009).

---

3. See Golubitsky and Stewart (2003) on the use of symmetry in designing neural networks to generate various types of animal gaits.

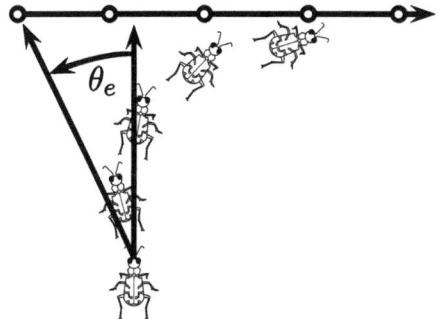

Figure 8.9
Tiger beetle pursuing a prey dummy based on the prey dummy's angular position, θ_e. The prey dummy moves horizontally from left to right. Initially the beetle turns to the left, but as the prey dummy passes by, the tiger beetle turns to the right (adapted from figure 1 in Haselsteiner, Gilbert, and Wang [2014]).

Haselsteiner, Gilbert, and Wang (2014) show that tiger beetles pursue prey with an angular velocity, ω_b, that is nearly proportional to the angular displacement, θ_e, between the beetle's present heading and the prey's position about 0.028 seconds in the past (see figure 8.9).[4] Since in the model ω_b depends on the angular position of the prey, not on how far away it is, the level curves of ω_b, as a function of the prey's position, are half-lines radiating out from the vertical line through the beetle's center. The level curves for several angular displacements, θ_e, are shown in the left panel of figure 8.10. The graph of ω_b in front of the beetle is a region of a surface known as a "helicoid." When the prey is directly in front of the beetle ($\theta_e = 0°$), the beetle does not turn, but as the prey moves to the left or right, the beetle turns in the same direction—more quickly, the further to the right or left the prey is—thereby producing pursuit behavior.

In figure 8.10, it can be seen that the helicoid turning function for the beetle (left panel) and the turning function for a Braitenberg vehicle (right panel) resemble each other. The kinematic rule for the tiger beetle can be well approximated in our model of Braitenberg vehicles for angular displacements less than 90°. Recall from section 2.2 that the central lune for a Braitenberg vehicle is the intersection of the two sensors' fields of view, that is, $\left(D_{(\ell,n)} \cap D_{(r,n)}\right)$. When the sensor weights of a Braitenberg vehicle are equal, the level curves of the turning function, $\bar{\omega}_n$, for $(X_n, Y_n)^T$ in the central lune are arcs of hyperbolas. The straight-line asymptotes of these arcs radiate from the vertical line through the midpoint $(0, N, 0)^T$ between the two sensors. Several level curves for $\bar{\omega}_n$ are shown in the right panels of figures 2.7 and 8.10. As can be seen in

4. See Collett and Land (1975) for similar results with hoverflies and Pequeño-Zurro, Shaikh, and Rañó (2019) for a version of Braitenberg vehicles based on the rate of change in the sensors' activations.

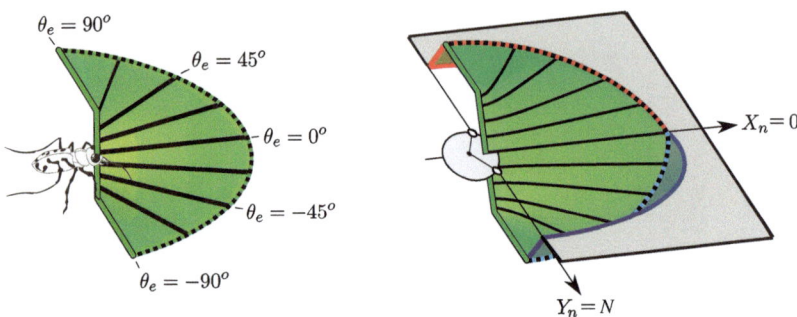

Figure 8.10
(left) The graph of the turning rate, ω_b, in the tiger beetle model, as a function of the prey's position in front of the beetle, is a region of a helicoid. Level curves are shown as black half lines. The beetle turns more quickly towards its prey when it is close to $90°$ or $–90°$. (right) The graph of the turning rate, $\bar{\omega}_n$, for a Braitenberg pursuer (with equal sensor weights) as a function of the other vehicle's position $(X_n, Y_n)^T$. The portion of the graph for $(X_n, Y_n)^T$ in the front half of the central lune is shaded green. This green surface, away from $(0, 0)^T$, approximates the helicoid region on the left.

the figures, these curves are close to straight, especially as we move away from the vertical line through $(0, N, 0)^T$. Thus, they are a good approximation for the helicoid of the tiger beetle model, which has perfectly straight level curves.

Recall from section 4.2 that states for vehicle n can be interpreted as representations of the position of the other vehicle $\neg n$. In our discussion of representational indeterminacy, we noted that a simpler representational scheme could have been used, with a one-dimensional agent space, in which agents only represent the relative angular position of one another. That is comparable to the framework Haselsteiner et al. use. Though the authors do not make use of the concept of a representation in their discussion, the internal states for their model of a tiger beetle could be interpreted as representations for the relative position of its prey. The resemblance between the turning functions for the Braitenberg vehicles and the tiger beetles also supports the idea that tiger beetle pursuits could be guided by representational processes similar to those performed by the Braitenberg vehicles, that is, guided not just by a representation of relative angular offset but also by a representation of the relative distance of the prey.

Epilogue

There are many directions for further research with this system, including additional study of the two-vehicle system, and ways of elaborating and extending it. There is, to begin with, more work to do studying this system. Within the four-dimensional weight space W_{total}, we have only explored a little bit beyond the union of the two-dimensional subspaces W_{same} and W_{rev}, what we described as a "paddle wheel" in section 2.5. Other bifurcations and additional forms of behavior occur outside of this region. Of particular interest is another two-dimensional subspace, $W_{opp} + W_{eq}$ (shown in figure 2.10), which is the set of all quadruples of weights in which both agents have bilaterally symmetric weights (for example, a pursuer with weights $(1, 1)$ and an avoider with weights $(-0.5, -0.5)$).

We discussed bilaterally symmetric vehicles in connection with W_{eq}. For bilaterally symmetric vehicles, in a single-file configuration, the value of the turning functions, $\bar{\omega}_n$, will be zero and the vehicles will continue to move in a single-file fashion. However, this motion may or may not be stable. We saw in sections 6.8, 6.10, 7.6, and 7.7 that within W_{eq}, when the two agents are both pursuers, one of them will "follow" the other towards a stable single-file configuration, but when they are both avoiders (within W_{eq}), this configuration is unstable and a small perturbation can make them stop following each other. Consider a curve that begins in the pursuer-pursuer half of W_{eq}, which leaves W_{eq}, enters $W_{opp} + W_{eq}$, and connects to a point in the avoider-avoider half of W_{eq}. Somewhere along this curve, a bifurcation must occur that changes the stability of the single-file configurations. Possibly there are other bifurcations that occur in $W_{opp} + W_{eq}$ as well.

The study of $W_{opp} + W_{eq}$ could be facilitated by using the total curvature of the periodic arcs, $\check{\kappa}$, in the physical paths (section 6.8). For weights directly on W_{eq}, the vehicles move in straight lines and their curvature is 0 everywhere. Within the pursuer-pursuer quadrant of W_{same}, the value of $\check{\kappa}$ for the physical paths approaches $1/2$ from one side of W_{eq} and $-1/2$ from the other side (i.e., the "half turn case" where each vehicle exits a periodic arc 180 degrees from the direction in which it entered, as shown in figure 6.17). In section 2.5, we saw an example of a pursuer-avoider pair with weights in $W_{opp} + W_{eq}$. In that example, the value of $\check{\kappa}$ was numerically close to 0 for their physical paths (i.e., the "no turn case" where each vehicle exits a periodic arc in the same direction as it entered, as shown in the right panel of figure 2.10). For weights very close to $W_{opp} + W_{eq}$, the value of $\check{\kappa}$ could take on values between $-1/2$ and $1/2$ (i.e. the "slight turn case" where each vehicle exits a periodic arc in almost the same direction as it entered). There could also be sign changes in $\check{\kappa}$ near $W_{opp} + W_{eq}$ as the weights move between the pursuer-purser and pursuer-avoider orthants. How the value of $\check{\kappa}$ varies could be insightful for understanding the bifurcations that occur near and inside of $W_{opp} + W_{eq}$. This in turn could be useful for understanding the turns taken by bilaterally symmetric animals in pursuit.

Another obvious direction for future research would be to consider more than two Braitenberg vehicles. The situation is comparable to work in celestial mechanics on systems with a finite number of point masses, that is, the "n-body problem." When $n = 2$, the situation is tractable. Newton found exact solutions for two bodies, that is, the "Kepler orbits" in which the paths of the two point masses are conic sections (Newton 1846). But when $n > 2$, the dynamics can be chaotic (which in this context means nonintegrable), making computation of the physical paths of the point masses difficult. The situation may be similar with Braitenberg vehicles. We have not observed chaos in two vehicles, although the system does have long transients, which can give the appearance of chaos. Given the analogy with celestial mechanics, it is natural to suppose that adding one more vehicle would result in chaotic behavior.

Swarm-like behaviors are possible with larger collections of Braitenberg vehicles, as has been observed in numerous computational and physical simulations (Mamduh et al. 2013; Dvoretskii et al. 2020; Kasmarik et al. 2020). Such systems can be mathematically intractable, making it necessary to rely more on numerical and empirical methods. Still, such studies could be informed by our detailed studies with two vehicles. In particular we showed in section 6.7 how an agential perspective could help understand the complicated dynamics of the total dynamical system for the two vehicles. For instance, when the agents' states were fixed, the agents revolved around each other, and when their states varied periodically, the agents were meandering around each other. An agential perspective could also help to understand emergent phenomena in multi-agent systems. Different types of paths in the agents' state spaces could correspond to different types of collective movement by a large number of vehicles. So the open dynamics of the agents' internal states could be a guide to how a swarm of Braitenberg vehicles behaves.

The agents and their environment could be changed in other ways, facilitating additional lines of study. Currently, the sensors are connected directly to the effectors. Inter-neurons could be added to support more complex behaviors. Variables representing degree of arousal and motivation could be introduced. The weights of the network could be updated using a learning rule. A natural way to study learning in these vehicles would be to consider how agents *learn* to pursue or avoid each other. The body of the agents could be changed. A third sensor could be added to a vehicle that would determine how it turns as it travels through three-dimensional space. The environment could be enriched. Features could also be added to the environment such as stationary objects, geographic features, and depletable resources. Vehicles could be evolved using genetic algorithms. These elaborations could be used to design real vehicles for use in physical environments, as has already been done to good effect (Lilienthal and Duckett 2004; Stolkin, Sheryll, and Hotaling 2007; Cruse et al. 2009; Rañó, Khamassi, and Wong-Lin 2021).

Another line of additional research is more philosophical, unpacking in greater detail the implications this kind of system has for debates in the philosophy of mind concerning the body and environment in cognition. The study of embodied cognition remains active, sometimes under the moniker of "4E Cognition" (the four "E's" correspond to embodied, embedded, enacted, and extended cognition). Two recent and extensive edited volumes on the topic are *The Oxford Handbook of 4E Cognition* (Newen, De Bruin, and Gallagher 2018), and *Embodiment, Enaction, and Culture* (Durt, Fuchs, and C. 2017). As of this writing (December 2022) there are more than 2,000 combined entries in the categories of "Extended Cognition" and "Embodiment and Situated Cognition" in the philpapers database (https://philpapers.org), with hundreds of new articles appearing since 2020. The approach is being applied to such topics as creativity, social cognition, personal identity, and love even as the old debates continue to play out between internalism and externalism (for example, see Aizawa [2018] in the Oxford anthology).

We believe our framework provides for a more balanced approach to these issues, allowing the insights of both sides of the debate to be recognized and applied in a common framework, and providing concrete tools for analyzing representational dynamics in environmentally coupled systems.

A Solutions to the Approximate Bifurcation Equation

In this appendix, we determine the precise conditions on the sensor weights (w_ℓ, w_r) in W_{same} for the existence of solutions to equation (6.3). This provides us with approximate conditions for the existence of solutions to the bifurcation equation (6.1).

We begin with some general considerations. $X_n = 0$ cannot be a solution to equation (6.3) since X_n appears in the denominator on the right-hand side. The regions $\mathcal{R}_r$, $\mathcal{R}_\ell$ are mirror symmetric about the diagonal $w_\ell = w_r$, so we mainly focus on $\mathcal{R}_r$ and right-turning revolving type relative equilibria. Right-turning revolving type relative equilibria correspond to positive solutions for X_n in equation (6.1). The positive X_n-axis is subdivided into four sub-intervals based on the range of an agent's two sensors (see figure 6.4). The interval farthest to the right is $[P + M, \infty)$, where the other agent is out of view of both sensors. So there is no solution with $X_n \in [P + M, \infty)$. We focus on the three remaining sub-intervals:

1. The interval $I_{(r,1)} = (0, M]$, where the other vehicle is between the vehicle's sensors.
2. The interval $I_{(r,2)} = [M, P - M]$, where the other vehicle is to the right of both sensors and still in view of both of them.
3. The interval $I_{(r,3)} = [P - M, P + M)$, where the other agent is only in view of the right sensor.

The corresponding subintervals for the negative X_n-axis are $I_{(\ell,j)} = -I_{(r,j)}$ for $j = 1, 2, 3$.

The function $\bar{\omega}(X_n, N)^T$ on the left-hand side of equation (6.3) is the restriction of the function $\bar{\omega}(X_n, Y_n)^T$ [defined in equation (2.8)] to the line through the sensors. This function is piecewise linear (examples are shown in figures 2.6 and 6.4). Each line segment in the graph is contained by a line that is the graph of a linear polynomial. The coefficients of these linear polynomials each have P in the denominator of a fraction. The linear polynomials can be simply expressed by letting P be a common denominator, putting each linear polynomial in the form $(AX_n + B)/P$ where A, B are functions of the vehicles' parameters. We obtain expressions for A and B for the line segments in the graph of $\bar{\omega}(X_n, N)^T$. The results are summarized in table A.1.

When $(w_\ell, w_r) = (0, 0)$, the graph of $\bar{\omega}(X_n, N)^T$ is identical to the X_n-axis and does not intersect the graph of $-2v/X_n$. As the weights move away from $(0, 0)$, there are two main ways the piecewise linear graph of $\bar{\omega}(X_n, N)^T$ can begin to intersect the hyperbolic graph of $-2v/X_n$:

	A	B
$I_{(\ell,3)}$	w_ℓ	$(P+M)w_\ell$
$I_{(\ell,2)}$	$w_\ell - w_r$	$(P+M)w_\ell - (P-M)w_r$
$I_{(\ell,1)} \cup I_{(r,1)}$	$-w_\ell - w_r$	$(P-M)w_\ell - (P-M)w_r$
$I_{(r,2)}$	$-w_\ell + w_r$	$(P-M)w_\ell - (P+M)w_r$
$I_{(r,3)}$	w_r	$-(P+M)w_r$

Table A.1
The values of A and B in the linear polynomial $(AX_n + B)/P$ in each interval.

1. **Tangentially:** A line segment in the graph of $\bar{\omega}(X_n, N)^T$ can become tangent to the graph of $-2v/X_n$ at a single point in the line segment's interior before any other point in the graph of $\bar{\omega}(X_n, N)^T$ reaches the graph of $-2v/X_n$.
2. **Non-tangentially:** A vertex in the graph of $\bar{\omega}(X_n, N)^T$ can reach the graph of $-2v/X_n$ before any other point in the graph of $\bar{\omega}(X_n, N)^T$ reaches the graph of $-2v/X_n$.

Since the branches of the hyperbolic graph of $-2v/X_n$ are strictly increasing functions, a necessary condition for a vertex to be an initial contact point is for the slope of the line segment to the left of the vertex to be no more than the slope of the tangent to the graph of $-2v/X_n$ at the contact point and the slope of the line segment to the right of the vertex to be no less than the slope of tangency.

The tangential and non-tangential initial contacts have common limiting cases that occur when a line segment on either side of a vertex in the graph of $\bar{\omega}(X_n, N)^T$ is tangent to the graph of $-2v/X_n$ at the contact point.

We will see that tangential intersections correspond to parabolic arcs in W_{same}, non-tangential intersections correspond to line segments in W_{same}, and the common limiting cases correspond to a common endpoint of a parabolic arc and line segment. Moreover, the parabolic arcs and line segments meet tangentially in W_{same}.

We obtain formulas for what A and B have to be so that a given value of X_n is the horizontal coordinate for the initial contact point. From table A.1 we see that A and B are invertible linear functions of w_l, w_r when X_n is in $I_{(r,1)} \cup I_{(r,2)}$. This allows us to obtain formulas for (w_l, w_r) so that $X_n \in I_{(r,1)} \cup I_{(r,2)}$ is the horizontal coordinate of the initial contact point. We will also show that the requirement $P > 3M$ prevents the graphs of the functions $\bar{\omega}(X_n, N)^T$ and $-2v/X_n$ from meeting tangentially for any $X_n \in I_{(r,3)}$. This is the reason for $3M$ in equation (2.2).

The condition that a line through a line segment in the graph of $\bar{\omega}(X_n, N)^T$ intersects the graph of $-2v/X_n$ is equivalent to a polynomial equation of degree two in X_n:

$$\frac{AX_n + B}{P} = -\frac{2v}{X_n} \quad \Longleftrightarrow \quad AX_n^2 + BX_n + 2vP = 0 \qquad (A.1)$$

since neither P nor X_n can be zero.

For a tangential intersection, we need the functions $\bar{\omega}(X_n, N)^T$ and $-2v/X_n$ to have the same derivative at the initial contact point. The derivative of $(AX_n + B)/P$ is A/P and the derivative of $-2v/X_n$ is $2v/X_n^2$. So the condition that the derivatives of the two functions are equal at X_n is also equivalent to a polynomial equation of degree two in X_n:

$$\frac{A}{P} = \frac{2v}{X_n^2} \iff A = 2vP/X_n^2 \iff AX_n^2 = 2vP.$$

This gives us the solution for A, and by substituting $AX_n^2 = 2vP$ into (A.1) we can express B in terms of X_n.

$$2vP + BX_n + 2vP = 0 \iff B = -\frac{4vP}{X_n}.$$

Thus A and B in terms of X_n is:

$$(A, B) = \frac{2vP}{X_n^2}(1, -2X_n). \tag{A.2}$$

Eliminating X_n from (A.2) gives us the discriminant of the quadratic polynomial in equation (A.1) which is:

$$B^2 - 8vPA = 0.$$

Thus, the discriminant locus is a parabola in the (A, B)-parameter plane. The image of a parabola under a linear map is a parabola. Since (w_ℓ, w_r) is the image of (A, B), under a linear map the condition for a tangential initial contact determines a parabola in W_{same}.

For a nontangential intersection, we let (A_-, B_-) be the value of (A, B) for the line segment in the graph of $\bar{\omega}(X_n, N)^T$ to the left of the vertex that meets the graph of $-2v/X_n$ and (A_+, B_+) be the value of (A, B) for the line segment to the right of the vertex.

The condition for a vertex to intersect the graph of $-2v/X_n$ comes from substituting the X_n coordinate of the vertex for the graph of $\bar{\omega}(X_n, N)^T$ into equation (A.1) and either (A_-, B_-) or (A_+, B_+) for (A, B). Since $\bar{\omega}(X_n, N)^T$ is continuous, the result will be the same equation for a line in W_{same}.

$$A_- X_n^2 + B_- X_n + 2vP = 0 \iff A_+ X_n^2 + B_+ X_n + 2vP = 0. \tag{A.3}$$

The slope of the tangent to the graph of $-2v/X_n$ is independent of the weights. The condition that the slope of the line segment to the left of the vertex is no more than the slope of tangency, and the slope of the line segment to the right of the vertex is no less than the slope of tangency is

$$\frac{A_-}{P} \le \frac{2v}{X_n^2} \le \frac{A_+}{P} \iff A_- X_n^2 \le 2vP \le A_+ X_n^2. \tag{A.4}$$

This determines a sector in W_{same}. The two edges of the sectors correspond to equality in (A.4). We will see that for the vertex at $X_n = M$, the intersection of the line in equation (A.3) with the sector in (A.4) is a finite line segment. For the vertex at $X_n = P - M$, the intersection of the line in equation (A.3) with the sector in equality (A.4) is a half-line.

The curve C_r in W_{same}

We proceed now from general considerations to specific cases of solutions to equation (6.3). We will successively consider solutions for $X_n \in I_{(r,1)}$, $X_n = M$, $X_n \in I_{(r,2)}$, $X_n = P - M$, and $X_n \in I_{(r,3)}$.

Case $X_n \in I_{(r,1)}$:

From table A.1 the linear function, and its inverse, for (A, B) in terms of (w_ℓ, w_r) on the interval $I_{(r,1)}$ is:

$$(A, B) = (w_\ell, w_r) \begin{pmatrix} -1 & P-M \\ -1 & -(P-M) \end{pmatrix}, \quad (w_\ell, w_r) = \frac{(A, B)}{2(P-M)} \begin{pmatrix} M-P & M-P \\ 1 & -1 \end{pmatrix}.$$

We compose this linear map for (w_ℓ, w_r) in terms of (A, B) with the right side of equation (A.2) and denote the resulting function by $Q_{(r,1)}$.

$$(w_\ell, w_r) = Q_{(r,1)}(X_n) = -\frac{vP}{(P-M)X_n^2} (P-M+2X_n, P-M-2X_n). \tag{A.5}$$

The values of w_ℓ and w_r go to infinity as X_n approaches 0, the left endpoint of $I_{(r,1)}$. The parabolic arc $Q_{(r,1)}(I_{r,1})$ has just the single endpoint.

$$\mathcal{C}_{(r,1)} = Q_{(r,1)}(M) = -\frac{vP}{M^2(P-M)} (P+M, P-3M). \tag{A.6}$$

Since we require $P > 3M$, this point is in the avoider-avoider quadrant of W_{same}.

Case $X_n = M$:

Substituting $X_n = M$ and the values of (A, B) for $I_{(r,1)}$ and $I_{(r,2)}$ from table A.1 into equation (A.3) gives:

$$M(P-2M) w_\ell - MP w_r + 2vP = 0 \tag{A.7}$$

and into (A.4) gives:

$$(-w_l - w_r)M^2 \leq 2vP \leq (-w_l + w_r)M^2. \tag{A.8}$$

The point $\mathcal{C}_{(r,1)}$ satisfies (A.7), brings about equality on the left side of (A.8), and satisfies the inequality on the right side of (A.8). The point

$$\mathcal{C}_{(r,2)} = -\frac{vP}{M^3} (P-M, P-3M) \tag{A.9}$$

satisfies (A.7), satisfies the inequality on the left side of (A.8), and brings about equality on the right side of (A.8). So the intersection of the line with equation (A.7) with the sector in (A.8) is the line segment $\overline{\mathcal{C}_{(r,1)}\mathcal{C}_{(r,2)}}$. It shares a common endpoint with the parabolic arc $Q_{(r,1)}(I_{(r,1)})$. Because $P > 3M$, the line segment $\overline{\mathcal{C}_{(r,1)}\mathcal{C}_{(r,2)}}$ is contained in the avoider-avoider quadrant of W_{same}.

A necessary condition for the initial contact of the graph of $\bar{\omega}(X_n, N)^T$ with the graph of $-2v/X_n$ to happen at the vertex $\bar{\omega}(M, N)$ is for $(w_\ell, w_r) \in \overline{\mathcal{C}_{(r,1)}\mathcal{C}_{(r,2)}}$. This condition is not sufficient, and we will show that in fact the initial contact cannot occur at this vertex.

Case $X_n \in I_{(r,2)}$:

From table A.1, the linear function, and its inverse, for (A, B) in terms of (w_ℓ, w_r) on the interval $I_{(r,2)}$ is

$$(A, B) = (w_\ell, w_r) \begin{pmatrix} -1 & P - M \\ 1 & -(P + M) \end{pmatrix}, \quad (w_\ell, w_r) = \frac{(A, B)}{2M} \begin{pmatrix} -(P + M) & -(P - M) \\ -1 & -1 \end{pmatrix}.$$

We compose this linear map for (w_ℓ, w_r) in terms of (A, B) with the right side of equation (A.2) and denote the resulting function by $Q_{(r,2)}$.

$$(w_\ell, w_r) = Q_{(r,2)}(X_n) = \frac{vP}{MX_n^2} \left(2X_n - (P + M),\ 2X_n - (P - M) \right) \tag{A.10}$$

$Q_{(r,2)}(I_{(r,2)})$ is a parabolic arc in W_{same}. $C_{(r,2)} = Q_{(r,2)}(M)$ is one endpoint and

$$C_{(r,3)} = Q_{(r,2)}(P - M) = \frac{vP}{M(P - M)^2} (P - 3M, P - M) \tag{A.11}$$

is the other endpoint. Since we require $P > 3M$, this point is in the pursuer-pursuer quadrant of W_{same}.

Case $X_n = P - M$:

Substituting $X_n = P - M$ and the values of (A, B) for $I_{(r,1)}$ and $I_{(r,2)}$ from table A.1 into equation (A.3) gives:

$$-2M(P - M)\, w_r + 2vP = 0 \quad \Longleftrightarrow \quad w_r = \frac{vP}{M(P - M)} \tag{A.12}$$

and into (A.4) gives:

$$(-w_l + w_r)(P - M)^2 \leq 2vP \leq w_r(P - M)^2. \tag{A.13}$$

The line for (A.12) is horizontal as is the edge of the sector corresponding to the inequality on the right side of (A.13). So long as $P > 3M$, line (A.12) is above the horizontal edge of the sector. Line (A.12) intersects the sector's other edge at the point $C_{(r,3)}$. So the intersection of line (A.12) with sector (A.13) is a half-line beginning at $C_{(r,3)}$. We let $C_{(r,4)}$ be any point on this line except for $C_{(r,3)}$. We denote the half-line by $\overrightarrow{C_{(r,3)}C_{(r,4)}}$.

A necessary condition for the initial contact of the graph of $\bar{\omega}(X_n, N)^T$ with the graph of $-2v/X_n$ to occur at the vertex $\bar{\omega}(P - M, N)$ is for $(w_\ell, w_r) \in \overrightarrow{C_{(r,3)}C_{(r,4)}}$.

Case $X_n \in I_{(r,3)}$:

From table A.1, we can see that on the interval $I_{(r,3)}$, the linear function for the parameters (A, B) is noninvertible, that is, (A, B) is independent of the value of w_ℓ. Substituting the values for A and B from table A.1 for $X_n \in I_{(r,3)}$ into (A.2) gives:

$$(w_r, -(P + M)w_r) = \frac{2vP}{X_n^2}(1,\ 2X_n) \quad \Longleftrightarrow \quad \begin{cases} w_r = 2vP/X_n^2 & \text{and} \\ w_r = -4vP/((P + M)X_n). \end{cases}$$

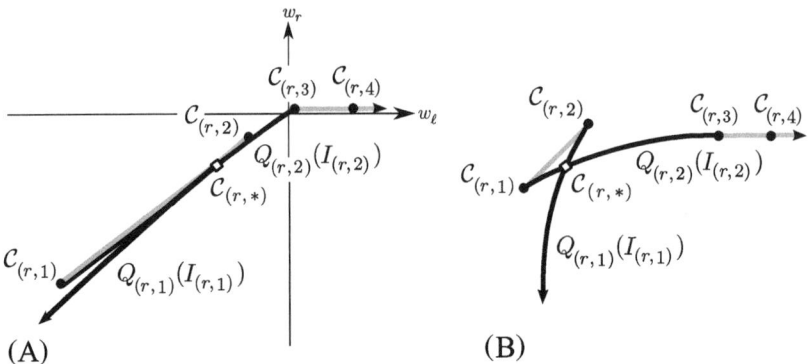

Figure A.1
The curve $\mathcal{C}_r$ in W_{same}. The parts of $\mathcal{C}_r$ can be hard to see because they tend to line up over the top of each other. The parabolic arcs $Q_{(r,1)}(I_{(r,1)})$ and $Q_{(r,2)}(I_{(r,2)})$ are shown in black. The line segment $\overline{\mathcal{C}_{(r,1)}\mathcal{C}_{(r,2)}}$ and half-line $\overrightarrow{\mathcal{C}_{(r,3)}\mathcal{C}_{(r,4)}}$ are shown in gray. The intersection point $\mathcal{C}_{(r,*)} = Q_{(r,1)}(I_{(r,1)}) \cap Q_{(r,2)}(I_{(r,2)})$ is at the white box. (A) A geometrically precise version of $\mathcal{C}_r$ and coordinate axes. The physical parameter values are $P = 8$, $\psi = \pi/2$, and $v = 1/500$. (B) A diffeomorphic transformation of $\mathcal{C}_r$ to make the line segments and parabolic arcs easier to see.

Eliminating w_r gives $X_n = (P + M)/2$. However $(P + M)/2 < P - M$ since we stipulate that $P > 3M$. Consequently, the line segment in the graph of $\bar{\omega}(X_n, N)^T$ over the interval $I_{(r,3)}$ never meets the graph of $-2v/X_n$ tangentially.

In summary, the graph of $\bar{\omega}(X_n, N)^T$ makes an initial contact with the right branch of the graph of $-2v/X_n$ as (w_ℓ, w_r) moves away from $(0, 0)$ when (w_ℓ, w_r) reaches either the parabolic arc $Q_{(r,1)}(I_{(r,1)})$, the line segment $\overline{\mathcal{C}_{(r,1)}\mathcal{C}_{(r,2)}}$, the parabolic arc $Q_{(r,2)}(I_{(r,2)})$, or the horizontal half-line $\overrightarrow{\mathcal{C}_{(r,3)}\mathcal{C}_{(r,4)}}$. We denote the union of these curves by $\mathcal{C}_r$.

$$\mathcal{C}_r = Q_{(r,1)}(I_{(r,1)}) \cup \overline{\mathcal{C}_{(r,1)}\mathcal{C}_{(r,2)}} \cup Q_{(r,2)}(I_{(r,2)}) \cup \overrightarrow{\mathcal{C}_{(r,3)}\mathcal{C}_{(r,4)}}.$$

An example of $\mathcal{C}_r$ for particular physical parameters is shown in figure A.1. The parabolic arc $Q_{(r,1)}(I_{(r,1)})$ goes from infinity to the point $\mathcal{C}_{(r,1)}$ where it meets an end point of $\overline{\mathcal{C}_{(r,1)}\mathcal{C}_{(r,2)}}$. The parabolic arc $Q_{(r,2)}(I_{(r,2)})$ and $\overline{\mathcal{C}_{(r,1)}\mathcal{C}_{(r,2)}}$ have the common end point $\mathcal{C}_{(r,2)}$. The parabolic arc $Q_{(r,2)}(I_{(r,2)})$ and $\overrightarrow{\mathcal{C}_{(r,3)}\mathcal{C}_{(r,4)}}$ have the common endpoint $\mathcal{C}_{(r,3)}$. The line segment $\overrightarrow{\mathcal{C}_{(r,3)}\mathcal{C}_{(r,4)}}$ goes from $\mathcal{C}_{(r,3)}$ to infinity. So $\mathcal{C}_r$ is the image of a continuous curve from $\mathbf{R}$ to W_{same}.

The Cross Point in $\mathcal{C}_r$

As (w_ℓ, w_r) is varied from $(0, 0)$, the first contact between the graph of the piecewise linear function $\bar{\omega}(X_n, N)^T$ and the right branch in the graph of $-2v/X_n$ must occur at some point on $\mathcal{C}_r$. If $\mathcal{C}_r$ did not cross itself then as (w_ℓ, w_r) is varied from $(0, 0)$ the first contact between the graph of $\bar{\omega}(X_n, N)^T$ and the right branch in the graph of $-2v/X_n$ could occur at any point in $\mathcal{C}_r$. However, as we will demonstrate, there is one crossing point in C_r (see figure A.1). So only a subset of C_r corresponds to an initial contact between the graph of $\bar{\omega}(X_n, N)^T$ and the right branch in the graph of $-2v/X_n$.

Clearly each of the parabolic arcs and line segments in C_r have no self-intersection points. So we only need to check if the different parabolic arcs or line segments in C_r intersect one another. There are four parts to C_r so there are $\binom{4}{2} = 6$ cases to check.

The first case we check is the two line segments. The line through $\overline{C_{(r,1)}C_{(r,2)}}$ does intersect the line through $\overrightarrow{C_{(r,3)}C_{(r,4)}}$, but because we stipulated that $P > 3M$, the line segment $\overline{C_{(r,1)}C_{(r,2)}}$ is below the w_r-axis while $\overrightarrow{C_{(r,3)}C_{(r,4)}}$ is above the w_r-axis [compare equations (A.6) and (A.9) with equation (A.12)]. So the line segments $\overline{C_{(r,1)}C_{(r,2)}}$ and $\overrightarrow{C_{(r,3)}C_{(r,4)}}$ do not intersect.

The next case we check is the parabolic arc $Q_{(r,1)}(I_{(r,1)})$ meeting the half-line $\overrightarrow{C_{(r,3)}C_{(r,4)}}$. A necessary condition for them to meet is that they have the same value for w_r. From equation (A.5) and (A.12), we get

$$-\frac{vP}{(P-M)X_n^2}(P-M-2X_n) = \frac{vP}{M(P-M)} \implies X_n^2 - 2MX_n + M(P-M) = 0.$$

The discriminant of this quadratic polynomial is $4M(2M-P)$, which is negative since we stipulated $P > 3M$. So this quadratic polynomial has no real roots, and the parabolic arc $Q_{(r,1)}(I_{(r,2)})$ and the half-line $\overrightarrow{C_{(r,3)}C_{(r,4)}}$ do not intersect.

For the next three cases, we make use of the theorem that when a line is tangent to a parabola the two curves only intersect at the point of tangency.

From equation (A.7), we see that the line segment $\overline{C_{(r,1)}C_{(r,2)}}$ is orthogonal to the vector $M(P-2M, -P)$. We compute a tangent vector to $Q_{(r,1)}(X_n)$ at $X_n = M$ from equation (A.5)

$$\frac{d}{dX_n}\bigg|_{X_n=M} Q_{(r,1)}(X_n) = -\frac{vP}{(P-M)X_n^3}(-2P+2M-2X_n, -2P+2M+2X_n)\big|_{X_n=M}$$

$$= \frac{2vP}{M^3(P-M)}(P, P-2M).$$

Since the dot product of this tangent vector with $M(P-2M, -P)$ is zero, the line segment $\overline{C_{(r,1)}C_{(r,2)}}$ is tangent to the parabolic arc $Q_{(r,1)}(I_{(r,1)})$ at their common endpoint where $X_n = M$, and that is their only intersection point.

We compute a tangent vector to $Q_{(r,2)}(X_n)$ at $X_n = M$ from equation (A.7)

$$\frac{d}{dX_n}\bigg|_{X_n=M} Q_{(r,2)}(X_n) = \frac{vP}{MX_n^3}(2P+2M-2X_n, 2P-2M-2X_n)\big|_{X_n=M}$$

$$= \frac{2vP}{M^4}(P, P-2M).$$

The dot product of this tangent vector with $M(P-2M, -P)$ is also zero. The line segment $\overline{C_{(r,1)}C_{(r,2)}}$ is tangent to the parabolic arc $Q_{(r,2)}(X_n)$ at their common endpoint where $X_n = M$ and that is their only intersection point.

Computing a tangent vector to $Q_{(r,2)}(X_n)$ at $X_n = P - M$ gives:

$$\frac{d}{dX_n}\bigg|_{X_n = P-M} Q_{(r,2)}(X_n) = \frac{vP}{MX_n^3}\left(2P + 2M - 2X_n, 2P - 2M - 2X_n\right)\big|_{X_n = P-M}$$

$$= \frac{4vP}{M(P-M)^3}(M, 0). \tag{A.14}$$

So the half-line $\overrightarrow{C_{(r,3)}C_{(r,4)}}$ is tangent to the parabolic arc $Q_{(r,2)}(X_n)$ at their common endpoint where $X_n = P - M$, and that is their only intersection point.

In the remaining case, the two parabolic arcs $Q_{(r,1)}(I_{(r,1)})$ and $Q_{(r,2)}(I_{(r,2)})$ do intersect. An intersection point in the parabolas $Q_{(r,1)}(X_n)$ and $Q_{(r,2)}(X_n)$ corresponds to the two lines through the line segments in the graph $\bar{\omega}(X_n, N)^T$ over the intervals $I_{(r,1)}$, $I_{(r,2)}$ simultaneously meeting the right branch in the graph of $-2v/X_n$ tangentially. We are particularly interested in when the line segments themselves simultaneously meet the graph of $-2v/X_n$ tangentially. This has to occur at different points in the graphs so that the X_n in $Q_{(r,1)}(X_n)$ is different from the X_n in $Q_{(r,2)}(X_n)$. To distinguish these two points in this section, we let $X_{(n,1)} \in I_{(r,1)}$ and $X_{(n,2)} \in I_{(r,2)}$.

The condition that the parabolas meet is $Q_{(r,1)}(X_{(n,1)}) = Q_{(r,2)}(X_{(n,2)})$, which expanded from equations (A.5) and (A.10) give us:

$$-\frac{vP}{(P-M)X_{(n,1)}^2}(P - M + 2X_{(n,1)}, \ P - M - 2X_{(n,1)})$$

$$= \frac{vP}{MX_{(n,2)}^2}(-P - M + 2X_{(n,2)}, \ -P + M + 2X_{(n,2)}),$$

which implies

$$-M\left((P-M)/X_{(n,1)}^2 + 2/X_{(n,1)}, \ (P-M)/X_{(n,1)}^2 - 2/X_{(n,1)}\right)$$

$$= (P-M)\left((-P-M)/X_{(n,2)}^2 + 2/X_{(n,2)}, \ (-P+M)/X_{(n,2)}^2 + 2/X_{(n,2)}\right).$$

We can simultaneously eliminate the terms containing $1/X_{(n,1)}^2$ and $1/X_{(n,2)}^2$ by subtracting the first component from the second component, and then we can express $X_{(n,1)}$ in terms of $X_{(n,2)}^2$:

$$-M(-4/X_{(n,1)}) = (P-M)(2M/X_{(n,2)}^2),$$

$$X_{(n,1)} = \frac{2X_{(n,2)}^2}{P-M}.$$

We can substitute this back into the top row, multiply through by $4X_{(n,2)}^4$, divide out by $P-M$, and rearrange to obtain a reducible cubic polynomial in $X_{(n,2)}$.

$$-M(P-M)\left(\frac{P-M}{2X_{(n,2)}^2}\right)^2 + 2M\left(\frac{P-M}{2X_{(n,2)}^2}\right) = -\frac{(P-M)^2}{X_{(n,2)}^2} + \frac{2(P-M)}{X_{(n,2)}},$$

$$8X_{(n,2)}^3 - 4PX_{(n,2)}^2 + M(P-M)^2 = 0,$$

$$(2X_{(n,2)} - (P-M))(4X_{(n,2)}^2 - 2MX_{(n,2)} - M(P-M)) = 0.$$

The discriminant to $4X_{(n,2)}^2 - 2MX_{(n,2)} - M(P-M)$ is $4(4MP-3M^2)$, which is positive since $P > 3M$. This gives us three[1] solutions for $X_{(n,2)}$:

$$X_{(n,2)} = \frac{P-M}{2},$$

$$X_{(n,2)} = \frac{M - \sqrt{4MP-3M^2}}{4},$$

$$X_{(n,2)} = \frac{M + \sqrt{4MP-3M^2}}{4}.$$

The solution $X_{(n,2)} = (P-M)/2$ is in the interval $I_{(r,2)}$, but

$$X_{(n,1)} = 2\left(\frac{P-M}{2}\right)^2 /(P-M) = \frac{P-M}{2}$$

is not in the interval $I_{(r,1)}$ since $P > 3M$, so the corresponding intersection point on the two parabolas $Q_{(r,1)}(X_{(n,1)})$, $Q_{(r,2)}(X_{(n,2)})$ does not lie on the parabolic arc $Q_{(r,1)}(I_{(r,1)})$. The value of $\sqrt{4MP-3M^2}$ is greater than that of M since $P > 3M$ so solution $X_{(n,2)} = (M - \sqrt{4MP-3M^2})/4$ is negative and not in the interval $I_{(r,2)}$. Therefore, the corresponding intersection point on the two parabolas $Q_{(r,1)}(X_{(n,1)})$, $Q_{(r,2)}(X_{(n,2)})$ does not lie on the parabolic arc $Q_{(r,2)}(I_{(r,2)})$.

The solution $X_{(n,2)} = (M + \sqrt{4MP-3M^2})/4$ does lie in the interval $I_{(r,2)}$. Clearly it is no smaller than M, and because $P > 3M$ it follows that

$$0 < 4(4P-7M)(P-M) \implies$$

$$4MP-3M^2 < (4P-5M)^2 \implies \frac{M + \sqrt{4MP-3M^2}}{4} < P-M.$$

---

1. Technically the two parabolas intersect at four points, one of which is $(0,0)$, but this point is not in the image of the parameterization $Q_{(r,1)}(X_{(n,1)})$, $Q_{(r,2)}(X_{(n,2)})$ for the parabolas. The point $(0,0)$ is only reached asymptotically by $Q_{(r,1)}(X_{(n,1)})$, $Q_{((r,2))}(X_{(n,2)})$ as $X_{(n,1)}, X_{(n,2)} \to \infty$. We already know there are no solutions to equation (6.3) when $(w_\ell, w_r) = (0,0)$.

We let $X_{(n,5)} = (M + \sqrt{4MP - 3M^2})/4$ and the corresponding value for $X_{(n,1)}$ be $X_{(n,4)} = 2X_{(n,2)}^2/(P-M)$. $X_{(n,4)}$ is positive and because $P > 3M$, it follows that

$$0 < 4(P-3M)(P-M) \implies 4MP - 3M^2 < (2P-3M)^2$$
$$\implies 2P - M + \sqrt{4MP - 3M^2} < 4P - 4M$$
$$\implies \frac{M}{4(P-M)}\left(2P - M + \sqrt{4MP - 3M^2}\right) < M.$$

so $X_{(n,4)} \in I_{(r,1)}$. The solution $(X_{(n,1)}, X_{(n,2)}) = (X_{(n,4)}, X_{(n,5)})$ corresponds to the only intersection point between the parabolic arcs $Q_{(r,1)}(I_{(r,1)})$, $Q_{(r,2)}(I_{(r,2)})$. Furthermore this is the only self-intersection point in the curve $\mathcal{C}_r$ (see figure A.1). We denote this intersection point by

$$\mathcal{C}_{(r,*)} = Q_{(r,1)}\left(X_{(n,4)}\right) = Q_{(r,2)}\left(X_{(n,5)}\right)$$

and define the intervals $I_{(r,4)} = (0, X_{(n,4)}]$, $I_{(r,5)} = [X_{(n,5)}, P-M]$, $I_{(\ell,4)} = -I_{(r,4)}$, and $I_{(\ell,5)} = I_{(r,5)}$.

The appearance of $\mathcal{C}_r$ can be deceptive. It is fairly clear that the common endpoint, $\mathcal{C}_{(r,1)}$, of the parabolic arc $Q_{(r,1)}(I_{(r,1)})$ and $\overline{\mathcal{C}_{(r,1)}\mathcal{C}_{(r,2)}}$ is a cusp in $\mathcal{C}_r$ and that the common endpoint, $\mathcal{C}_{(r,2)}$, of the parabolic arc $Q_{(r,2)}(I_{(r,2)})$ and $\overline{\mathcal{C}_{(r,1)}\mathcal{C}_{(r,2)}}$ is another cusp in $\mathcal{C}_r$ [see figures A.1(A) and A.2(A)]. What is less clear in the figures is that the parabolic arc $Q_{(r,2)}(I_{(r,2)})$ meets the half-line $\overrightarrow{\mathcal{C}_{(r,3)}\mathcal{C}_{(r,4)}}$ smoothly at the point $\mathcal{C}_{(r,3)}$. A large discontinuity in the curvature of $\mathcal{C}_r$ at $\mathcal{C}_{(r,3)}$ makes it appear that they do not meet smoothly [see figure A.1(A)]. On the other hand, the parabolic arcs $Q_{(r,1)}(I_{(r,4)})$ and $Q_{(r,2)}(I_{(r,5)})$ do not meet smoothly at the cross point $\mathcal{C}_{(r,*)}$ but the angle between the tangents to $Q_{(r,1)}(I_{(r,4)})$ and $Q_{(r,2)}(I_{(r,5)})$ at $\mathcal{C}_{(r,*)}$ is so small that it can appear as though they do meet smoothly [see figures A.1(A) and A.2].

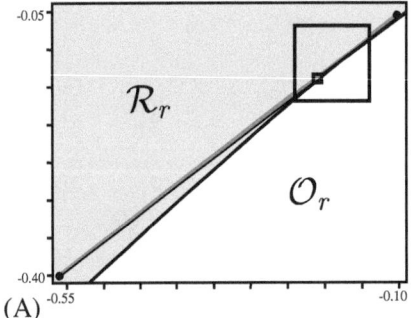
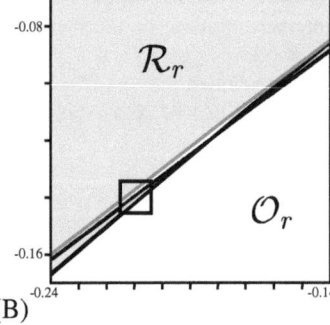

(A)

(B)

Figure A.2
Magnifications of the W_{same} subspace showing $\mathcal{S}_r$ as a thin white region bounded by gray and black curves. $\mathcal{S}_r$ is a subset of $\mathcal{R}_r$. The region shown in (B) is a magnification of the outer "rubberbanded" region shown in (A). The inner "rubberbanded" region shown in (A) also appears in (B). This region is further magnified in figure A.3(A). The physical parameters are $P = 8$, $\psi = 0.95(\pi/2)$, and $\nu = 1/100$.

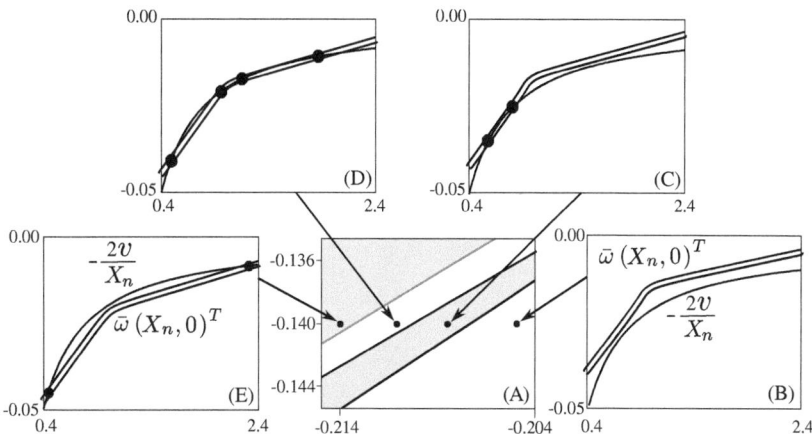

Figure A.3
(A) The magnification of the "rubber banded" region of the W_{same} in figure A.2(B). In figures A.3(B–E), the weight w_r is fixed at -0.1400 while w_ℓ takes on the four values -0.2050, -0.2085, -0.2110, and -0.2140. In each panel, the graphs of $\bar{\omega}(X_n, 0)^T$ are shown as thick white curves with black outlines and the graphs of $-2v/X_n$ are shown as thin black curves. (B) At $w_\ell = -0.2050$ (inside $\mathcal{O}_r$) the two graphs do not intersect. (C) At $w_\ell = -0.2085$ (inside $\mathcal{R}_r \backslash \mathcal{S}_r$), the two graphs intersect at two points. (D) At $w_\ell = -0.2110$ (inside $\mathcal{S}_r$), the two graphs intersect at four points. (E) At $w_\ell = -0.2140$ (back inside $\mathcal{R}_r \backslash \mathcal{S}_r$), the two graphs intersect at two points.

The interior of the thin planar region bounded by the arc of the parabola $Q_{(r,2)}$ from the point $\mathcal{C}_{(r,*)}$ to the point $\mathcal{C}_{(r,1)}$, the line segment $\overline{\mathcal{C}_{(r,1)}\mathcal{C}_{(r,2)}}$, and the arc of the parabola $Q_{(r,1)}$) from $\mathcal{C}_{(r,2)}$ back to the point $\mathcal{C}_{(r,*)}$ will by denoted by $\mathcal{S}_r$. This is the region where there are four solutions to equation (6.3).

Note that the self-intersection of $\mathcal{C}_r$ at $\mathcal{C}_{(r,*)}$ is not an artifact from approximating the solutions to (6.1) with solutions to (6.3). When $\psi = \pi/2$, equations (6.1) and (6.3) coincide. As ψ is lowered from $\pi/2$, the solutions to equations (6.1) and (6.3) separate but as long as ψ remains close enough to $\pi/2$, the region $\mathcal{S}_r$ is not empty. There will be a region in W_{same} such that the graphs of $\bar{\omega}(X_n, 0)^T$ and $-2v/X_n$ intersect at four distinct points in the open interval $(0, P + M)$. Figure A.3 shows the graphs of $\bar{\omega}(X_n, 0)^T$ and $-2v/X_n$ as (w_ℓ, w_r) is varied through $\mathcal{S}_r$.

B   Computing the Matrix $\mathbf{J}_{(r,*)}$

The partial derivatives for F_1', F_2', F_4', and F_5' in $\mathbf{J}_{(r,*)}$ are easily computed; their entries are mostly zeros. To compute the partial derivatives for F_3' and F_6' in $\mathbf{J}_{(r,*)}$, we make use of equation (2.8) and the chain rule:

$$\nabla F_j' = \left(\frac{\partial \bar{\omega}_n}{\partial X_j} \quad \frac{\partial \bar{\omega}_n}{\partial Y_j} \right) \begin{pmatrix} \nabla X_j \\ \nabla Y_j \end{pmatrix},$$

where $j = 3, 6$ and ∇ stands for the differential operator:

$$\left(\frac{\partial}{\partial x_1'}, \frac{\partial}{\partial y_1'}, \frac{\partial}{\partial \theta_1'}, \frac{\partial}{\partial x_2'}, \frac{\partial}{\partial y_2'}, \frac{\partial}{\partial \theta_2'} \right).$$

that is, a hextuple of partial derivative operators that acts on a scalar function.

We compute the second factor of $\nabla F_j'$ first, that is, $(\nabla X_n, \nabla Y_n)^T$. To do this, we can take advantage of the invariance of $\mathcal{C}_n$ (the coordinate transform that takes a state of the total system to the body frame for agent n) under proper congruences, as stated in in equation (5.13). We can take the derivative of $\mathcal{C}_n$ using equation (4.2) for $n = 1, 2$.

Since $\mathcal{C}_n$ is a function from a six-dimensional space to a two-dimensional space, its derivative $D\mathcal{C}_n$ is a 2×6 matrix. After performing some simplifications, we get:

$$D\mathcal{C}_1(s') = \begin{pmatrix} \nabla X_1(x_1', y_1', \theta_1', x_2', y_2', \theta_2') \\ \nabla Y_1(x_1', y_1', \theta_1', x_2', y_2', \theta_2') \end{pmatrix}$$

$$= \begin{pmatrix} -\sin(\theta_1') & \cos(\theta_1') & Y_1 & \sin(\theta_1') & -\cos(\theta_1') & 0 \\ -\cos(\theta_1') & -\sin(\theta_1') & -X_1 & \cos(\theta_1') & \sin(\theta_1') & 0 \end{pmatrix}$$

and

$$D\mathcal{C}_2(s') = \begin{pmatrix} \nabla X_2(x_1', y_1', x_2', y_2', \theta_1', \theta_2') \\ \nabla Y_2(x_1', y_1', x_2', y_2', \theta_1', \theta_2') \end{pmatrix}$$

$$= \begin{pmatrix} \sin(\theta_2') & -\cos(\theta_2') & 0 & -\sin(\theta_2') & \cos(\theta_2') & Y_2 \\ \cos(\theta_2') & \sin(\theta_2') & 0 & -\cos(\theta_2') & -\sin(\theta_2') & -X_2 \end{pmatrix}.$$

Evaluating these derivatives at $s_{(r,*)}$ gives:

$$\begin{pmatrix} \nabla X_1 \\ \nabla Y_1 \end{pmatrix} = \begin{pmatrix} -1 & 0 & 0 & 1 & 0 & 0 \\ 0 & -1 & -X_{(r,*)} & 0 & 1 & 0 \end{pmatrix}$$

and

$$\begin{pmatrix} \nabla X_2 \\ \nabla Y_2 \end{pmatrix} = \begin{pmatrix} -1 & 0 & 0 & 1 & 0 & 0 \\ 0 & -1 & 0 & 0 & 1 & -X_{(r,*)} \end{pmatrix}.$$

We now turn to the first factor of $\nabla F'_j$, that is, $(\partial \bar{\omega}_n / \partial X_n, \partial \bar{\omega}_n / \partial Y_n)$, which we express in short-hand form using the quantities $\nu_{(r,*)}$ and $\mu_{(r,*)}$ introduced in equation (6.9).

$$\left(\frac{\partial \bar{\omega}_n}{\partial X_n}, \frac{\partial \bar{\omega}_n}{\partial Y_n} \right) = -\left(\nu_{(r,*)}, \mu_{(r,*)} \right).$$

The minus sign is for convenience in calculations. The values for $\nu_{(r,*)}$ and $\mu_{(r,*)}$ can be computed from equation (2.8).

$$\mu_{(r,*)} = N \left(w_{(\ell,n)} \frac{f'(d_{(\ell,*)})}{d_{(\ell,*)}} - w_{(r,n)} \frac{f'(d_{(r,*)})}{d_{(r,*)}} \right)$$

$$\nu_{(r,*)} = w_{(r,n)} \left(X_{(r,*)} - M \right) \frac{f'(d_{(r,*)})}{d_{(r,*)}} - w_{(\ell,n)} \left(X_{(r,*)} + M \right) \frac{f'(d_{\ell,*})}{d_{(\ell,*)}},$$

(B.1)

where

$$d_{(\ell,*)} = \sqrt{\left(X_{(r,*)} + M \right)^2 + N^2} \quad \text{and} \quad d_{(r,*)} = \sqrt{\left(X_{(r,*)} - M \right)^2 + N^2}.$$

We can now multiply the two factors from the chain rule together to get $\nabla F'_3$ and $\nabla F'_6$:

$$\nabla F'_3(s_{(r,*)}) = (\nu_{(r,*)}, \mu_{(r,*)}, \mathcal{T}_{(r,*)}, -\nu_{(r,*)}, -\mu_{(r,*)}, 0),$$

$$\nabla F'_6(s_{(r,*)}) = (\nu_{(r,*)}, \mu_{(r,*)}, 0, -\nu_{(r,*)}, -\mu_{(r,*)}, \mathcal{T}_{(r,*)}).$$

The whole matrix is shown in equation (6.10).

C

The Eigenvalues and Eigenvectors for $\mathbf{J}_{(r,*)}$ and $\mathbf{J}_T$

The matrix $\mathbf{J}_{(r,*)}$ and its eigenvalues, as functions of the parameters of the total dynamical system, were presented in section 6.5 on the linear stability analysis of revolving type relative equilibria. The matrix $\mathbf{J}_T$ and its eigenvalues, as functions of the parameters of the total dynamical system, were presented in section 7.5 on the linear stability analysis of side-by-side relative equilibria.

In this appendix, we outline a relatively simple strategy to confirm that these are the eigenvalues for $\mathbf{J}_{(r,*)}$ and $\mathbf{J}_T$. The eigenvalues for $\mathbf{J}_{(\ell,*)}$ can be obtained from the eigenvalues for $\mathbf{J}_{(r,*)}$ by switching the weights $w_{(\ell,n)}$ and $w_{(r,n)}$ in the formulas. The confirmation is achieved by multiplying the matrices for $\mathbf{J}_{(r,*)}$ and $\mathbf{J}_T$ times vectors that we present here, a task we leave to the reader. The results from these matrix multiplications are presented in equations (C.1) through (C.8). We also present an eigenvector for each eigenvalue. This comes down to performing basic arithmetic operations using the components of the matrices and vectors. For a pair of complex conjugate eigenvalues, we can sidestep several complex arithmetic operations by presenting a basis for the two-dimensional real eigenspace of the conjugate eigenvalues. The eigenvalues can then be confirmed with a few subsequent complex arithmetic operations, which we perform in this appendix.

This outline is organized into six steps:

Step 1. Show that $\pm i\,\Omega_{(r,*)}$ are eigenvalues for $\mathbf{J}_{(r,*)}$ and that 0 is an eigenvalue for $\mathbf{J}_T$ with a multiplicity of at least two.

Step 2. Show that $\mathcal{T}_{(r,*)}$ is an eigenvalue for $\mathbf{J}_{(r,*)}$ and $\mathcal{T}_T$ is an eigenvalue for $\mathbf{J}_T$.

Step 3. Show that 0 is an eigenvalue for $\mathbf{J}_{(r,*)}$.

Step 4. Show that $\mathcal{S}_{(r,*)}^{\pm}$ are eigenvalues for $\mathbf{J}_{(r,*)}$.

Step 5. Show that 0 is an eigenvalue for $\mathbf{J}_T$ with a multiplicity of at least three.

Step 6. Show that $\mathcal{S}_T^{\pm}$ are eigenvalues for $\mathbf{J}_T$.

Each of $\mathbf{J}_{(r,*)}$ and $\mathbf{J}_T$ act on the six-dimensional tangent space to S_T, which may suggest that we would need to multiply each of the matrices $\mathbf{J}_{(r,*)}$ and $\mathbf{J}_T$ with six distinct vectors. However, three of the eigenvectors for $\mathbf{J}_{(r,*)}$ and for $\mathbf{J}_T$ have the same form. The exact form for the vectors is presented in each step. To concisely make use of these eigenvectors in steps 1 and 2, we employ a single form for the matrices that covers both cases, $\mathbf{J}_{(r,*)}$ and $\mathbf{J}_T$. We combine the two cases by simply dropping the subscripts:

$$\mathbf{J} = \begin{pmatrix} 0 & \Omega & -\nu & 0 & 0 & 0 \\ -\Omega & 0 & 0 & 0 & 0 & 0 \\ \nu & \mu & \mathfrak{T} & -\nu & -\mu & 0 \\ 0 & 0 & 0 & 0 & \Omega & \nu \\ 0 & 0 & 0 & -\Omega & 0 & 0 \\ \nu & \mu & 0 & -\nu & -\mu & \mathfrak{T} \end{pmatrix}.$$

By setting $\Omega = \Omega_{(r,*)}$, $\nu = \nu_{(r,*)}$, $\mu = \mu_{(r,*)}$, and $\mathfrak{T} = \mathfrak{T}_{(r,*)}$ we get $\mathbf{J} = \mathbf{J}_{(r,*)}$. By setting $\Omega = 0$, $\nu = \nu_T$, $\mu = \mu_T$, and $\mathfrak{T} = \mathfrak{T}_T$, we get $\mathbf{J} = \mathbf{J}_T$.

The names we will use for the vectors are as follows:

$$\mathbf{e}_1, \ \mathbf{e}_2, \ \mathbf{e}_3, \qquad \mathbf{e}_{(0,r,*)}, \ \mathbf{e}_{(4,r,*)}, \ \mathbf{e}_{(5,r,*)}, \qquad \mathbf{e}_{(0,T)}, \ \mathbf{e}_{(4,T)}, \ \mathbf{e}_{(5,T)}.$$

Steps 1 and 2 will use the vectors $\mathbf{e}_1$, $\mathbf{e}_2$, $\mathbf{e}_3$ to confirm eigenvalues for both $\mathbf{J}_{(r,*)}$ and $\mathbf{J}_T$. Steps 3 and 4 will use the vectors $\mathbf{e}_{(0,r,*)}$, $\mathbf{e}_{(4,r,*)}$, $\mathbf{e}_{(5,r,*)}$ to confirm the remaining eigenvalues for $\mathbf{J}_{(r,*)}$. Steps 5 and 6 will use the vectors $\mathbf{e}_{(0,T)}$, $\mathbf{e}_{(4,T)}$, $\mathbf{e}_{(5,T)}$ to confirm the remaining eigenvalues for $\mathbf{J}_T$.

Step 1. Show that $\pm i\,\Omega_{(r,*)}$ are eigenvalues for $\mathbf{J}_{(r,*)}$ and that 0 is an eigenvalue for $\mathbf{J}_T$ with a multiplicity of at least two. A basis for the corresponding eigenspace is:

$$\mathbf{e}_1 = (1, 0, 0, 1, 0, 0)^T \qquad \mathbf{e}_2 = (0, 1, 0, 0, 1, 0)^T.$$

It is easy to check that

$$\mathbf{J}\,\mathbf{e}_1 = -\Omega\,\mathbf{e}_2 \qquad \text{and} \qquad \mathbf{J}\,\mathbf{e}_2 = \Omega\,\mathbf{e}_1. \tag{C.1}$$

This implies that $\mathbf{J}(\mathbf{e}_1 \pm i\,\mathbf{e}_2) = -\Omega\,\mathbf{e}_2 \pm i\,\Omega\mathbf{e}_1 = \pm i\,\Omega\,(\mathbf{e}_1 \pm i\,\mathbf{e}_2)$ so that $\pm i\,\Omega$ are complex conjugate eigenvalues of $\mathbf{J}$ with corresponding complex conjugate eigenvectors $(\mathbf{e}_1 \pm i\,\mathbf{e}_2)$. For $\mathbf{J} = \mathbf{J}_{(r,*)}$ the eigenvalues are $\pm i\,\Omega_{(r,*)}$, and for $\mathbf{J} = \mathbf{J}_T$ the eigenvalue is 0 with a multiplicity of at least two.

Step 2. Show that $\mathfrak{T}_{(r,*)}$ is an eigenvalue for $\mathbf{J}_{(r,*)}$ and $\mathfrak{T}_T$ is an eigenvalue for $\mathbf{J}_T$. A corresponding eigenvector is

$$\mathbf{e}_3 = \left(-\nu\mathfrak{T}, \ \nu\Omega, \ \mathfrak{T}^2 + \Omega^2, \ -\nu\mathfrak{T}, \ \nu\Omega, \ -\mathfrak{T}^2 - \Omega^2 \right)^T.$$

It is not very difficult to check that

$$\mathbf{J}\,\mathbf{e}_3 = \mathfrak{T}\,\mathbf{e}_3. \tag{C.2}$$

So $\mathfrak{T}_{(r,*)}$ is an eigenvalue for $\mathbf{J}_{(r,*)}$ and $\mathfrak{T}_T$ is an eigenvalue for $\mathbf{J}_T$.

Step 3. Show that 0 is an eigenvalue for $\mathbf{J}_{(r,*)}$. A corresponding eigenvector is

$$\mathbf{e}_{(0,r,*)} = \left(0, \ -X_{(r,*)}, 2, 0, \ X_{(r,*)}, 2 \right)^T.$$

Using the fact that $\Omega_{(r,*)} = -2\nu/X_{(r,*)}$ it is easy to check that

$$\mathbf{J}_{(r,*)}\,\mathbf{e}_{(0,r,*)} = (0,0,0,0,0,0)^T. \tag{C.3}$$

Step 4. Show that $\mathcal{S}_{(r,*)}^{\pm}$ are eigenvalues for $\mathbf{J}_{(r,*)}$. We can re-express the salient eigenvalues defined in equation (6.12) more concisely by using the non-negative real quantity:

$$\zeta_{(r,*)} = \sqrt{\left|\mathcal{T}_{(r,*)}^2 - 4\left(\Omega_{(r,*)}^2 + 2\nu\nu_{(r,*)}\right)\right|}.$$

There are two cases to consider according to the relative sizes of $\mathcal{T}_{(r,*)}^2$ and $4\left(\Omega_{(r,*)}^2 + 2\nu\nu_{(r,*)}\right)$.

If $\mathcal{T}_{(r,*)}^2 \geq 4\left(\Omega_{(r,*)}^2 + 2\nu\nu_{(r,*)}\right)$, then the salient eigenvalues are real:

$$\mathcal{S}_{(r,*)}^{\pm} = \frac{\mathcal{T}_{(r,*)} \pm \zeta_{(r,*)}}{2}.$$

If $\mathcal{T}_{(r,*)}^2 < 4\left(\Omega_{(r,*)}^2 + 2\nu\nu_{(r,*)}\right)$, then the salient eigenvalues are complex conjugates:

$$\mathcal{S}_{(r,*)}^{\pm} = \frac{\mathcal{T}_{(r,*)} \pm i\zeta_{(r,*)}}{2}.$$

Either way, as we will see, a basis for the corresponding real eigenspace is:

$$\mathbf{e}_{(4,r,*)} = \left(-\nu\,\mathcal{T}_{(r,*)},\ 2\nu\Omega_{(r,*)},\ \mathcal{T}_{(r,*)}^2 - 4\nu\nu_{(r,*)},\ \nu\mathcal{T}_{(r,*)},\ -2\nu\Omega_{(r,*)},\ \mathcal{T}_{(r,*)}^2 - 4\nu\nu_{(r,*)}\right)^T,$$

$$\mathbf{e}_{(5,r,*)} = \zeta_{(r,*)}\left(-\nu,\ 0,\ \mathcal{T}_{(r,*)},\ \nu,\ 0,\ \mathcal{T}_{(r,*)}\right)^T.$$

It is not very difficult to check that for $\mathcal{T}_{(r,*)}^2 \geq 4\left(\Omega_{(r,*)}^2 + 2\nu\nu_{(r,*)}\right)$

$$\mathbf{J}_{(r,*)}\left(\mathbf{e}_{(4,r,*)}\right) = \frac{\mathcal{T}_{(r,*)}\,\mathbf{e}_{(4,r,*)} + \zeta_{(r,*)}\,\mathbf{e}_{(5,r,*)}}{2},$$

$$\mathbf{J}_{(r,*)}\left(\mathbf{e}_{(5,r,*)}\right) = \frac{\zeta_{(r,*)}\,\mathbf{e}_{(4,r,*)} + \mathcal{T}_{(r,*)}\,\mathbf{e}_{(5,r,*)}}{2}. \tag{C.4}$$

So the span of $\mathbf{e}_{(4,r,*)}$ and $\mathbf{e}_{(5,r,*)}$ is indeed an invariant subspace for $\mathbf{J}_{(r,*)}$. Furthermore, equation (C.4) implies that:

$$\mathbf{J}_{(r,*)}\left(\mathbf{e}_{(4,r,*)} \pm \mathbf{e}_{(5,r,*)}\right) = \mathbf{J}_{(r,*)}(\mathbf{e}_{(4,r,*)}) \pm \mathbf{J}_{(r,*)}(\mathbf{e}_{(5,r,*)})$$

$$= \left(\frac{\mathcal{T}_{(r,*)}\,\mathbf{e}_{(4,r,*)} + \zeta_{(r,*)}\,\mathbf{e}_{(5,r,*)}}{2}\right) \pm \left(\frac{\zeta_{(r,*)}\,\mathbf{e}_{(4,r,*)} + \mathcal{T}_{(r,*)}\,\mathbf{e}_{(5,r,*)}}{2}\right)$$

$$= \left(\frac{\mathcal{T}_{(r,*)} \pm \zeta_{(r,*)}}{2}\right)\mathbf{e}_{(4,r,*)} \pm \left(\frac{\mathcal{T}_{(r,*)} \pm \zeta_{(r,*)}}{2}\right)\mathbf{e}_{(5,r,*)}$$

$$= \mathcal{S}_{(r,*)}^{\pm}\left(\mathbf{e}_{(4,r,*)} \pm \mathbf{e}_{(5,r,*)}\right).$$

So $\mathcal{S}_{(r,*)}^{+}$ is a real eigenvalue of $\mathbf{J}_{(r,*)}$ with real eigenvector $\left(\mathbf{e}_{(4,r,*)} + \mathbf{e}_{(5,r,*)}\right)$, and $\mathcal{S}_{(r,*)}^{-}$ is a real eigenvalue of $\mathbf{J}_{(r,*)}$ with real eigenvector $\left(\mathbf{e}_{(4,r,*)} - \mathbf{e}_{(5,r,*)}\right)$ whenever $\mathcal{T}_{(r,*)}^2 \geq 4\left(\Omega_{(r,*)}^2 + 2\nu\nu_{(r,*)}\right)$.

We can check for the inequality $\mathcal{T}^2_{(r,*)} < 4\left(\Omega^2_{(r,*)} + 2\nu\nu_{(r,*)}\right)$ in a similar way:

$$\mathbf{J}_{(r,*)}\left(\mathbf{e}_{(4,r,*)}\right) = \frac{\mathcal{T}_{(r,*)}\,\mathbf{e}_{(4,r,*)} - \zeta_{(r,*)}\,\mathbf{e}_{(5,r,*)}}{2},$$

$$\mathbf{J}_{(r,*)}\left(\mathbf{e}_{(5,r,*)}\right) = \frac{\zeta_{(r,*)}\,\mathbf{e}_{(4,r,*)} + \mathcal{T}_{(r,*)}\,\mathbf{e}_{(5,r,*)}}{2}. \tag{C.5}$$

So the span of $\mathbf{e}_{(4,r,*)}$ and $\mathbf{e}_{(5,r,*)}$ is again an invariant subspace for $\mathbf{J}_{(r,*)}$. Furthermore equation (C.5) implies:

$$\mathbf{J}_{(r,*)}\left(\mathbf{e}_{(4,r,*)} \pm i\,\mathbf{e}_{(5,r,*)}\right) = \mathbf{J}_{(r,*)}(\mathbf{e}_{(4,r,*)}) \pm i\,\mathbf{J}_{(r,*)}(\mathbf{e}_{(5,r,*)})$$

$$= \left(\frac{\mathcal{T}_{(r,*)}\,\mathbf{e}_{(4,r,*)} - \zeta_{(r,*)}\,\mathbf{e}_{(5,r,*)}}{2}\right) \pm i\left(\frac{\zeta_{(r,*)}\,\mathbf{e}_{(4,r,*)} + \mathcal{T}_{(r,*)}\,\mathbf{e}_{(5,r,*)}}{2}\right)$$

$$= \left(\frac{\mathcal{T}_{(r,*)} \pm i\,\zeta_{(r,*)}}{2}\right)\mathbf{e}_{(4,r,*)} \pm i\left(\frac{\mathcal{T}_{(r,*)} \pm i\,\zeta_{(r,*)}}{2}\right)\mathbf{e}_{(5,r,*)}$$

$$= 8^{\pm}_{(r,*)}\left(\mathbf{e}_{(4,r,*)} \pm i\,\mathbf{e}_{(5,r,*)}\right).$$

So $8^{\pm}_{(r,*)}$ are complex conjugate eigenvalues of $\mathbf{J}_{(r,*)}$ with complex conjugate eigenvectors $\left(\mathbf{e}_{(4,r,*)} \pm i\,\mathbf{e}_{(5,r,*)}\right)$ whenever $\mathcal{T}^2_{(r,*)} < 4\left(\Omega^2_{(r,*)} + 2\nu\nu_{(r,*)}\right)$.

Step 5. Show that 0 is an eigenvalue for $\mathbf{J}_T$ with a multiplicity of at least three. We showed in step 1 that 0 is an eigenvalue for $\mathbf{J}_T$ with a multiplicity of two. So we just show that there is one more eigenvector for the eigenvalue 0. One such eigenvector is:

$$\mathbf{e}_{(0,T)} = (\mu_T,\ -\nu_T,\ 0,\ -\mu_T,\ \nu_T,\ 0)^T.$$

It is easy to check that:

$$\mathbf{J}_T\left(\mathbf{e}_{(0,T)}\right) = (0,0,0,0,0,0)^T. \tag{C.6}$$

Step 6. Show that $8^{\pm}_T$ are eigenvalues for $\mathbf{J}_T$. This step works like step 4 except the basis vectors for the corresponding eigenspaces are different. We can re-express the salient eigenvalues defined in equation (7.10) more concisely by using the non-negative real quantity:

$$\zeta_T = \sqrt{\left|\mathcal{T}^2_T - 8\nu\nu_T\right|}.$$

There are two cases to consider according to the relative sizes of $\mathcal{T}^2_T$ and $8\nu\nu_T$. If $\mathcal{T}^2_T \geq 8\nu\nu_T$ then the salient eigenvalues are real:

$$8^{\pm}_T = \frac{\mathcal{T}_T \pm \zeta_T}{2}.$$

If $\mathcal{T}^2_T < 8\nu\nu_T$ then the salient eigenvalues are complex conjugates:

$$8^{\pm}_T = \frac{\mathcal{T}_T \pm i\,\zeta_T}{2}.$$

Either way, as we will see, a basis for the corresponding real eigenspace is:

$$\mathbf{e}_{(4,T)} = (-\mathcal{T}_T,\ 0,\ 4\nu_T,\ \mathcal{T}_T,\ 0,\ 4\nu_T)^T\,,$$

$$\mathbf{e}_{(5,T)} = \zeta_T\,(1,\ 0,\ 0,\ -1,\ 0,\ 0)^T\,.$$

It is not very difficult to check that for $\mathcal{T}_T^2 \geq 8\nu\nu_T$:

$$\mathbf{J}_T\left(\mathbf{e}_{(4,T)}\right) = \frac{\mathcal{T}_T\,\mathbf{e}_{(4,T)} + \zeta_T\,\mathbf{e}_{(5,T)}}{2}\,,$$

$$\mathbf{J}_T\left(\mathbf{e}_{(5,T)}\right) = \frac{\zeta_T\,\mathbf{e}_{(4,T)} + \mathcal{T}_T\,\mathbf{e}_{(5,T)}}{2}\,. \tag{C.7}$$

So the span of $\mathbf{e}_{(4,T)}$ and $\mathbf{e}_{(5,T)}$ is an invariant subspace for $\mathbf{J}_T$, and by the same reasoning as in step 4:

$$\mathbf{J}_T\left(\mathbf{e}_{(4,T)} \pm \mathbf{e}_{(5,T)}\right) = \mathcal{S}_T^{\pm}\left(\mathbf{e}_{(4,T)} \pm \mathbf{e}_{(5,T)}\right)\,.$$

So $\mathcal{S}_T^+$ is a real eigenvalue of $\mathbf{J}_T$ with real eigenvector $\left(\mathbf{e}_{(4,T)} + \mathbf{e}_{(5,T)}\right)$, and $\mathcal{S}_T^-$ is a real eigenvalue of $\mathbf{J}_T$ with real eigenvector $\left(\mathbf{e}_{(4,T)} - \mathbf{e}_{(5,T)}\right)$ whenever $\mathcal{T}_T^2 \geq 8\nu\nu_T$.

We can check for $\mathcal{T}_T^2 < 8\nu\nu_T$ in a similar way:

$$\mathbf{J}_T\left(\mathbf{e}_{(4,T)}\right) = \frac{\mathcal{T}_T\,\mathbf{e}_{(4,T)} - \zeta_T\,\mathbf{e}_{(5,T)}}{2}\,,$$

$$\mathbf{J}_T\left(\mathbf{e}_{(5,T)}\right) = \frac{\zeta_T\,\mathbf{e}_{(4,T)} + \mathcal{T}_T\,\mathbf{e}_{(5,T)}}{2}\,. \tag{C.8}$$

So the span of $\mathbf{e}_{(4,T)}$ and $\mathbf{e}_{(5,T)}$ is an invariant subspace for $\mathbf{J}_T$, and by the same reasoning as in step 4:

$$\mathbf{J}_T\left(\mathbf{e}_{(4,T)} \pm i\,\mathbf{e}_{(5,T)}\right) = \mathcal{S}_T^{\pm}\left(\mathbf{e}_{(4,T)} \pm i\,\mathbf{e}_{(5,T)}\right)\,.$$

So $\mathcal{S}_T^{\pm}$ are complex conjugate eigenvalues of $\mathbf{J}_T$ with complex conjugate eigenvectors $\left(\mathbf{e}_{(4,T)} \pm i\,\mathbf{e}_{(5,T)}\right)$ whenever $\mathcal{T}_T^2 < 8\nu\nu_T$.

The verification of equations (C.1) through (C.8) completes the confirmation process. The set of eigenvalues for $\mathbf{J}_{(r,*)}$ is $\left\{0,\ \pm i\Omega_{(r,*)},\ \mathcal{T}_{(r,*)},\ \mathcal{S}_{(r,*)}^{\pm}\right\}$, and the set of eigenvalues for $\mathbf{J}_T$ is $\left\{0,\ \mathcal{T}_T,\ \mathcal{S}_T^{\pm}\right\}$ where 0 has a multiplicity of at least three.

We can also see now that the eigenvalues for $\mathbf{J}_T$ can be obtained from the eigenvalues for $\mathbf{J}_{(r,*)}$ by the artifice of replacing $X_{(r,*)}$ with X_T, $\mu_{(r,*)}$ with μ_T, $\nu_{(r,*)}$ with ν_T, and $\Omega_{(r,*)}$ with 0. It might seem that we could have used the same trick to get the eigenvectors. However, we do not get the vectors $\mathbf{e}_{(0,T)}$, $\mathbf{e}_{(4,T)}$, or $\mathbf{e}_{(5,T)}$ by performing this kind of substitution on the vectors $\mathbf{e}_{(0,r,*)}$, $\mathbf{e}_{(4,r,*)}$, or $\mathbf{e}_{(5,r,*)}$.

List of Symbols

Symbols not defined in the main text are given definitions here and are not associated to any page.

α	A subscript used to indicate that a variable refers to an agent system. It can have its own subscript 1, 2, or n indicating which agent it refers to.	p. 67
α_s	A function from the special Euclidean group to the total state space.	p. 101
β	A parameter to continuously deform φ_{τ_n} to ϕ_τ.	p. 86
Δm_*	The difference between the slopes of $\bar\omega(X_n, 0)^T$ and $-2v/X_n$ at the intersection point X_*.	p. 127
∇	The gradient operator.	p. 219
$\zeta_{(r,*)}$	An intermediate quantity used to compute the eigenvalues of the matrix $\mathbf{J}_{(r,*)}$.	p. 223
ζ_T	An intermediate quantity used to compute the eigenvalues of the matrix $\mathbf{J}_T$.	p. 224
Θ	An arbitrary angle in radians.	p. 15
θ_n	The heading of vehicle n in radians.	p. 13
$\kappa(t)$	The local curvature of a planar curve parameterized by t.	p. 142
$\breve{\kappa}$	The total curvature of a periodic arc for a planar curve with periodically varying curvature.	p. 143
λ	The parameter for the example dynamical systems in chapter 3.	p. 41
$\mu_{(r,*)}$	One of the entries of the matrix $\mathbf{J}_{(r,*)}$.	p. 124
μ_T	One of the entries of the matrix $\mathbf{J}_T$.	p. 170
$\nu_{(r,*)}$	One of the entries of the matrix $\mathbf{J}_{(r,*)}$.	p. 124
ν_T	One of the entries of the matrix $\mathbf{J}_T$.	p. 170
π	A continuous surjection from the total state space to the agent state space.	p. 68
π_n	The continuous surjection from the total state space to agent n's state space.	p. 71
σ	A continuous injection of an agent's state space into the total state space.	p. 68
σ_n	The continuous injection of agent n's state space into the total state space.	p. 71

τ	A subscript used to indicate that a variable refers to the total system. (also the dummy variable of an integral).	p. 67
ϕ	A generic dynamical system.	p. 25
ϕ_α	A generic agent dynamical system of an open dynamical system.	p. 71
ϕ_{α_n}	An agent dynamical system for agent n.	p. 71
ϕ_τ	A generic total dynamical system of an open dynamical system.	p. 68
	and the total dynamical system for a pair Braitenberg vehicles.	p. 85
φ_τ	A dynamical system on the total space that relates the total and agent dynamical systems.	p. 69
φ_{τ_n}	A dynamical system on the total space that relates the total dynamical system and the dynamical system for agent n.	p. 84
ψ	The angular offset of the vehicle's sensors from its heading.	p. 11
Ω_*	The angular velocity of the Braitenberg vehicles when the total dynamical system is in the revolving type relative equilibrium $\mathcal{E}_*$. The numbers $\pm i\,\Omega_*$ are tangential eigenvalues for $\mathcal{E}_*$.	p. 120
ω_n	Turning function (angular speed) for vehicle n in the rest frame. Takes six arguments corresponding to the state of the total system.	p. 17
$\bar{\omega}_n$	Turning function (angular speed) for vehicle n in its body frame. Takes two arguments corresponding to the location of agent $\neg n$ in agent n's body frame.	p. 17
$A, A_\pm$	A coefficient in the linear polynomials for $\bar{\omega}_n$ restricted to the line through the sensors.	p. 207
$\bar{a}$	The maximum possible activation of one sensor when the other sensor's activation is at its maximum value.	p. 72
$\begin{pmatrix} a_{(\ell,n)} \\ a_{(r,n)} \end{pmatrix}$	A pair of activations for agent n's left and right sensors.	p. 9
$B, B_\pm$	A coefficient in the linear polynomials for $\bar{\omega}_n$ restricted to the line through the sensors.	p. 207
C_r	An extension of the boundary curve $\partial\mathcal{R}_r$ inside of $\mathcal{R}_r$.	p. 117
$C_{(r,*)}$	The self-intersection point of C_r.	p. 117
$C_{(r,n)}$	Four points of the curve C_r, $n = 1, 2, 3, 4$.	p. 210
$\mathcal{C}_n$	The continuous surjection from the rest frame to the body frame of agent n. Used to compute the continuous surjection π.	p. 76
$\mathcal{C}_n^{-1}$	The continuous right inverse of $\mathcal{C}_n$ used to compute the continuous injection σ_n.	p. 84
D	The total derivative operator.	p. 33

$\mathcal{D}$	The same function as $\widetilde{\mathcal{D}}$ but with its range restricted to make it surjective. Used to compute the continuous surjection π.	p. 76
$\mathcal{D}^{-1}$	The continuous right inverse of $\mathcal{D}$ used to compute the continuous injection σ_n.	p. 83
$\widetilde{\mathcal{D}}$	The continuous function from an agent's body frame to the pair of distances of the other agent from its sensors.	p. 76
$\begin{pmatrix} d_{(\ell,n)} \\ d_{(r,n)} \end{pmatrix}$	The pair of distances from agent n's left and right sensors to the other agent.	p. 14
$\mathcal{E}$	The relative equilibrium for the example in section 3.6.	p. 60
$\mathcal{E}_{(\ell,*)}$	A left-turning relative equilibrium.	p. 114
$\mathcal{E}_{(r,*)}$	A right-turning relative equilibrium.	p. 114
$\mathcal{E}_T$	A side-by-side relative equilibrium.	p. 164
$\mathbf{E}(2)$	The Euclidean group, the full congruence group for two-dimensional Euclidean geometry.	p. 96
$\mathbf{e}_*$	Eigenvectors for $\mathbf{J}_{(r,*)}$ or for $\mathbf{J}_T$.	p. 222
F	The function for the differential equation used to define the total dynamical system. The functions F_n are its six components.	p. 85
F'	The function F converted to a rotating frame. The functions F'_n are its six components.	p. 124
F''	The function F converted to a translating frame.	p. 169
$\mathcal{F}$	The same function as $\widetilde{\mathcal{F}}$ but with its range restricted to make it surjective. Used to compute the continuous surjection π.	p. 78
$\mathcal{F}^{-1}$	The continuous right inverse of $\mathcal{F}$ used to compute the continuous injection σ_n.	p. 82
$\widetilde{\mathcal{F}}$	A continuous function that maps pairs of distances to pairs of sensor activations.	p. 78
f	The scaling function that maps distances to activation levels (also used for an example of a dynamical system within section 3.1).	p. 16
G	An isomorphism from the special Euclidean group of the plane to a group of transformations of the total state space.	p. 98
g	An arbitrary proper congruence of the Euclidean plane.	p. 95
$\mathcal{H}$	The semi-infinite strip consisting of all pairs of distances from an agent's sensors to the other agent.	p. 77
h	A one-parameter subgroup of the special Euclidean group.	p. 96
I_*	Intervals of the X_n-axis of the body frame for agent n.	p. 207
$\mathbf{J}$	The Jacobian (total derivative) for the example dynamical systems in chapter 3.	p. 33
$\mathbf{J}_{(r,*)}$	The Jacobian (total derivative) for the relative equilibrium $\mathcal{E}_{(r,*)}$.	p. 125
$\mathbf{J}_T$	The Jacobian (total derivative) for the relative equilibrium $\mathcal{E}_T$.	p. 170
L	Distance of the sensors from the center of the vehicle.	p. 11

M	Half the distance between a vehicle's sensors.	p. 12
$\mathcal{M}_\ell$	The region in W_{same} where the Braitenberg vehicles engage in "left-turning" meander.	p. 144
$\mathcal{M}_r$	The region in W_{same} where the Braitenberg vehicles engage in "right-turning" meander.	p. 132
$\mathcal{M}_T$	The region of W_{rev} where the Braitenberg vehicles engage in "linear meander".	p. 160
N	How far the sensors are in front of an agent's location.	p. 12
$\mathcal{O}_\ell$	The regions of the weight spaces where left-turning relative equilibria do not exist.	p. 116
$\mathcal{O}_r$	The regions of the weight spaces where right-turning relative equilibria do not exist.	p. 116
P	Range of the sensors (radius within which the other agent produces a non-zero sensor activation).	p. 11
$\mathcal{P}_C$	A counter-rotating "relative periodic orbit."	p. 174
$\mathcal{P}_{(\ell,*)}$	An inner or outer "relative periodic orbit" that emerges from $\mathcal{E}_{(\ell,*)}$ in a Hopf-like bifurcation.	p. 115
$\mathcal{P}_{(r,*)}$	An inner or outer "relative periodic orbit" that emerges from $\mathcal{E}_{(r,*)}$ in a Hopf-like bifurcation.	p. 115
$\mathcal{P}_T$	A "relative periodic orbit" that emerges from $\mathcal{E}_T$ in a Hopf-like bifurcation.	p. 159
$\mathcal{P}_{(\ell,\pm)}$	A pair of "left-turning relative periodic orbits." that appear to emerge together in a saddle node-like bifurcation.	p. 153
$\mathcal{Q}_{\text{eq}}$	An invariant set that occurs when the sensor weights are all equal.	p. 155
$Q_{(r,n)}$	Quadratic polynomials for parabolic arcs that approximate portions of the curve $\mathcal{C}_r$,	p. 210
R_Θ	Rotation matrix for angle Θ.	p. 15
$\mathcal{R}_\ell$	The region in W_{same} where left-turning relative equilibria exist.	p. 116
$\mathcal{R}_r$	The region in W_{same} where right-turning relative equilibria exist.	p. 116
$\mathcal{R}_T$	The region in W_{rev} where side-by-side relative equilibria exist.	p. 159
$\mathbf{R}$	The set of all real numbers.	
$\mathbf{R}_{\geq 0}$	The set of nonnegative real numbers.	
S	A generic state space for a dynamical system.	p. 26
S_α	A generic agent state space for an open dynamical system.	p. 67
S_{α_n}	The agent state space for vehicle n.	p. 71
S_τ	A generic total state space for an open dynamical system and the total state space for a pair of Braitenberg vehicles.	p. 67 p. 70
$\mathcal{S}_r$	The region of $\mathcal{R}_r$ where there are four right turning solutions to equation (6.3).	p. 217
$\mathcal{S}^\pm_{(r,*)}$	The two salient eigenvalues for the relative equilibrium $\mathcal{E}_{(r,*)}$.	p. 125
$\mathcal{S}^\pm_T$	The two salient eigenvalues for the relative equilibrium $\mathcal{E}_T$.	p. 171
$\mathbf{SO}(2)$	Special orthogonal group (all rotations about the origin of $\mathbf{R}^2$).	p. 61

SE(2)	Special Euclidean group (all translations and rotations of $\mathbf{R}^2$).	p. 89
s	An arbitrary state of the total state space (also used for the arc length of a curve, within section 6.2).	p. 25
T	A generic time space for a dynamical system and the minimal period of the curvature of a curve.	p. 26 p. 142
$T_{(-vt)}$	The one-parameter subgroup of translations of $\mathbf{R}^2$ generated by the vector $(0, -v)^T$.	p. 169
$\mathfrak{T}_{(r,*)}$	One of the normal eigenvalues of the relative equilibrium $\mathcal{E}_{(r,*)}$.	p. 125
$\mathfrak{T}_T$	One of the normal eigenvalues of the relative equilibrium $\mathcal{E}_T$.	p. 170
U	An intermediate quantity used to compute the location of the agent $\neg n$ using its distance from the sensors of agent n.	p. 83
v_n	The speed of vehicle n.	p. 11
W_{eq}	The subspace of the total weight space where all four weights are equal.	p. 20
W_{opp}	The subspace of the total weight space where the weights of the two agents have opposite signs from each other.	p. 22
W_{rev}	The subspace of the total weight space where the agents' weights are reversed from each other.	p. 19
W_{same}	The subspace of the total weight space where the agents have the same weights as each other.	p. 18
W_{total}	The total weight space.	p. 17
$\begin{pmatrix} w_\ell \\ w_r \end{pmatrix}$	Coordinates for W_{same}.	p. 113
	or coordinates for W_{rev}.	p. 157
$X_{(n,*)}$	A point in the interval I_*.	p. 214
$X_{(r,*)}$	A right-turning solution to bifurcation equation (6.1)	p. 121
X_T	A solution to bifurcation equation (7.5)	p. 166
$\begin{pmatrix} X_n \\ Y_n \end{pmatrix}$	The location of vehicle $\neg n$ in vehicle n's body frame.	p. 14
$\begin{pmatrix} x_n \\ y_n \end{pmatrix}$	The location of vehicle n in the rest frame.	p. 8
Z	The set of all integers.	
$\mathbf{Z}_2$	The transformation group of $\mathbf{R}^2$ consisting of the identity map and the reflection about the y-axis.	p. 40
$\mathbf{Z}_{\geq 0}$	The set of nonnegative integers.	

References

Abraham, F., R. H. Abraham, and C. D. Shaw. 1990. *A visual introduction to dynamical systems theory for psychology.* Santa Cruz, CA: Aerial Press.

Abraham, R. H., and C. D. Shaw. 1992. *Dynamics–The geometry of behavior 2nd ed.* Redwood City, CA: Addison–Wesley Publishing Company.

Ahmed, N., and W. J. Teahan. 2021. "Using compression to discover interesting behaviours in a hybrid Braitenberg vehicle." *IEEE Access* 9:11316–11327.

Aizawa, K. 2018. "Critical note." In *The Oxford handbook of 4E cognition,* 117. Oxford: Oxford University Press.

Aranson, I. S. ed. 2016. *Physical models of cell motility.* International Publishing: Springer.

Arnold, L., I. Chueshov, and G. Ochs. 2005. "Random dynamical systems methods in ship stability: A case study." In *Interacting Stochastic Systems,* 409–433. Berlin: Springer.

Barack, D. L., and J. W. Krakauer. 2021. "Two views on the cognitive brain." *Nature Reviews Neuroscience* 22 (6): 359–371.

Barkley, D. 1992. "Linear stability analysis of rotating spiral waves in excitable media." *Physical Review Letters* 68 (13): 2090–2093.

Barkley, D. 1994. "Euclidean symmetry and the dynamics of rotating spiral waves." *Physical Review Letters* 72 (1): 164–168.

Barkley, D., and I. G. Kevrekidis. 1994. "A dynamical systems approach to spiral wave dynamics." *Chaos* 4 (3):453–460.

Barnsley, M. F., and S. Demko. 1985. "Iterated function systems and the global construction of fractals." *Proceedings of the Royal Society of London. A. Mathematical and Physical Sciences* 399 (1817): 243–275.

Belousov, B. P. 1985. "A periodic reaction and its mechanism." In *Oscillations and traveling waves in chemical systems,* edited by Richard J. Field and Maria Burger. New York: John Wiley & Sons.

Bhalla, A. P. S., B. E. Griffith, and N. A. Patankar. 2013. "A forced damped oscillation framework for undulatory swimming provides new insights into how propulsion arises in active and passive swimming." *PLoS Computational Biology* 9 (6): e1003097.

Bicho, E., and G. Schöner. 1997. "The dynamics approach to autonomous robotics demonstrated on a low-level vehicle platform." *Robotics and Autonomous Systems* 21:23–35.

Birkhoff, G. D. 1927. "Dynamical systems." In *Colloquium Publications,* vol. 9. Providence, RI: American Mathematical Society.

Bisazza, A., L. J. Rogers, and G. Vallortigara. 1998. "The origins of cerebral asymmetry: A review of evidence of behavioural and brain lateralization in fishes, reptiles and amphibians." *Neuroscience and Biobehavior Reviews* 22 (3): 411–426.

Braitenberg, V. 1984. *Vehicles, experiments in synthetic psychology.* Cambridge, MA: MIT Press.

Bredon, G. E. 1972. "Introduction to compact transformation groups." In *Pure and Applied Mathematics,* vol. 46. New York: Academic Press.

Brun, P. T., B. Audoly, N. M. Ribe, T. S. Eaves, and J. R. Lister. 2015. "Liquid ropes: A geometrical model for thin viscous jet instabilities." *Physical Review Letters, 114* 17:174501.

Carne, T. K. 2012. *Geometry and groups.* Cambridge: Cambridge University Press.

Chen, I. C., O. Kuksenok, V. V. Yashin, R. M. Moslin, A. C. Balazs, and K. J. Van Vliet. 2011. "Shape-and size-dependent patterns in self-oscillating polymer gels." *Soft Matter* 7 (7): 3141–3146.

Chossat, P., and M. Golubitsky. 1988. "Iterates of maps with symmetry." *SIAM Journal on Mathematical Analysis* 19 (6): 1259–1270.

Collett, T. S., and M. F. Land. 1975. "Visual control of flight behaviour in the hoverfly *Syritta pipiens* L J." *Journal of Comparative Physiology* 99 (1): 1–66.

Cooper, R., N. Nudo, J. M. González, S. B. Vinson, and H. Liang. 2011. "Side-dominance of *Periplaneta americana* persists through antenna amputation." *J. Insect Behavior* 24 (3): 175–185.

Cruse, H., V. Dürr, M. Schilling, and J. Schmitz. 2009. "Principles of insect locomotion." In *Spatial temporal patterns for action-oriented perception in roving robots,* 43–96. Berlin, Heidelberg: Springer.

Cushman, R. 1983. "Geometry of the energy momentum mapping of the spherical pendulum." *Centrum voor Wiskunde en Informatica Newsletter* 1:4–18.

Diaconis, P., and D. Freedman. 1999. "Iterated random functions." *SIAM Review, 41 (1),* 45–76.

Doubrovinski, K., and K. Kruse. 2008. "Cytoskeletal waves in the absence of molecular motors." *Europhysics Letters* 83 (1):18003.

Dreher, A., I. S. Aranson, and K. Kruse. 2014. "Spiral actin-polymerization waves can generate amoeboidal cell crawling." *New Journal of Physics* 16:055007.

Durt, C., T. Fuchs, and Tewes C., eds. 2017. *Embodiment, enaction, and culture: Investigating the constitution of the shared world.* Cambridge, MA: MIT Press.

Dvoretskii, S., Z. Gong, A. Gupta, J. Parent, and B. Alicea. 2020. "Braitenberg vehicles as developmental neurosimulation." ArXiv preprint. arXiv: 2003.07689.

Efimov, I. R., V. I. Krinsky, and J. Jalife. 1995. "Dynamics of rotating vortices in the Beeler-Reuter model of cardiac tissue." *Chaos Solitons and Fractals* 5 (3):513–526.

Epstein, I. R., and K. Showalter. 1996. "Nonlinear chemical dynamics: Oscillations, patterns, and chaos." *Journal of Physical Chemistry* 100 (31):13132–13147.

Euclid. 1956. *Elements*. New York: Dover.

Euler, L. 1776. *Formulae generales pro translatione quacunque corporum rigidorum*. Vol. 478. Euler Archive - All Works. https://scholarlycommons.pacific.edu/euler-works /478.

Farey, J. 1827. *A treatise on the steam engine: Historical, practical, and descriptive*. London: Longman, Rees, Orme, Brown, & Green.

Fenichel, N. 1972. "Persistence and smoothness of invariant manifolds for flows." *Indiana Univ. Math. J.* 21 (3): 193–226.

Field, M. J. 1980. "Equivariant dynamical systems." *Transactions of the American Mathematical Society* 259 (1): 185–205.

Field, M. J. 2020. *Lectures on bifurcations, dynamics and symmetry*. Boca Raton, FL: CRC Press.

Field, R. J., E. Körös, and R. M. Noyes. 1972. "Oscillations in chemical systems. II. Thorough analysis of temporal oscillation in the bromate-cerium-malonic acid system." *Journal of the American Chemical Society* 94 (25): 8649–8664.

Field, R. J., and R. M. Noyes. 1974. "Oscillations in chemical systems IV. Limit cycle behavior in a model of a real chemical reaction." *The Journal of Chemical Physics* 60 (5): 1877–1884.

Frasnelli, E. 2013. "Brain and behavioral lateralization in invertebrates." *Frontiers in Psychology* 4:939.

Ghose, K., T. K. Horiuchi, P. S. Krishnaprasad, and C. F. Moss. 2006. "Echolocating bats use a nearly time-optimal strategy to intercept prey." *PLoS Biology* 4 (5): e108.

Golubitsky, M., V. G. LeBlanc, and I. Melbourne. 1997. "Meandering of the spiral tip: An alternative approach." *Journal of Nonlinear Science* 7:557–586.

Golubitsky, M., D. Schaefer, and I. Stewart. 1988. "Singularities and groups in bifurcation theory, II." In *Applied Mathematical Sciences,* vol. 69. New York: Springer-Verlag.

Golubitsky, M., and I. Stewart. 2003. *The symmetry perspective: From equilibrium to chaos in phase space and physical space*. 200. Basel–Boston–Berlin: Birkhäuser Verlag.

Gouin, E., M. D. Welch, and P. Cossart. 2005. "Actin-based motility of intracellular pathogens." *Current Opinion in Microbiology* 8 (1): 35–45.

Guckenheimer, J., and P. Holmes. 1983. "Nonlinear oscillations, dynamical systems, and bifurcations of vector fields." In *Applied Mathematical Sciences,* vol. 42. New York: Springer-Verlag.

Hand, D. J. 2008. "Random dynamical systems: Theory and applications by Rabi Bhattacharya, Mukul Majumdar." *International Statistical Review* 76 (1): 143–144.

Haselsteiner, Andreas F., Cole Gilbert, and Z. Jane Wang. 2014. "Tiger beetles pursue prey using a proportional control law with a delay of one half-stride." *Journal of the Royal Society Interface* 11:20140216.

Hasselblatt, B., and A. Katok, eds. 2002. *Handbook of dynamical systems, Volume 1A.* North Holland: Elsevier.

Herz-Fischler, R. 1987. *A mathematical history of division in extreme and mean ratio.* Waterloo, Ontario: Wilfrid Laurier Univ. Press.

Hiraiwa, T., M. Y. Matsuo, T. Ohkuma, T. Ohta, and M. Sano. 2010. "Dynamics of a deformable self-propelled domain." *Europhysics Letters* 91 (2): 20001.

Hirsch, M. W., C. C. Pugh, and M. Shub. 1970. "Invariant manifolds." *Bulletin of the American Mathematical Society* 76 (5): 1015–1019.

Hirsch, M. W., C. C. Pugh, and M. Shub. 1977. "Invariant manifolds." In *Lecture Notes in Mathematics,* vol. 583. Berlin: Springer-Verlag.

Hirsch, M. W., S. Smale, and R. L. Devaney. 2012. *Differential equations, dynamical systems, and an introduction to chaos.* San Diego: Academic Press.

Hotton, S. 2010. "A dynamical systems approach to actin-based motility in *Listeria monocytogenes.*" *Europhysics Letters* 92:30005.

Hotton, S. 2016. "A geometric invariant for the study of planar curves and its application to spiral tip meander." ArXiv preprint. arXiv: 1602.07758.

Hotton, S., and J. Yoshimi. 2010. "The dynamics of embodied cognition." *International Journal of Bifurcation and Chaos* 20 (4): 943–972.

Hotton, S., and J. Yoshimi. 2011. "Extending dynamical systems theory to model embodied cognition." *Cognitive Science* 35 (3): 444–479.

Hunt, E. R., T. O'Shea-Wheller, G. F. Albery, T. H. Bridger, M. Gumn, and N. R. Franks. 2014. "Ants show a leftward turning bias when exploring unknown nest sites." *Biology Letters* 10:20140945.

Ichino, T., T. Asahi, H. Kitahata, N. Magome, K. Agladze, and K. Yoshikawa. 2007. "Microfreight delivered by chemical waves." *Journal of Physical Chemistry C* 112:3032–3035.

Ishihara, S. 2013. "Cell migration model with multiple chemical compasses." ArXiv preprint. arXiv: 1301.6466.

Izhikevich, E. M. 2007. *Dynamical systems in neuroscience.* Cambridge, MA: MIT Press.

Jahnke, W., W. E. Skaggs, and A. T. Winfree. 1989. "Chemical vortex dynamics in the Belousov-Zhabotinskii reaction and in the two-variable Oregonator model." *Journal of Physical Chemistry* 93 (2): 740–749.

Kasmarik, K., S. Abpeikar, M. M. Khan, N. Khattab, M. Barlow, and M. Garratt. 2020. "Autonomous recognition of collective behaviour in robot swarms." In *Australasian Joint Conference on Artificial Intelligence,* 281–293.

Keener, James P, and John J Tyson. 1986. "Spiral waves in the Belousov-Zhabotinskii reaction." *Physica D: Nonlinear Phenomena* 21 (2-3): 307–324.

Kheowan, O. U., V. Gáspár, V. S. Zykov, and S. C. Müller. 2001. "Measurements of kinematical parameters of spiral waves in media of low excitability." *Physical Chemistry Chemical Physics* 3 (21): 4747–4752.

Kiss, I. Z., J. H. Merkin, S. K. Scott, and P. L. Simon. 2003. "Travelling waves in the Oregonator model for the BZ reaction." *Physical Chemistry Chemical Physics* 5 (24): 5448–5453.

Klein, F. 1893. "A comparative review of recent researches in geometry." *Bulletin of the American Mathematical Society* 2 (10): 215–249.

Klein, F. 1897. *Mathematical theory of the top: Lectures delivered on the occasion of the sesquicentennial celebration of Princeton University.* New York: Charles Scribner's Sons.

Klein, F., and A. Sommerfeld. 2012. *The theory of the top, volume III: Perturbations. Astronomical and geophysical applications (1910).* New York: Springer Science & Business Media.

Krupa, M. 1990. "Bifurcations from relative equilibria." *SIAM J. Math. Anal* 21:1453–1486.

Levin, I., R. Deegan, and E. Sharon. 2020. "Self-oscillating membranes: Chemomechanical sheets show autonomous periodic shape transformation." *Physical Review Letters* 125 (17): 178001.

Li, Ge, Qi Ouyang, Valery Petrov, and Harry L. Swinney. 1996. "Transition from simple rotating chemical spirals to meandering and traveling spirals." *Physical Review Letters* 77 (10): 2105.

Lilienthal, A. J., and T. Duckett. 2004. "Experimental analysis of gas-sensitive Braitenberg vehicles." *Advanced Robotics* 18 (8): 817–834.

Lohmann, A. C., A. J. Corcoran, and T. L. Hedrick. 2019. "Dragonflies use underdamped pursuit to chase conspecifics." *Journal of Experimental Biology* 222 (11): jeb190884.

Long, John H., Adam C. Lammert, Charles A. Pell, Mathieu Kemp, James A. Strother, Hugh C. Crenshaw, and Matthew J. McHenry. 2004. "A navigational primitive: Biorobotic implementation of cycloptic helical klinotaxis in planar motion." *IEEE Journal of Oceanic Engineering* 29 (3): 795–806.

Lotka, A. J. 1956. *Elements of mathematical biology.* Mineola, NY: Dover Publications.

Maeda, S., Y. Hara, R. Yoshida, and S. Hashimoto. 2008. "Peristaltic motion of polymer gels." *Angewandte Chemie International Edition* 47 (35): 6690–6693.

Mamduh, S. M., K. Kamarudin, S. M. Saad, A. Y. M. Shakaff, A. Zakaria, and A. H. Abdullah. 2013. "Braitenberg swarm vehicles for odour plume tracking in laminar airflow." In *IEEE Symposium on Computers & Informatics (ISCI),* 1–6. IEEE.

Mane, R. 1978. "Persistent manifolds are normally hyperbolic." *Transactions of the American Mathematical Society* 246:261–283.

Manz, N., B. T. Ginn, and O. Steinbock. 2003. "Meandering spiral waves in the 1,4-cyclohexanedione Belousov-Zhabotinsky System catalyzed by Fe[batho(SO$_3$)$_2$]$_3^{4-/3-}$." *The Journal of Physical Chemistry A* 107 (50): 11008–11012.

Markman, A. B., and E. Dietrich. 2000. "In defense of representation." *Cognitive Psychology* 40 (2): 138–171.

Matsui, E. T. 2001. "Bifurcation and stability of relative equilibria with isotropy in Lagrangian systems with symmetry." PhD diss., University of California, Santa Cruz.

Maxwell, J. C. 1868. "I. On governors." *Proceedings of the Royal Society of London*, no. 16, 270–283.

McHenry, Matthew J. 2001. "Mechanisms of helical swimming: Asymmetries in the morphology, movement and mechanics of larvae of the ascidian *Distaplia occidentalis*." *Journal of Experimental Biology* 204 (17): 2959–2973.

Meiss, J. D. 2007. *Differential dynamical systems.* Philadelphia: Society for Industrial & Applied Mathematics.

Milnor, J. W. 1950. "On the total curvature of knots." *Annals of Mathematics* 52 (2): 248–257.

Newen, A., L. De Bruin, and S. Gallagher. 2018. *The Oxford handbook of 4E cognition.* Oxford University Press.

Newton, I. 1846. *The mathematical principles of natural philosophy (1687).* New York: Daniel Adee.

Noest, R. M., and Z. J. Wang. 2017. "A tiger beetle's pursuit of prey depends on distance." *Physical Biology* 14 (2): 026004.

Ogawa, N., H. Oku, K. Hashimoto, and M. Ishikawa. 2006. "A physical model for galvanotaxis of *Paramecium* cell." *Journal of Theoretical Biology* 242:314–328.

Ohta, T. 2017. "Dynamics of deformable active particles." *Journal of the Physical Society of Japan* 86 (7): 072001.

Pequeño-Zurro, A., D. Shaikh, and I. Rañó. 2019. "Temporal changes in stimulus perception improve bio-inspired source seeking." ArXiv preprint. arXiv: 1903.10279.

Perko, L. 2001. "Differential equations and dynamical systems (third ed.)" In *Texts in Applied Mathematics,* vol. 7. New York: Springer-Verlag.

Philippides, A., B. Baddeley, P. Husbands, and P. Graham. 2012. "How can embodiment simplify the problem of view-based navigation?" In *Biomimetic and Biohybrid Systems: First International Conference, Living Machines,* 216–227. Barcelona: Springer.

Poinsot, L. 1851. *Théorie nouvelle de la rotation des corps (No. 143).* Paris: Bachelier.

Port, R., and T. Van Gelder, eds. 1995. *Mind as motion: Explorations in the dynamics of cognition.* Cambridge, MA: MIT Press.

Prenowitz, W., and H. Swain. 1966. "*Congruence and motion in geometry.*" In *Thinking with Mathematics Series,* vol. 12. Boston: D.C. Heath & Company.

Pugh, C., and M. Shub. 1970. "The Ω–stability theorem for flows." *Inventiones Mathematicae* 11 (2): 150–158.

Rañó, I. 2009a. "A steering taxis model and the qualitative analysis of its trajectories." *Adaptive Behavior* 17 (3): 197–211.

Rañó, I. 2009b. "Hanging around and wandering on mobile robots with a unique controller." In *Proc. of the 4th European Conf. on Mobile Robots,* 135–140.

Rañó, I. 2010. "An empirical evidence of Braitenberg vehicle 2b behaving as a billiard ball." *Lecture Notes in Computer Science* 6226:293–302.

Rañó, I. 2011. "On the convergence of Braitenberg vehicle 3a immersed in parabolic stimuli." In *2011 IEEE/RSJ International Conference on Intelligent Robots and Systems,* 2346–2351.

Rañó, I. 2012a. "A model and formal analysis of braitenberg vehicles 2 and 3." In *IEEE International Conference on Robotics and Automation,* 910–915. IEEE.

Rañó, I. 2012b. "On the systematic analysis of the Braitenberg vehicle 2b for point-like stimulus sources." *Bioinspiration and Biomimetics* 7:036015 (14pp).

Rañó, I., M. Khamassi, and K. Wong-Lin. 2021. "Stability analysis of bio-inspired source seeking with noisy sensors." In *2021 European Control Conference (ECC),* 341–346. IEEE.

Ren, L., L. Wang, Q. Gao, R. Teng, Z. Xu, J. Wang, C. Pan, and I. R. Epstein. 2020. "Programmed locomotion of an active gel driven by spiral waves." *Angewandte Chemie* 132 (18): 7172–7178.

Reufer, M., R. Besseling, J. Schwarz-Linek, V. A. Martinez, A. N. Morozov, J. Arlt, D. Trubitsyn, F. B. Ward, and W. C. K. Poon. 2014. "Switching of swimming modes in *Magnetospirillium gryphiswaldense.*" *Biophysics Journal* 106 (1): 37–46.

Robinson, C. 1995. *Dynamical systems: Stability, symbolic dynamics, and chaos.* Boca Raton, FL: CRC Press.

Ruelle, D. 1973. "Bifurcations in the presence of a symmetry group." *Archive for Rational Mechanics and Analysis* 51 (2): 136–152.

Rutenberg, A. D., and M. Grant. 2001. "Curved tails in polymerization-based bacterial motility." *Physical Review E* 64 (2): 021904.

Sacker, Robert John. 1964. *On invariant surfaces and bifurcation of periodic solutions of ordinary differential equations.* New York: New York University.

Salomon, R. 1999. "Evolving and optimizing Braitenberg vehicles by means of evolution strategies." *International Journal of Smart Engineering System Design* 2 (1): 69–77.

Salumäe, T., I. Rañó, O. Akanyeti, and M. Kruusmaa. 2012. "Against the flow: A Braitenberg controller for a fish robot." In *2012 IEEE International Conference on Robotics and Automation,* 4210–4215. IEEE.

Sandstede, B., A. Scheel, and C. Wulff. 1999. "Bifurcations and dynamics of spiral waves." *Journal of Nonlinear Science* 9 (4): 439–478.

Schmidt, B., and S. C. Müller. 1997. "Forced parallel drift of spiral waves in the Belousov-Zhabotinsky reaction." *Physical Review E* 55 (4): 4390–4393.

Sfondrini, A. 2013. "An introduction to universality and renormalization group techniques." ArXiv preprint. arXiv: 1210.2262.

Shaikh, D., and I. Rañó. 2020. "Braitenberg vehicles as computational tools for research in neuroscience." *Frontiers in Bioengineering and Biotechnology* 8:565963.

Shenoy, V. B., D. T. Tambe, A. Prasad, and J. A. Theriot. 2007. "A kinematic description of the trajectories of *Listeria monocytogenes* propelled by actin comet tails." *Proceedings of the National Academy Sciences* 104 (20): 8229–8234.

Skinner, G. S., and H. L. Swinney. 1991. "Periodic to quasiperiodic transition of chemical spiral rotation." *Physica D: Nonlinear Phenomena* 48 (1): 1–16.

Smale, S. 1967. "Differentiable dynamical systems." *Bulletin of the American Mathematical Society* 73 (6): 747–817.

Soo, F. S., and J. A. Theriot. 2005. "Large-scale quantitative analysis of sources of variation in the actin polymerization-based movement of *Listeria monocytogenes*." *Biophysical Journal* 89 (1): 703–723.

Stillwell, J. 2005. "The four pillars of geometry." In *Undergraduate Texts in Mathematics*. New York: Springer-Verlag.

Stolkin, R., R. Sheryll, and L. Hotaling. 2007. "Braitenbergian experiments with simple aquatic robots." In *OCEANS 2007,* 1–7. IEEE.

Strogatz, S. H. 2018. *Nonlinear dynamics and chaos: with applications to physics, biology, chemistry, and engineering.* Boca Raton, FL: CRC press.

Tanaka, R., T. Nomoto, T. Toyota, H. Kitahata, and M. Fujinami. 2013. "Delayed response of interfacial tension in propagating chemical waves of the Belousov–Zhabotinsky Reaction without stirring." *The Journal of Physical Chemistry B* 117 (44): 13893–13898.

Tarama, M. 2017. "Swinging motion of active deformable particles in Poiseuille flow." *Physical Review E* 96 (2): 022602.

Templeton, J. J., D. J. Mountjoy, S. R. Pryke, and S. C. Griffith. 2012. "In the eye of the beholder: Visual mate choice lateralization in a polymorphic songbird." *Biology Letters* 8:924–927.

Thelen, E., and L. B. Smith. 1994. *A dynamic systems approach to the development of cognition and action.* Cambridge, MA: MIT Press.

Tilney, L. G., and D. A. Portnoy. 1989. "Actin filaments and the growth, movement, and spread of the intracellular bacterial parasite *Listeria monocytogenes*." *The Journal of Cell Biology* 109 (4): 1597–1608.

Van Gelder, T. 1995. "What might cognition be, if not computation?" *The Journal of Philosophy* 92 (7): 345–381.

Vanderbauwhede, A., M. Krupa, and M. Golubitsky. 1989. "Secondary bifurcations in symmetric systems." In *Lecture notes in pure and applied mathematics,* 118:709–716. New York: Marcel Dekker.

Vicker, M. G. 2000. "Reaction–diffusion waves of actin filament polymerization/depolymerization in Dictyostelium pseudopodium extension and cell locomotion." *Biophysical Chemistry* 84 (2): 87–98.

Walter, W. G. 1950. "An imitation of life." *Scientific American* 182 (5): 42–45.

Weiner, O. D., W. A. Marganski, L. F. Wu, S. J. Altschuler, and M. W. Kirschner. 2007. "An actin-based wave generator organizes cell motility." *PLoS Biology* 5 (9): 2053–2063.

Wen, F. L., K. T. Leung, and H. Y. Chen. 2012a. "Curved trajectories of actin-based motility in two dimensions." *Europhysics Letters* 98:38005.

Wen, F. L., K. T. Leung, and H. Y. Chen. 2012b. "Trajectories of Listeria-type motility in two dimensions." *Physical Review E* 86:061902.

Wen, F. L., K. T. Leung, and H. Y. Chen. 2016. "Spontaneous symmetry breaking for geometrical trajectories of actin-based motility in three dimensions." *Physical Review E* 94:012401.

Whitelam, S., T. Bretschneider, and N. J. Burroughs. 2009. "Transformation from spots to waves in a model of actin pattern formation." *Physical Review Letters* 102 (19): 198103.

Whitney, H. 1937. "On regular closed curves in the plane." *Compositio Math* 4:276–284.

Wiener, N. 1954. *The human use of human beings.* London: Da Capo Press.

Wiggins, S. 1994. "Normally hyperbolic invariant manifolds in dynamical systems," vol. 105. Berlin: Springer Science & Business Media.

Wilson, R., and L. Foglia. 2021. "Embodied cognition." In *The Stanford encyclopedia of philosophy (Winter 2021 Edition),* edited by Edward N. Zalta. https://plato.stanford.edu/archives/win2021/entries/embodied-cognition.

Winfree, A. T. 1972. "Spiral waves of chemical activity." *Science* 175 (4022): 634–636.

Winfree, A. T. 1973. "Scroll-shaped waves of chemical activity in three dimensions." *Science* 181:937–939.

Winfree, A. T. 1991. "Varieties of spiral wave behavior: An experimentalist's approach to the theory of excitable media." *Chaos* 1 (3): 303–334.

Workman, L., and R. J. Andrew. 1986. "Asymmetries of eye use in birds." *Animal Behavior* 34 (5): 1582–1584.

Yoshimi, J. 2012. "Active internalism and open dynamical systems." *Philosophical Psychology* 25 (1): 1–24.

Young, Y. N., M. J. Shelley, and D. B. Stein. 2021. "The many behaviors of deformable active droplets." *Mathematical Biosciences and Engineering* 18 (3): 2849–2881.

Zaikin, A. N., and A. M. Zhabotinsky. 1970. "Concentration wave propagation in two-dimensional liquid-phase self-oscillating system." *Nature* 225 (5232): 535–537.

Zhabotinsky, A. M. 1964. "Periodical oxidation of malonic acid in solution (a study of the Belousov reaction kinetics)." *Biofizika* 9:306–311.

Zhabotinsky, A. M. 2007. "Belousov-Zhabotinsky reaction." *Scholarpedia* 2 (9): 1435, revision #91050.

Zhabotinsky, A. M., F. Buchholtz, A. B. Kiyatkin, and I. R. Epstein. 1993. "Oscillations and waves in metal-ion-catalyzed bromate oscillating reactions in highly oxidized states." *Journal of Physical Chemistry* 97 (29): 7578–7584.

Zhabotinsky, A. M., and A. N. Zaikin. 1973. "Autowave processes in a distributed chemical system." *Journal of Theoretical Biology* 40 (1): 45–61.

Ziebert, F., and I. S. Aranson. 2016. "Computational approaches to substrate-based cell motility." *npj Computational Materials* 2 (1): 1–16.

Ziepke, A., I. Maryshev, I. S. Aranson, and E. Frey. 2022. "Multi-scale organization in communicating active matter." *Nature Communications* 13 (1): 6727.

Zykov, V. S. 1986. "Cycloid circulation of spiral waves in an excitable medium." *Biophysics* 31 (5): 940–944.

Index